Spatial Justice

There can be no justice that is not spatial. Against a recent tendency to despatialise law, matter, bodies and even space itself, this book insists on spatialising them, arguing that there can be neither law nor justice that are not articulated through and in space.

Spatial Justice presents a new theory and a radical application of the material connection between space – in the geographical as well as sociological and philosophical sense – and the law – in the broadest sense that includes written and oral law, but also embodied social and political norms. More specifically, it argues that spatial justice is the struggle of various bodies – human, natural, non-organic, technological – to occupy a certain space at a certain time. Seen in this way, spatial justice is the most radical offspring of the spatial turn, since, as this book demonstrates, spatial justice can be found at the core of most contemporary legal and political issues – issues such as geopolitical conflicts, environmental issues, animality, colonisation, droning, the cyberspace and so on. In order to argue this, the book employs the *lawscape*, as the tautology between law and space, and the concept of *atmosphere* in its geological, political, aesthetic, legal and biological dimension.

Written by a leading theorist in the area, *Spatial Justice: Body, Lawscape, Atmosphere* forges a new interdisciplinary understanding of space and law, while offering a fresh approach to current geopolitical, spatiolegal and ecological issues.

Andreas Philippopoulos-Mihalopoulos is Professor of Law & Theory at Westminster University, London

Space, Materiality and the Normative
Series Editors: Andreas Philippopoulos-Mihalopoulos and Christian Borch

Space, Materiality and the Normative presents new ways of thinking about the connections between space and materiality from a normative perspective. At the interface of law, social theory, politics, architecture, geography and urban studies, the series is concerned with addressing the use, regulation and experience of space and materiality, broadly understood, and in particular with exploring their links and the challenges they raise for law, politics and normativity.

Books in this series:

Spatial Justice
Body, Lawscape, Atmosphere
Andreas Philippopoulos-Mihalopoulos

Forthcoming:

Urban Commons
Rethinking the City
Christian Borch and Martin Kornberger

A Jurisprudence of Movement
Common Law, Walking, Unsettling Place
Olivia Barr

Placing International Law
Authority, Jurisdiction, Technique
Fleur Johns, Shaun McVeigh, Sundhya Pahuja, Thomas Skouteris, and Robert Wai

Spatial Justice

Body, Lawscape, Atmosphere

Andreas Philippopoulos-Mihalopoulos

First published 2015
by Routledge
2 Park Square, Milton Park, Abingdon, Oxon, OX14 4RN

and by Routledge
711 Third Avenue, New York, NY 10017

a GlassHouse Book

Routledge is an imprint of the Taylor & Francis Group, an informa business

© 2015 Andreas Philippopoulos-Mihalopoulos

The right of Andreas Philippopoulos-Mihalopoulos to be identified as author of this work has been asserted by him in accordance with sections 77 and 78 of the Copyright, Designs and Patents Act 1988.

All rights reserved. No part of this book may be reprinted or reproduced or utilised in any form or by any electronic, mechanical, or other means, now known or hereafter invented, including photocopying and recording, or in any information storage or retrieval system, without permission in writing from the publishers.

Trademark notice: Product or corporate names may be trademarks or registered trademarks, and are used only for identification and explanation without intent to infringe.

British Library Cataloguing in Publication Data
A catalogue record for this book is available from the British Library

Library of Congress Cataloging-in-Publication Data
Philippopoulos-Mihalopoulos, Andreas, author.
Spatial justice : body, lawscape, atmosphere / Andreas Philippopoulos-Mihalopoulos.
pages cm. -- (Space, materiality, and the normative)
Includes bibliographical references.
ISBN 978-1-138-01738-2 -- ISBN 978-1-315-78052-8 1. Justice. 2. Justice, Administration of--Philosophy. 3. Space perception. 4. Geographical perception. I. Title.
K240.P45 2015
340'.114--dc23
2014020910

ISBN: 978-1-138-01738-2 (hbk)
ISBN: 978-1-315-78052-8 (ebk)

Typeset in Baskerville by
Fish Books

To Elias (the other author)

Contents

Acknowledgements — ix

Introduction — 1

1 **Law's spatial turn** — 15
   1.1 *Points of turning* 15
   1.2 *False turns* 23
   1.3 *Abstractions beyond metaphors* 28

2 **Welcome to the lawscape** — 38
   2.1 *Emerging spaces, emerging bodies, emerging law* 39
   2.2 *Posthuman epistemology* 59
   2.3 *The lawscape* 65
   2.4 *Posthuman, immanent, fractal: one lawscape* 79
   2.5 *The repeated time of the lawscape* 87
   2.6 *Walking the lawscape* 94

3 **From lawscape to atmosphere: affects, bodies, air** — 107
   3.1 *Affects: senses, emotions, symbols* 110
   3.2 *Atmosphere* 122
   3.3 *Engineering and perpetuating an atmosphere* 139
   3.4 *Coda: back in the room* 145

4 **A change of air: the posthuman atmosphere** — 151
   4.1 *The Earth that moves* 153
   4.2 *Ruptures in the service of atmosphere* 163
   4.3 *'I don't know'* 166

5	**The rupture of spatial justice**	**174**
	5.1 An aspatial spatial justice 175	
	5.2 The desire to move, the desire to stand still 184	
	5.3 A rupture in the continuum 192	
	5.4 Withdrawal 198	
6	**The islands**	**220**
	6.1 The first island 221	
	6.2 Are we there yet? The second island 228	
	6.3 Repetition: the double island 232	

Bibliography   237
Index   259

Acknowledgements

It was not supposed to be difficult to write this book. It was meant to be a consolidation of my spatiolegal thinking of the past few years. It was supposed to be my final word on the spatiolegal, which would allow me to move on to issues of materiality and elementarity that, although spatial, are different from space. Yet this book turned out to be so much more than just a consolidation.

First of all, I did not manage to keep my thinking from running wild while writing the book. In other words, this book failed to be a consolidation. It has proven to be a deep rethinking of my previous ideas and an emergence of new concepts and connections that occasionally took me by surprise. This means that both the law and the spatiality employed here are fully material, and often elemental. In the course of the book I have come to accept that the text has its own materiality and way of evolving and I would not want to stand in its way. So this book is a line of flight, a joyful torture to write, and a tortuous line to ride. But it took me somewhere new.

Second, the whole process of thinking and writing this book brought me in discussion with numerous people whose humblingly generous feedback has determined the core of the book. This feedback often took the form of articulate comments during the discussion that followed my lectures; other times, meaningful smiles and vehement head movements (yay or nay) during these lectures; and more often than not, talks and walks and coffees around invitations to think together. Feedback also regularly comes through my students' generous engagement with my ideas and our discussions in and outside the classroom, which makes us all sometimes dizzy with delight and wonder. Finally, feedback comes through people who have thought and used my concepts in their own work, and the way they have given new life to them, whether critical or affirmative (or both). All this has determined the book.

It so happens that some people and some spaces manage to make your line of thought kink, go this rather than that way, push your limits in a different direction – in short, have an *encounter* with something that exceeds you. I have learnt to heed these encounters and the way they affect me because they pierce my skin in unsettling yet tremendously decisive ways. These encounters, some long some short, some continuing some singular, but all of them deeply affecting, I would like to list and thank here. My first thanks, appropriately, go to the spaces of writing: Venice and Caffé Brasília by the Fenice; Copenhagen and Riccos at Nørrebro; Sydney and Lot 19 at Elizabeth Bay; London and the Bethnal Green café culture; Suffolk and Tony's horse farm; Thessaloniki and its quiet folds. I would also like to thank Roswitha Gerlitz and Tony Macintosh for the space to think; the Department of Politics, Management and Philosophy at Copenhagen Business School for their always thorough engagement with my work and financial and personal support; the Department of Planning and Complex Environments at the IUAV University in Venice for funding a part of this project and inviting me as Guest Professor; the School of Law, University of Technology Sydney for their generous invitation as Distinguished Research Fellow and for facilitating my work in most gentle ways; the Melbourne Law School, Criminology Department, and the Institute of International Law and the Humanities, University of Melbourne, for exposing me to so many new ways of thinking about space; and above all my own institution, the Westminster Law School, for their unfaltering support and even tolerance of my off the wall ideas.

Discussions that have changed the line of my thought and life (not an exaggeration) I have had with many. With apologies to the ones I forget, I would like to thank here Anne Bottomley and Nathan Moore for their constant becoming; Andrea Brighenti for pushing and pushing; Olivia Barr for all our mutual folding; Andrea Pavoni for our always frenetic agreements; Danilo Mandic, Caterina Nirta, Victoria Brooks, Pravin Jerayaj, Hans von Rettig, Debdatta Chowdhury, Kay Lalor for being so much more than just doctoral students; Chris Butler, Illan Wall and that inspired bunch of exciting people on Strathmore Island, inebriated with the way ideas were exploding before us; Stewart Williams and another inspiring bunch of people in Tasmania; Jan Hogan and Adam Laurence for showing me what matter can do; Leopold Lambert for showing me the joy of funambulism; Emily Grabham and Rasmus Johnsen for showing me time; Frances Restuccia for showing me the Messiah; Yoriko Otomo and Ed Mussawir for showing me animals; Laurent De Sutter and Mark Halsey for showing me Deleuze; Radha d'Souza for showing me other geographies; Alison Young and Peter Rush for showing me space and family and food; Christian Borch and Timon Beyes for showing me atmosphere; Daphne Bidstrup Hjorth for showing me thought without thesis; Oren Ben Dor for showing me friendship; Anna Grear for showing me vulnerability; Amy Kulper for showing me air; Augusto Cusinato for showing me the landscape; Matilda Arvidsson for showing me garden; Chiara Mazzoleni and Rick Mohr for showing me city; Antonio Riello for showing me Sunday happiness; Marlene Klein for showing me the water in the evening; Adriano Cancellieri and Elena

Ostanel for opening up spatial justice to new fields; Penny Adamopoulou, Sirena Chalhoub and Angela Condello for their fearlessly disrupting theatrics during my talks; my editor-for-life, Colin Perrin, who encouraged me to write this book, saying the rather delightful 'you do not need theory, you *are* theory'; my sister Christina for being my guru; Elias for writing this book in his way as I was writing it in mine; and my students, undergraduate and graduate, who have never stopped making me question what I think I know.

My deepest thanks for reading parts or even the whole of this book, and often for so much more than that, to Sharron FitzGerald, Anna Grear, Miriam Tedeschi and Francesca Anasaloni, Thanos Zartaloudis, Hans Lindahl, Peter Goodrich, Alison Young, Linda Mulchahy, Andrea Pavoni. This book has been rehearsed in one way or another in various institutions that invited me to talk about my work. I would like to thank Westminster University's School of Architecture and Built Environment; the Westminster Centre for the Study of Democracy; and the Westminster Law School; Goldsmiths London; the LSE; UCL; King's College London; Brunel University; University of Oxford; University of Southampton; University of Sussex; University of Warwick; University of Kent in Canterbury; IUAV Venice; Ca Foscari Venice; Venice International University; Politecnico di Milano; University of Oslo, Norway; Universität Bielefeld, Germany; Université du Sud, Toulon, France; IISL, Onati, Spain; University of Lund, Sweden; Kungliga Tekniska Högskolan, Stockholm Sweden; University of Tilburg, the Netherlands; The Berg Institute for Law and History, Tel Aviv, Israel; Al-Quds University, East Jerusalem, Palestinian Occupied Territories; UTAS, Hobart, Tasmania; University of Technology, Sydney; Griffith Law School, Brisbane; Australian National University, Canberra and specifically Des Manderson for the invitation to speak at one of the most inspiringly buzzing audience I have ever encountered at the Australasian Law and Humanities and Law and Literature Annual Conference.

My thanks finally to the publishers of the following texts, extracts of which have been reworked in this book with permission: 'Atmospheres of Law: Senses, Affects, Lawscapes' in *Emotion, Space and Society* 7(1), 35–44, 2013; 'Mapping the Lawscape: Spatial Law and the Body', in Z. Bankowski and M. Del Mar and P. Maharg (eds), *The Arts and the Legal Academy: Beyond Text in Legal Education*, Aldershot: Ashgate, 2012; 'Law, Space, Bodies: The Emergence of Spatial Justice', in L. De Sutter and K. McGee (eds), *Deleuze and Law*, Edinburgh: Edinburgh University Press, 2012; 'Law's Spatial Turn: Geography, Justice and a Certain Fear of Space', *Law, Culture and Humanities*, 7(2), 1–16, 2011; 'Spatial Justice: Law and the Geography of Withdrawal', *International Journal of Law in Context*, 6(3), 1–16, 2010.

Introduction

There is no outside! But we forget this...How lovely it is that we forget![1]

I.I There is no outside

But we need an outside. This is the cry of this book. The cry is spatialised, legalised, rarefied, embodied, ruptured, dissimulated, but underneath all this, it remains a cry. A particularly echoing one too: in the box of the Anthropocene, where humans are both everywhere and decentralised, and in which all material bodies are clustered, breathing space is limited, future is closing in, human extinction is a possible reality. One stands on one's toes and looks for the outside.

This book does not offer an outside. Instead, it offers small consolations in the form of ruptures. Since I cannot 'forget' that there is no outside, I dwell fully in the inside. The inside is a continuum. I try to make the best of it by partiotioning it, sometimes even creating illusions of interiors and exteriors. I have previously called these ruptures 'immanent transcendence'.[2] You will not find much transcendence in this book, even immanently. You will, however, find lots of folds in which to cover up, glass to hide behind, spaces on which to slide back and withdraw, scales of invisibilisation to play with, illusions to reflect on. Is this transcendence? Yes, if transcendence is to take shelter from the violence of immanence. The ultimate purpose of it all is to fight a repeating, constantly reconfigured and contingent fight, not in order to find an outside but in order to reorient the inside in a way that bodies can fit in better with each other. This is my hope, and the name I have chosen for it is *spatial justice*. I need to qualify it though: there is nothing grand, apocalyptic or messianic about it. There is nothing of a solution, blueprint, or end goal in it. There is instead a deep-seated desire to eradicate the desires (it is all about desire

1 Nietzsche, 2005: 175
2 Philippopoulos-Mihalopoulos, 2011b

against desire) that have placed us central and made us vacuous effigies of human preponderance. Spatial justice emerges as withdrawal, namely a body's moving away from its desire to carry on with the comfort offered by supposedly free choices, power structures or even by fate. Withdrawal simultaneously moves away from desire while following another desire for spatial justice. Yet, there is nowhere to withdraw but deeper inside, a little further this or that side. Withdrawal is not an escape but a shaking up of desire in order to orient the legal space differently. To withdraw is to fight.

This fight needs tools, strategies, or even weapons. All of them operate as *ruptures*, namely cracks of the continuum that allow renegotiation and differential movement. The main rupture is *withdrawal*, which allows for the emergence of a space of renegotiation and reorientation. This is closely followed by the *lawscape*, a concept that has haunted my texts for the last few years, reflecting the inherent yet tortuous way in which law and space come together. Yet another is *dissimulation*, itself a rupture but with the specific result of having bodies presenting themselves in different positions than the ones they actually occupy. Another is *invisibilisation* and *visibilisation*, a video game of ontological presence rather than phenomenological visuality that makes the lawscape more or less legal, more or less spatial. Even this is a rupture of sorts, better hidden, indeed better dissimulated. The other weapon is *atmosphere*, an enclosure of desire that brims with the illusion of spatial justice. Finally, and paradoxically, the last weapon is that of the *continuum*, which is used in plural forms, multiplying as lawscape or atmosphere, both bringing us back from the illusions of rupture and enabling ruptures to take place. Two things need to be clarified: these notions ebb and flow, sometimes overlapping, mostly withdrawing from full definition and from each other. Their definitions are never unitary but always contextual, drawing from the assemblage in which they are found. For many of them I offer various definitions, adding or substracting something from the previous one. The point of this is the desire not to *fix* things, ethically or otherwise. This is the second thing that I must clarify: while there is a strong and I hope obvious ethical direction in this book, I resist falling into easy moralising traps. No concept, practice, or ideology can claim an a priori ethical orientation. None of them is above strategising, and none guarantees a specific outcome, whether positive or negative – not even the main concept of the book, spatial justice. Everything is correlative to its assemblage. This does not mean that I subscribe to cultural relativism. Rather, I am sensitised to the way concepts, practices and ideologies become operationalised and fixed in specific spatiolegal conditions that allow them to maintain the illusion of just, fair, proper and so on.

So, there is no outside. This is the guiding position, indeed the *thesis* of the book. Faithful to its etymology, the thesis (=position, placement) that negates the outside informs the way bodies, the lawscape, atmosphere, and spatial justice position themselves in relation to each other, in a connection of movement, yet at the same time withdrawn from each other, ontologically different and never fully present. Emphatically, I am not dwelling on the negative of the thesis but its positive: there is a continuum that contains everything, including its own ruptures, conflicts, invisibilisations, dissimulations. This continuum is not some anything-goes well-wishing culturally-relative flat ontology but a tilted, power-structured surface, on which bodies move and rest and position themselves, thus affecting the tilt while being affected by it. It is all about how bodies position themselves. This 'thesis' comes from my engagement with the writings of Spinoza, Nietzsche, Deleuze, Luhmann, Grosz, Braidotti, Gatens, Negarestani, Morton, Serres, Guattari, Lyotard, Derrida, object oriented ontologies as well as environmental and ecological issues, posthumanist feminist theory, embodiment, questions of identity and emplacement, architecture and issues of materiality. While these and other theorists and theories appear below, this text does not subscribe to one school of thought. It is neither Deleuzian (despite appearances), nor Luhmannian, materialist, critical or sociolegal. Nor does it engage in critique, in the sense of appraising existing literature and commenting on its shortfalls (with the exception of two instances where I felt I had to do this in order to defend the *need* for my radically differentiated position). This text flows along the various theories, respectfully (but not always) picking up flows of thought and applications, without however making a meal out of theoretical loyalty or opposition. If anything, this book attempts to be consistent with itself more than with any specific theory. Not an easy feat, since the way I understand consistency is as a surface of flows and confluences, but also ruptures in the form of contradictions, openings, enclosures, incompleteness, singularities that do not fit in, as well as the concept of continuum that mediates everything.

The main thrust of the book is the concept and practice of *spatial justice*. I define spatial justice as *the conflict between bodies that are moved by a desire to occupy the same space at the same time*. This is neither merely distributive justice, nor regional democracy, but an embodied desire that presents itself ontologically. My goal is twofold: first, to think of the concept in a spatial, corporeal, and generally material way. This is a reaction to most current theorisations of spatial justice that, surprisingly, are too aspatial, with the result that the concept becomes deprived of the radical potential offered by understandings of space not only in geography but also architecture, philosophy, quantum physics, feminism, ecological thought and so on. The second goal is to think of the concept of spatial justice in a legal context. The majority of theorisations focus on a broad, political understanding of justice, ignoring its most obvious connection with law. This connection, as I show especially in Chapters 5 and 6, contextualises spatial justice as a process of legal reorientation rather than solution. I hasten to add that the law I employ here is a spatialised law that does not dwell on the textual (that too) but expands on the space and bodies that incorporate it and act it out.

This law is the *lawscape*. Law and space cannot be separated from each other. They are constantly conditioned by each other, allowing one to emerge from within its connection to the other. Yet, this connection is paradoxical, interfolding and excessive. It is not a connection of dialectical interdependence but, as I show consistently throughout the book, a 'folding' of one into the other, a connection of non-causal enveloping that allows each one to emerge depending on the conditions. *The lawscape is the way the ontological tautology between law and space unfolds as difference.* This takes place as a play of visibilisation and invisibilisation. A no-smoking sign in a public space is a visibilisation of the law, but the free, open space of an art gallery performs an invisibilisation of the law. In the first case, space becomes a little more *obviously* regulated and thus cedes priority to the law. In the second, space retains a façade of ambling and seemingly unconstrained unconstrained movement, free from legal presence. So at various points a lawscape appears more or less legal, or more or less spatial. This interplay is what allows a body to negotiate its position in the lawscape, to affect the lawscape and ultimately to reorient it. Counter-intuitively, in/visibilisation does not operate on a visual, phenomenological level, nor does it assume a viewer. In/visibilisation is the ontological condition of the lawscape, which goes beyond an epistemological/disciplinary connection between law and geography.

In the lawscape, neither law nor space vanish completely. The lawscape remains both law and space at all times, thus allowing a manoeuvring space for the various bodies that constitute the lawscape to affect it in various ways. Yet at points, the lawscape dissimulates its presence for as long as it can, enlisting every trick in the book in order to do so. At these points, the lawscape becomes *atmosphere*. Although the difference between lawscape and atmosphere is a matter of degrees, since both belong to the same spatiolegal continuum, the turning into atmosphere is a spectacular display of smoke and mirrors. A cosy atmosphere at home is about the materiality of bodies coming together around a dinner table, fully dissimulating the fact that the private remains a lawscape. An oppressive atmosphere in a prison is about the law's intensified presence that directs bodies in corridors of movement, fully dissimulating the materiality of the lawscape with its possibilities of resistance and negotiation. In both cases, the lawscape has become an enclosure that is no longer conditioned by mere in/visibilisation but by an intense *dissimulation*. Again, dissimulation is not a phenomenological concept but an ontological one. It emerges as a rupture in the continuum of the lawscape and operates on the level of ontological withdrawal of the lawscape: it is self-dissimulation. It allows the lawscape to dissimulate itself and become an atmosphere. *Atmosphere is the lawscape's wet dream.* Every lawscape wants to become an atmosphere when it grows up.

But growing up entails fixety, stability, consistency. An atmosphere is an enclosure of affects that spread through affective imitation between bodies. I define atmosphere as *the affective ontology of excess between, through and against bodies*. These bodies produce the atmosphere and its excess, and find themselves situated, trapped or liberated in it.

When I refer to bodies, I do not mean human bodies only. The *bodies* of the title are human, nonhuman, technological, natural, immaterial, material, elemental, systemic. In this I follow Spinoza, and after him Deleuze, as well as object oriented ontologies and new materialisms, in order to decentre the human and see her in its interconnectedness with other bodies. I must say right from the outset that for this book, my main focus is the human body. My conceptualisation of it though is *posthuman*, that is to say a decentred extension of the human that finds itself in assemblages with other bodies that are not just human. This is conditioned by a human affective preponderance that has taken over the planet: we are the only species whose presence is now irreversibly documented on the texture of the earth. This, as I have already indicated, is the much-debated geological era of the Anthropocene, which in some discussions risks reinserting anthropocentrism from the back door. Instead, I conceptualise the Anthropocene as a space of human responsibility. This responsibility emerges from the ontological *indistinction* (which is not the same as equality) between human and nonhuman bodies.

On the basis of lawscape, atmosphere and bodies, the concept of spatial justice emerges. As I have already mentioned, *spatial justice is the conflict between bodies that are moved by a desire to occupy the same space at the same time.* My definition is deliberately open. Spatial justice is not the answer to the geopolitical, geophysical, property-based or population displacement conflicts, but a spatiotemporally positioned question: not what *can* happen, but what *happens* when bodies claim the same space at the same time. Spatial justice is the emergence of a negotiation amongst bodies. It is not a Habermasian dialogue, or a levelling of the playing field; nor however is it an already foreclosed outcome of structural power imbalances. It is a corporeal and spatial movement that includes but is not limited to linguistic communication. It is not an intrahuman movement but a posthuman: when a human body moves, a whole assemblage of material and immaterial bodies moves along. When a fishing boat moves into a fish-depleted area, a whole assemblage of material and immaterial bodies moves along, that includes human bodies, politics, law, finance; but also biodiversity, climate change, iron and wood and water, fish and plankton, histories and jurisdictional boundaries. Finally, it is not about equally strong bodies on a flat ontology but about unequal bodies on a tilted surface that by dint of the assemblage of which they are part, they might have an ability to pose the question of emplacement in different parameters than the ones that have thus far determined the power imbalances.

Spatial justice emerges from a movement of withdrawal from the atmosphere. We are all inside atmospheres. We are all part of the earth's atmosphere, co-producing it along other bodies. Yet we are not aware of the atmosphere. An atmosphere is the perfect enclosure. Withdrawing from the atmosphere is a movement that allows a momentary rupture: during the length of a retained, withdrawn breath, the importance of air becomes asphyxiating. While difficult to withdraw from earth's atmosphere,[3] it is only marginally less difficult to withdraw from a legally engineered atmosphere. We are all part of its emergence. We perpetuate it with our positioning, political choices, legal embodiments, body functions. Bodies *desire* atmosphere. Think of shopping malls and consumerism. Withdrawing from an atmosphere is a withdrawal from the desire of the body itself. Removing the body from the atmosphere is not enough. One needs to remove one's body from the body of its desire.

Withdrawal from atmosphere does not guarantee spatial justice. It has only one possible effect: the re-emergence of the lawscape. Breaking the glass enclosure of the atmosphere at best recalls the lawscape from within. This is better than it sounds: a body cannot negotiate when atmospherically conditioned. Atmospheres are affective events that cannot be fought headlong because there is nowhere from where to fight them. 'We are always inside an object.'[4] The lawscape, being the negotiation of in/visibilisation between law and space, allows for a certain amount of corporeal manoeuvring. It is not a matter of consciousness or awareness. Just like atmosphere, the lawscape is preconscious and posthuman. The lawscape itself determines its in/visibilisation depending on its self-perpetuating needs. Within the lawscape's struggle for self-perpetuation, human agency becomes redefined as material, assemblic agency that is as much about nonhuman bodies as it is about the human bodies that contribute to the emergence of agency. But at least there is that possibility of manoeuvring space. In atmosphere, everything is desire for the autopoietic perpetuation of the very same atmosphere.

Spatial justice, therefore, is not a prescribed avenue but merely the possible reorientation of the lawscape according to the bodies that have withdrawn. The one who withdraws has the power. This is not a passive withdrawal, it has nothing to do with removing oneself from the conflict or sacrificing oneself to a moral priority of the Other. If anything, spatial justice is a space without the Other. Withdrawal is an arduous ontological movement: a *nomadic* fight for *monadic* bodies. It is *nomadic* because it requires a spatial movement *away* from existing structures. The bodies of

---

3   Although see Negarestani, 2008, on how we are moving underneath the surface of the earth in a cthuloid movement of the war against the sun; and Colebrook, 2014, on the death of the posthuman as the ultimate withdrawal from the atmosphere.

4   Morton, 2013: 17

its emergence are *monadic* in the sense that they are ontologically singular yet in a continuum of indistinction. These paradoxical formulations originate in my reading of autopoietic theory in combination with object oriented ontologies.[5] The crux of this book's ontological take is this: while always part of assemblages with other bodies, every body withdraws. Whether this takes place through self-dissimulation, self-in/visibilisation, or strategic planning, it remains an ontological fact and even necessity, as I show in Chapter 5. Bodies are Leibnizian monads, houses without windows, autopoietic units that remain ontologically withdrawn. No body ever presents itself fully ontologically. Every body is a vastness, open to the actual and virtual combinations of assemblages, contourless, amoeba-like. A part of the vastness always withdraws. Think of human psychology, a coin, the global financial system, your mobile telephone, climate change, life, death. All are bodies, all are withdrawn. Access to a causal explanation, heads or tails, a bank account, a touch-screen, a rainy July, a breath, or a skin cold to the touch are mere epistemological concessions that offer the illusion of knowing. The more we know about how bodies operate, the less we know about bodies. Withdrawn bodies move and rest, differentiating one another on the basis of their speed, nomadically clustering together in temporary or less temporary assemblages. They associate with each other and can orchestrate their withdrawal collectively or individually. But a withdrawal is never an individual move. It always drags the world along.

1.2 Quick dip inside

Withdrawal is multidirectional. Just like all concepts in this book, it does not have an a priori moral content but a contextualised, assemblage-related, situated ethical content.[6] I have talked about withdrawal as a movement away from a legally engineered atmosphere, its potential for the emergence of spatial justice, and as an ontological feature of all bodies. The latter has two consequences: first, that the affective desire for spatial justice is not a cultural phenomenon or an outcome of the recent rather trendy spatial turn of legal scholarship, but a *conative* characteristic of the body. Popularised by Jane Bennett,[7] the term comes from Spinoza (who in turn was inspired by Hobbes) and refers to the *conatus*, namely the will of

---

5 This book does not engage with autopoietic theory for the purposes of withdrawal but builds on my previous analyses, 2009 and 2013a.
6 The difference between morality and ethics has been put most clearly by Luhmann, 2004 and 1987. For Luhmann, ethics is an internalised, second-order observation of morality that is interested in internal consistency rather than external fitting-in. Morality, on the other hand, has a colonising effect that demands social uniformity. A similar distinction has been made by Spinoza, 2000, for whom ethics is always situated and yet mediated by the univocal continuum that for Spinoza is god.
7 Bennett, 2001 and 2010

each body to carry on being and becoming, 'the endeavour or struggle to persist in being'.[8] It can be also put in autopoietic terms, as the body's necessary operation to generate itself and its own elements in order to carry on its being and becoming:[9] in that sense, every body is autopoietic. It can also be put in the way Gilles Deleuze has put it, namely as 'the right of the existing mode'.[10] There are many ways in which it can be expressed but in every case, as I explain in the text, spatial justice is a matter of survival for all bodies involved because withdrawal (as the space of unfolding of spatial justice) is an ontological necessity. Second, withdrawal does not fit in with theories such as Actor Networks Theory as established by Bruno Latour, or broadly Science and Technology Studies, which do not reserve a space for non-connection.[11] In this book, an important role is reserved for ruptures, whether they are withdrawals or dissimulations or anything else that might crop up. The point is that not everything connects, despite their assemblage emplacement. On the contrary, *because* of the assemblage emplacement, every body withdraws. For this reason, my theoretical affiliations with the above theories, and Latourian theory specifically, which otherwise could have claimed a prime place in what I write and think, are less pronounced than references to new materialisms, object oriented ontologies and to some extent speculative realism.

This comes together with what I have already referred to as the continuum. All bodies are part of the continuum, which I understand as a surface of assemblages that, significantly, has no outside. This emanates directly from a Spinozan univocity, the one substance, which for Spinoza is god or nature, and which mediates between all bodies.[12] It has then been elaborated and enriched with Nietzsche's eternal return by Deleuze and Guattari in the form of the *plane of immanence* or *consistency*.[13] Amongst others, Moira Gatens has taken up the term in explicitly Spinozan ways,

---

8   Gatens and Lloyd, 1999: 26
9   Luhmann, 2012, but also Maturana and Varela, 1972
10 Deleuze, 1988: 102
11 This might be an unfair reading since ruptures and disconnections can be found in Latour as well. However, withdrawal is not just that, but a radical movement that bridges the ontological and epistemological in that it can be strategically employed. See Harman, 2009, who has a similar objection to Quentin Meillassoux's, 2008, *correlationism*, namely that although bodies (in the present vocabulary) cannot exist without each other, this does not exhaust the bodies. At the same time, I am interested in the emotional aspects of affect, always from an ontological point of view, but as part of a broader understanding of the affective turn – again, something that as things stand, is lacking from Latour. The final reason is Latour's understanding of the law as narrowly defined, and whose materiality is confined to the operations of legal institutions. See Latour, 2009, and Pottage's, 2012, reading of it.
12 Spinoza, 2000
13 Deleuze and Guattari, 1988 and 1994

adding feminist and ethnographic insights.[14] With such rich offerings, there is not a great deal I can offer to the continuum except for the following four points: first, *the continuum is one yet several, and often overlapping.* This is a feature of most concepts in this book, namely a singularity and a plurality at the same time. Lawscape as continuum means that there is only one lawscape formed by space and law; yet the continuum fractalises into a flat surface of multiple lawscapes, various cities, homes, rooms, jungles, planets, all repeating the mechanism of in/visibilisation. Atmosphere is another continuum which, however, becomes-other atmosphere in the case of rupture with the previous atmosphere. Finally, bodies are a continuum of indistinction in terms of inherent value or moral importance. Yet, and this is the second thing I add to existing theoretical continua: *each body retains its singularity through and despite its assemblage position.* As I have already mentioned, a body is not just relations. Although always in assemblages with other human/nonhuman bodies, a body remains monadic, withdrawn from the ontological continuum, either because of its multiple simultaneous presence in various assemblages; or because of the contourless nature of the body that is determined as much by assemblages as by its own power; or by means of identity differentiation in the form of *haecceity* which is an assemblage singularity as I show in Chapter 2; or finally because the body uses the ontological feature of withdrawal strategically. A body always withdraws.

The third reason for which the present continuum differs is *the inclusion of rupture.* Since the continuum has no outside ('there is no edge! We can't jump out of the universe'[15]), ruptures are ontologically included in the continuum in the form of folds (that is, connections), in/visibilisations, withdrawals, atmospheric enclosures that rupture the continuum. Ruptures contribute to the continuity of the continuum by allowing it to gain momentum and carry on spreading spatially and temporally. But why are they needed? No outside means a vast inside that cannot be understood, handled, colonised, manipulated. This inside is untenable for humans and nonhumans alike. The continuum must be ruptured. And so we do: we split it into rooms, property, territories, packs, relations, time. We split it into human and nonhuman, races and genders, spaces of withdrawal and those of visibilisation. These are Spinozan fictions, extensions of imagination that are necessary for understanding.[16] We dissimulate the immanence of the continuum by hooking on the figurative nails on the wall the Cubists used to paint on their paintings so that people could understand them and hook on something when immersed in those overlapping surfaces of fragmen-

---

14 Gatens, 1996a and b; but see also Bennett, 2010; Hardt and Negri, 2004.
15 Morton, 2013: 17
16 Spinoza, 2009 and 2000

tation. The continuum is too much. It pushes us to withdraw from ontology and dwell in the relatively comfort of epistemology.[17] We choose to *forget* that there is no outside, as Nietzsche sings at the beginning of this introduction. Moira Gatens and Genevieve Lloyd refer to it as 'the capacity to feign'.[18] So we construct material and immaterial boundaries that separate bodies from each other; we elevate skin into a severing screen; we exclude future generations from our present actions; we put distance between us and our waste; we cover it up with our hind legs, hastily, patchily, unconvincingly. This is often adequate. We take recourse to what Teresa Brennan calls the foundational fantasy of the difference between the self and the environment,[19] or what Timothy Morton calls the rift between ontology and epistemology.[20] In this way, we carry on with the world, 'forever taking leave'.[21] *Ontology and epistemology unfold in parallel while folded together.* They remain *parallel* to each other in the Spinozan manner of non-connection yet co-emergence and co-development.[22] Spinoza's concept of parallelism informs a great deal of this book. Moira Gatens and Genevieve Lloyd have put it very clearly: 'Nothing that happens in the order of thought depends causally on anything that happens in the order of material things, or vice versa. But the 'order of thought' and the 'order of things' – again precisely because of the status of each as complete expression of Substance – are mapped onto one another in a relation of correspondence'.[23] Parallelism, just like folding, is not just togetherness but also distance. In this book, distance can only come through (the necessary illusion of) rupture. In that sense, epistemology ruptures ontology and becomes a necessary dissimulation; ontology ruptures epistemology through the eternal return to the continuum (Spinoza's *substance*). This means only one thing: that there is no difference between illusion and necessity. All ruptures are necessary; all ruptures are illusionary. I say this in full awareness of the fact that spatial justice has been described here as a rupture. This does not denigrate the concept; rather, it makes its intervention even more decisive.

In the continuum of this book, there is no space from which a judgment on these practices can be passed. They are neither wrong nor problematic as such but can be found so only with regards to their context, their own ruptured continuum, and in relation to the responsibility carried in the particular situated instance. Nor do they have to be bridged, mended or closed. They remain excessive. In that sense, I am not claiming that 'my'

17 Hubbard, 2012
18 Gatens and Lloyd, 1999: 34
19 Brennan, 2004
20 Morton, 2013
21 Rilke, 1995: 381
22 Spinoza, 2000
23 Gatens and Lloyd, 1999: 3

ontology is better than 'yours'. My access to the ontological plane is not greater than anyone else's.[24] I can only fold into it my own epistemological ruptures and see how they work, how their excess affects the rest of the plane, and how the plane becomes-other through my ruptures. At the same time, the continuum is not the same as a flat ontology. The continuum ruptures because of its excessive production. Excess is politically important and cannot be accommodated by a neutral, seemingly egalitarian flat continuum. Ruptures give rise to excess. Current posthuman discourses often situate themselves in an unproblematically flat surface, rupture-less, without withdrawals, enclosures, dissimulations. Whether this is the *life* of vitalist writing, embodiment of posthuman corporeality, flatness of new ontology, the earth of new ecologies, flow and movement of Deleuzians, even Gaia in Latour's recent work or Sloterdijk's glasshouse, none of these continua allows for rupture, conflict, partitioning, distance, withdrawal.

With this, we reach the final point of difference, to which I have already referred: *this continuum is tilted*, its flatness leaning sideways. Flat horizontality cannot miraculously flatten political distinctions, unequal spatiotemporal conditions, power imbalances. Moving beyond dialectics does not guarantee flat ontology. This is something bodies still have to fight for, and the fight is not levelled. Some bodies are stonger than others, weigh more, pull that side of the surface down and make other, weaker bodies circulate in predetermined ways. Humans and farmed animals, a pack of wolves and an unarmed human, global warming and low-lying islands, drought and bamboos, capitalist finance and the urban poor: encounters between unequal bodies in terms of power cannot be resolved through a normative flatness, but a strategic rupture. This is the point of withdrawal as a tool for spatial justice.

1.3 The chapters

The book follows a straightforward structure: the first two chapters are on law and space, with a description, explanation and empirical case on the lawscape. Chapters 3 and 4 follow the same pattern but for atmosphere. And Chapters 5 and 6 do the same but this time for spatial justice. It is a bit of an internal joke the fact that, although I keep on saying that spatial justice is not a messianic or transcendental justice, and does not demand waiting, the chapter order actually makes the reader wait. Indeed, at least explicitly, spatial justice only emerges in the last two chapters. However, the journey is precipitous, the spaces are continuous and the moves carry throughout the pages of the book. Spatial justice can only take place as a withdrawal from an atmosphere, and atmosphere is the way the lawscape dissimulates itself. At the same time, spatial justice can only emerge from within the lawscape. It is a full, yet excessive, circle. Interspersed in the text are discussions on the spatial responsibility of indistinction that take the form of situatedness, which is particularly evident in the warped temporality and spatiality of the Anthropocene. There are also discussions on time and repetition, objects as bodies, dissimulation as strategy, and the posthuman as reality. All these emerge within the ruptured continuum of this book.

Another clarification before I describe briefly the chapters. While I flirt with law and geography, I do not necessarily consider this book to be part of this literature. There are others who have done this much more systematically than I could ever do.[25] The reason is that my focus is not geography but *space*. I am not trying to establish a disciplinary dialogue but a postdisciplinary questioning, in an era where posthuman, anthropocene, assemblages, objects and bodies have entered the quotidian vocabulary. At the same time, this is not a philosophical work, despite the references and occasional tenor of the text. The reason is that I am interested in the applied and empirical side of concepts as much as their conceptualisation. While what I call empirical here would not dare compare to the 'real' empirical studies on law and space out there, it attempts to ground concepts on non-conceptual situations. The empirical fields I use here are the London lawscape, the Triveneto shepherd atmospherics, and Robinson Crusoe's literary island. Each field grounds, confirms, occasionally contradicts but always extends the prior conceptual discussion respectively on the lawscape, atmosphere and spatial justice, and pushes the discussion onto the following step.

Every chapter opens up a different lawscape. Chapter 1 is still in some respects introductory, and begins with its feet firmly on the ground of current disciplinary attempts to reflect on space and law, but begins to swirl dangerously when the question of the law's spatial turn arises. I question the ability of the law to turn and what it would mean not just for the discipline but also for space. I try this out, first by pointing at important 'micro-turns' that have happened in areas connected to law and space, and which remain a deep inspiration for my work here; I carry on with a brief critique of some of the existing literature for its rather simplified understanding of what space is and what spatiality can mean for the law; and then by thinking about metaphorical uses of space, which although not problematic in themselves, do not advance a material understanding of either law or space. Materiality, however, is not the opposite of abstraction. In the final section of the chapter, I deal with abstraction as a necessary component of law and space theorisation on the basis of its political and strategic relevance.

---

25 Delaney, 2010, is the obvious candidate for this. See also Braverman *et al.*, 2013.

Chapter 2 welcomes you to the lawscape. Bathrooms and airports open the discussion, which moves on to emergent conceptualisations of space. It quickly becomes obvious, however, that any attempt at isolating space from either law or bodies is an epistemological crutch which, at this stage, I find unnecessary. I plunge therefore in various assemblages of law, space and bodies, roughly clustered around the term *manifold*, that include feminist readings of labyrinths, material agencies, psychoanalytic readings of marble, Australian indigenous songlines, Deleuzian laws and spaces, and so on. Oddly but intentionally, I have folded the methodological choice of this book in this part of Chapter 2. The reason is that from this point onwards, the concepts emerge fully and follow closely what I describe as a *posthuman* epistemology. The chapter moves on to a discussion of the lawscape, with extended explanations of what I understand as law and space for the purposes of the lawscape. The definition of the lawscape is based on the interplay of in/visibilisation, which, however, is not visuality. I am keen to avoid this association, since the lawscape's in/visibilisation has to do with ontological presence and not phenomenological and generally experiential vision. Section 4 focuses on the three qualities of the lawscape, namely the fractal lawscape, the immanent lawscape and the posthuman lawscape. Section 5 brings time in the discussion and spatialises it as repetition that produces difference across the lawscape continuum.[26] Finally, the last section takes a walk in London with my students, as part of an attempt to have an 'encounter' with the ontology of the lawscape through the epistemological avenue of walking.

Chapter 3 opens up a comfortable room for you to sit in and read this text, complete with music, aromas and games. This room puts forward the concept of affect as the symbolic, sensorial and emotional flow between bodies. Affects generate atmosphere. An atmosphere is described here as an affective ontology of excess that takes place in, through and against bodies. Atmosphere is lawscape dissimulated, brimming with enclosure, illusionary constructions of outside and a velocity that surpasses that of individual bodies. Atmosphere is elemental, engulfing while partitioning air. This is the focus of Section 2 of the chapter. Atmospheres are not just cosy events; they can be oppressive as well as conflictual spaces. In all cases, atmospheres can be engineered in order to look spontaneous and rhizomatic, rather than directed and predetermined. In Section 3, I list the main

---

26 Although the focus here is mainly on spatial considerations for emphatic reasons of balance redressing, there is no denying that time must be taken into account in an equally radical manner. Doreen Massey, 2005, has performed this balancing act admirably, but I find that, at least in law, there is a greater need for a strong statement on spatial relevance and a temporary withdrawal of time from the scene, until space manages to present its case. This does not discount the temporal arguments, and indeed I dedicate parts of the discussion to the temporality of the various concepts. It is just that they will have to wait for a little while longer.

steps of engineering an atmosphere and in Section 4, the coda, I turn the room back into the lawscape.

Chapter 4 finds us in the valleys of Triveneto mountains following the local transhumant shepherds and flocks, namely animal/human assemblages that are fully nomadic without even a semi-permanent base. This is a unique phenomenon in the world and the perfect site for extending the discussion on atmosphere. The assemblage moves while standing still, creating its own atmospherics of pause, dissimulation and withdrawal, removed from the intense local regulatory regime of private and public property, zoning, highways, as well as other flocks. In Section 2, I discuss this atmospheric withdrawal while looking at issues of space, territory, property and identity. In Section 3, the possibility of plural atmospherics is discussed, namely whether atmospheres can come into conflict with each other. The final fourth section focuses, focuses on how an atmosphere can be at the same time an engineered event *and* an emergence between and through bodies, and the answer is, once again, dissimulation.

Chapter 5 begins with a rather banal instance of imaginary conflict in the space of a concert hall. In the context, the concept of spatial justice is defined as the question behind the desire that makes bodies occupy a particular space at a particular time. Before that, Section 1 looks into some of the current literature on spatial justice and offers a critique on the basis of its aspatiality. Section 2 builds on the discussions on spatiality in this book in order to sketch movement and pause as a non-dialectical fold in which spatial justice emerges. Section 3 describes spatial justice as a rupture in the continuum and looks into ways in which this rupture takes place, notably as collectivity, as well as illusion or conative necessity. Section 4 goes through the motions of spatial justice as withdrawal (always a rupture) and returns to the lawscape of the concert hall.

Chapter 6 is a cluster of superimposed islands: initially Robinson Crusoe's, then Friday's and finally an elementary, solar island. For this chapter I am reading Michel Tournier's novel *Vendredi*, which is a rewriting of Defoe's Robinson Crusoe. I am putting it together with a text Deleuze wrote on this very novel. From these two texts, I extract a lawscape that keeps on becoming reorientated by the various bodies populating the island. The first section looks at the obsessive lawscaping of the newly shipwrecked Robinson. Section 2 stays briefly on the second island that is generated by a strange turn in the events. Section 3 begins the repetition of the island as difference, leading to a withdrawal which leads to spatial justice, but also a withdrawal *of* spatial justice for another, elementary, meteoric justice.

Chapter 1

Law's spatial turn

1.1 Points of turning

In the last couple of decades, the law has been discovering its own spatial credentials. This can mean many things. One is that the law understands its spatial relevance. Another might be that the law understands its spatial provenance. On another level, but related to the above, the law as a discipline has been constructing its own space, carving out chunks of disciplinary materiality and piecing them together to simultaneously become more relevant and more interdisciplinary. This means that 'the law' has fleshed out its multiplicity, making the whole endeavour of talking about 'the law' an impossible feat. Through space, the law has discovered its own situatedness, and with this its own fragmented terrain. These are much more than simple metaphors, although they are often perceived as just that. The law now constructs itself as a location in a biopolitical net of spaces, awakening to what Michel Foucault in his oft-quoted 1960s lecture *Of Other Spaces*, called the 'relations of proximity between points'.[1] Legal theory is progressively more comfortable with concepts such as mapping, scale, territory, boundary, and other geographical terms. More specifically, sociolegal and critical legal scholarship have turned to concepts and practices of emplacement, consideration of local conditions, geographical peculiarities of the case in question, and so on. We have reached a point where we can comfortably say that *the law has turned spatial.*

This last sentence sounds deliberately grand. If it has grabbed your attention, then it has achieved its goal. We can now proceed to the sobering caveats: *first*, that the law *can* indeed turn; *second*, that the space towards which law has turned remains adequately spatialised; and *third*, that this might mean something, not just for legal theory but for law's connection to actual spaces, in their material and embodied dimensions. I shall refrain from passing judgment on the first caveat until later; but let me say here that for the law to turn, an understanding of law as spatial through-and-

1   Foucault, 1986

through is required. This understanding must take place both internally, that is in the way law sees itself, and externally, that is in the way others see law. Nothing short of a U-turn, in other words. Whether this has taken place in law, it remains to be seen. Let me also reserve opinion on the third caveat, namely whether law's spatial turn might have an actual impact on spaces, objects and bodies, rather than its being a merely theoretical advancement. An answer to this can only be given in the course of this book, and only in relation to what I consider the most important yet most neglected aspect of law's spatial turn, namely spatial justice.

What I can do here, however, is focus on the second caveat, namely whether law's spatial turn is sufficiently spatial. In order to answer, it is important to take into account the genealogy of law's spatial turning, and examine, however briefly, some vantage points through which the connection between law and space has been introduced. Emphatically, this is neither a quest for the origin of legal spatiality, nor an exhaustive history of the turn.[2] On the contrary, it is an attempt to show that there is no single origin, no single turn, and that the turns are refracted in space. It is very important, I think, not to essentialise the connection and to resist treating the below as originary moments for law's spatiality. One way to avoid doing this is by presenting the various vantage points in an unprioritised order. Indeed, I do not consider any one more decisive than others, even if some have appeared temporally prior to others. Another way is to offer a non-exhaustive list, which is indeed what the below is intended to be. Resisting awarding originary or exhaustive status to the below is faithful to the understanding of space that this book adopts. While I return to this, it is relevant here to mention, somewhat axiomatically, that *the space of law's spatial turn is non-Euclidean, non-measurable, non-directional, non-unitary, non-linear and non-metaphorical.* This does not mean, however, that it is a negative space. This space is perfectly actual, walkable, equally available for picnics and imprisonment, appropriations and exclusions, colonisations and decolonisations. It is a manifold space open to opportunities and controlled by compulsions, arranged by emotions and constrained by senses; it contains various orientations but not always prescribed directions; it is relational but accommodates differentiated power balances between the various bodies that circulate in it; it is not fully relational in that it allows for bodies to withdraw ontologically; it is reversible yet cannot easily be controlled; and it is fully material, confident with the use of spatial metaphors but going deeper than them.

Before we begin, let me offer a brief contextualisation. Although instances of space and law's mutual production have been consistently even if not always explicitly recognised, they have been historically kept apart. There are several reasons for this: the prioritisation of time over space, largely due

---

2   A thorough and amost exhaustive treatment can be found in Braverman *et al.*, 2013

to the pure abstraction of the latter;[3] the not-so-well concealed necessity to dissimulate the specific geography of the origin of the law, for otherwise the law could not claim its placeless universal applicability; or, conversely, the need to disengage law from any spatial geographical (imperialistic, nationalistic, colonialist, supremacist, capitalist) bias. But here, in this book, I begin in bidisciplinary trauma. Now (and *here*) we know that law and space are consistently folded into one another resulting in hybrid epistemologies,[4] simultaneously unfolding onto ontological positions. It is not an exaggeration to say that, epistemologically speaking, law's spatial turn is now an established mainstream sociolegal, interdisciplinary and critical field of research, branching out on a multiplicity of applied and theoretical perspectives, and constantly evaluating the law from various viewpoints that are linked to critical and human geography, embodiment, globalisation and localisation, ecology, population movements, climate change economies, asylum seekers, human rights geographies, natural resources and environmental justice, rights to the city, monuments and memorials, and so on. In short, law's spatial turn has managed to involve a considerable volume of sociolegal, critical and generally contextualised legal theory. A final caveat before I go through some of these points: the criterion for inclusion is that they are movements *in after* space, contained yet both incorporating and exceeding the space of their unfolding. None of the below, at least at its better manifestation, unfolds itself with space as mere *background* or *container*. Rather, they all engage with space in more or less successful ways, and attempt to *perform* spatiality by producing their own spatial positionings. From these, I shall be extracting some notions and movements that in turn inform a great deal of this book.

Arguably the most immediately relevant vantage point is the feminist understanding and use of urban space as an instance of critique of a specifically gendered sexual reality.[5] The relation between space, place, bodies and the law has been explored by feminism as part of a greater identity project that reverses the usual prioritisations of the male, the mind, the public domain, time and reality, in favour of new semiologisations of their relation with the female, the body, the private, space, place and the imaginary. The feminist vantage point has contributed to law's spatial turn a long list of valuable possibilities for thought and action. Not without a certain productive hesitation, I would suggest that one of the most important and personally formative element is the bringing forth of ontological vulnerability.[6] Both as self-awareness and as a condition of existence, ontological vulnerability opens up a new legal space that transcends the distinction

---

3 Harvey, 1990: 214
4 Pue, 1990; Cooper, 1998; Blomley, 2002
5 Indicatively, Rose, 1993; Cooper, 1998; Butler, 1993; Grosz, 1995; Little, 1994; Greed, 1994
6 See especially Grear, 2011 and the whole journal issue there; Albertson *et al.*, 2013.

between private and public and establishes a demand for the law to face and protect the vulnerable. Although originating in and intimately connected to gender studies, it is not an exclusive female predicament, nor necessarily connected to gender differences. For, at the same time, the space of ontological vulnerability places the awareness of human acentrality and the subsequent species vulnerability squarely within the legal space of turning. Vulnerability is what in new materialist terms has been called *fragility of things*.[7] This is the beginning of the *posthuman*, with immediate connections to a new understanding of space, denuded from its usual familiar quality of measurability.

Intimately connected to the feminist interweaving of law and space is the queer theoretical and applied understanding that privatises and/or sexualises the public in ways that radically challenge the law and its own conceptual and geographic boundaries.[8] While queer spatial reading has also contributed to the bringing forth of vulnerability, what perhaps has been the single most important contribution to law's spatial turn is the process, or perhaps the play, of *dissimulation*. Queer spatialities begun with layering and mapping the infravisible routes of desire, affording the reader access to spatial codes and corridors that were constructed precisely to escape the heteronormative gaze. This necessity of dissimulation has often been performed as a playful spatial enactment of desire in queer spatialities. The movement has been one of *withdrawal*: move underground, in public toilets, dark rooms, municipal baths.[9] Or withdraw deeper, in folds of the private in which law would not reach: indeed withdraw from the law that stopped the acting out of desire, and into a different space of a different legality. As we shall see later, the leap is a short one, from dissimulation of the performance of desire, to dissimulation as a strategy for law's spatial turn; and withdrawal for another law, to withdrawal as a gesture that leads to spatial justice.

A spatialised biopolitical understanding of identity is another route to the spatiolegal interweaving, either through a phenomenology of (limitation and facilitation of) movement,[10] a sensualisation of the everyday of law,[11] excretal, carnal, aural and other interconnection with the human,[12] aesthetic engagement,[13] a criminological analysis of space,[14] or an exploration of the 'cognitive unconscious'[15] both in legal and ethical terms as

7 Connolly, 2013
8 Moran, 2002
9 E.g., Prior and Cusack, 2008 and Prior *et al.* 2013
10 Sennett, 1970; Finnegan, 2002; Butler and Parr, 1999
11 Lash, 2001; Valverde, 2003; Turner and Manderson, 2007
12 Longhurst, 2000; Hyde, 1997
13 Blomley, 1998
14 Moran *et al.*, 2003

spatial *conatus*, namely the Spinozan force that pushes life towards its maintenance, affirmation and even excess.[16] Such routes are particularly relevant to law's spatial turning, because they have the potential to distract the quest from its more usual identity focus, as it remains safely couched in its well-rehearsed sameness/difference discourse, and refocus instead on monadological yet contextualised conceptualisations of identity, that open up to a temporal and spatial *situatedness*.

A slightly indirect point of connection between law and space is the political appropriation of the legal. This theoretical exploration begins with social participation *à la Situationiste*,[17] and suggests ways of mobilising the political potential of space, thereby recalling basic notions of the triangulation between the spatial, the legal and the political, such as the city as *polis* and the ideal conversion from *urbs* to *civitas*. Politics here stands also as a symbol for the engagement of the spatial with society at large, especially as an actor that transcends nation-state sovereignty, such as the example of the Refugee Cities project,[18] or theoretical instantiations of the legal autonomy of *polis*.[19] Another connection between law and space has been through law's exceptionalism, either in the form of critique of such spatiolegal exceptions as the 'war on terror' and Guantanamo Bay,[20] or in more everyday manifestations of deprivation and displacement, often following a methodological frame inspired by Carl Schmitt/Giorgio Agamben, which naturally connects to questions and critiques of sovereignty.[21] The political is regularly used as a mediator between the legal and the spatial. While this reveals the political character of both law and space, it also poses a difficulty when it comes to circumventing the political in order to focus on the legality of the *polis*. It would seem that a conceptual tying down of the law is harder than the equivalent move embodied in the distinction between *the political* and *politics*. This difficulty paves the way for a reimagination of the spatiolegal as *tautology* between law and space, rather than a spatialised distinction.

Connected to the last point is the way the spatiolegal has begun appearing in land law/property law issues. Recent literature on this displays an impressive array of both consistency and variety: Anne Bottomley's work on

---

15 Lakoff and Johnson, 1999
16 The excess comes from Deleuze's 1988, understanding of conatus.
17 The 'unitary city', as Debord, Baudrillard, Lefebvre and others were advocating. See McDonough, 2004, for the original Situationist texts.
18 Derrida, 2001
19 Nancy, 2006: 53: 'The *polis* rests firstly on the fact that it gives itself its own law [*loi*]. It can invoke a prescription or a divine guarantee for this law; but it is to the *polis* itself that the determined establishment, formulation, observation and improvement of law belongs.' See also Bonomi and Abruzzese, 2004.
20 Strawson, 2002; Motha, 2006
21 Wall, 2012; Joyce, 2013

co-operatives and the shopping mall,[22] Sarah Keenan's work on Australian aboriginals and settlers' law,[23] Sandy Kedar's work on settlers states and legalised dispossession,[24] Chris Butler's Lefebvrian analysis of property,[25] Anna Grear's work on human rights and property,[26] Irus Braverman's work on Israel and Palestine,[27] Antonia Layard's work on issues of property and right to the city,[28] Sarah Blandy on gated communities,[29] Helen Carr's work on social housing,[30] Olivia Barr's work on movement to place:[31] these are examples of the richness of approaches that have managed to supplement some solid examples of existing literature, such as the works of Davina Cooper and Nick Blomley amongst others,[32] and together leave behind the old property law adage, to be found in every good and semi-good property law textbook, that land is not a thing but relations over the thing. This is now being rethought through a new legal materiality that unfolds in those spaces and through those bodies that traditional property law has managed to circumvent. Why might the law have felt the need to circumvent these? We have just touched upon law's most established self-dissimulations, and one that has ferociously resisted spatial turning: namely, that law's universal applicability is the only guarantee to achieve justice, itself equally distributed. Fortunately, this is not always taken on face value. TWAIL (Third World Approaches to International Law) and postcolonial scholars for example have seen through this façade and have understood all too well that universality does not mean justice. Yet the allure is too strong and even critique is often addressed to the *lack* of universality of the law. Recently, Luis Eslava and Sundhya Pahuja have done away with this façade too, by calling on postcolonial legal scholars to look at international law as material, spatial and agonistic.[33] But this is not easy. Space stops law from flying free from geographical shackles, and turning makes justice's blindfolded head swim. Isn't this against the law's self-proclaimed universal application? Isn't this a problem for the rule of law? But here, rhetorical dreams and drab realities are conflated. So what if the aggrieved party in a property law dispute is a woman? We have gender equality, and it would be patronising to differentiate: 'the conferment of special protection on married women, differentiating them from other sureties, is unacceptably

22 Bottomley, 2009 and 2007b
23 Keenan, 2010 and Keenan *et al.*, 2013
24 Kedar, 2003
25 Butler, 2013a
26 Grear, 2012
27 Braverman, 2007 and 2012
28 Layard, 2012
29 Blandy, 2011
30 Carr, 2013
31 Barr, 2010
32 Cooper, 1998; Blomley, 1998 and 2004
33 Eslava and Pahuja, 2012

patronising and wholly inconsistent with modern notions of the status of women'.³⁴ Self-aggrandising legal rhetoric or law-undermining embodied reality? Law as universal and abstract has achieved gender equality. Law as emplaced and embodied is far from achieving anything like this. The rhetoric carries on turning away from the uncomfortable materiality of space.

Possibly because it is still early days, or because the spatial turn has already done the groundwork, materiality has breezed in legal theory with much less bruhaha than spatiality. Bruno Latour's work on the *Conseil d'État*, although not directly spatial, delimits a judicial space in which legal materiality is observed and criticised as lacking.³⁵ Emilie Cloatre's work on law's material dimensions has introduced this very specific genre of contingency that comes from the ineffability of matter from within the law, and which adds to the spatial contingency.³⁵ Alain Pottage adds to this materiality by showing how Latour has not quite got it right: the law has always been material and indeed much more material than even the law itself allows to admit.³⁷ This is not a theoretical point. The regular moot point of whether the law is disconnected from material reality, also seen on an epistemological level as legal closure, is largely misplaced: in practice, law has never been more than an interdisciplinary or even postdisciplinary snapshot of a heady mix including geography, history, psychology, chemistry, physics, economics, the media, religion and so on. Even a simple case in a lower court necessitates a disciplinary outing. It is not, therefore, surprising that legal materiality has also been connected to environmental issues,³⁸ consumerism,³⁹ surveillance,⁴⁰ and legal agency,⁴¹ all of which presuppose a spatiality that is intimately acquainted with law.

Although not always understood explicitly in terms of spatiality and law, postcolonial literature has allowed law to turn spatial in distinct ways. Perhaps most relevantly, it allows the law to come to terms with a situated history and the processes in which history is being imprinted on the space of conflict. Nicolas Wey Gomez's expansive work on *The Tropics of Empire* has shown how space emerges already constructed even before its official 'discovery'.⁴² Through an archival research that spans centuries, Wey Gomez traces Columbus's desire (a desire that appears faithful to and representative of contemporary spatial and social thinking) to establish heaven on

---

34   Battersby, 1995: 36
35   Latour, 2009
36   Cloatre, 2008; Cloatre and Wright, 2012
37   Pottage, 2011; Visman, 2008
38   E.g., Grear, 2011
39   E.g., Layard, 2010
40   Bigot, 2006
41   Pottage, 2012; Philippopoulos-Mihalopoulos, 2014
42   Wey Gomez, 2008

earth on what was considered the 'torrid zones' of the globe, which relied on the simultaneous assertion of moral inferiority of the indigenous. While these forms of colonisation remain, though, for example, conditions of ghettoisation in urban slums across the globe, postmodernity has been characterised by a slightly different move: that of flatness. Tayyab Mahmud shows how the idea of a flat world trammelled by flows of global information marginalises both dominance and resurgence.[43] For Mahmud, this is the way geography, and law's turn to it, has been safely despatialised, eliminating any grounded element of difference. This is a very contemporary problem, as I show in the following section and plays on various levels of dissimulation. Valerie Kerruisch's work, for example, critiques the judiciary for implicitly asserting still the doctrine of terra nullius while acting from a political context which, at least formally, has condemned the doctrine.[44] Colonialism already marks the places of its contact and its arrival, as Brenna Bhandar shows.[45] But these markings cannot be captured by current dealings of sovereignty – perhaps the main engagement of legal postcolonial theory. Sovereignty, it would seem, is already part of the problem. We see instead the rise of a new use of téchne in the sense of new legal, post-sovereignty technologies that allow the complexity and assemblage-like nature of spatial claims to come forth. What is more, these new technologies manage to deal with the lingering colonial inscriptions on space. For colonialism also marks the spaces of its departure. Imperialism remains present in the postcolonial project of 'development', as Radha d' Souza writes, and the tools used to assert colonial dominance must now be adopted in order to overturn the very dominance: 'law, science and technology, and space as conceptual categories that are integral to understanding geo-historically constituted social structures.'[46] This is not a simple invitation to contextualised thinking. D' Souza points out the complexity of such an endeavour both from within the postcolonial space and from without, further complicated by the various layers of interpretation, exoticisation, orientalisation and so on. Rather, d' Souza writes: '"seeking" in the Indian/Eastern tradition, far from a mystical enterprise that "Western" traditions sometimes make it out to be, has a distinctive *methodology*. The methodology may be described as *"following the question"'*.[47]

As emphasised at the beginning of this section, these are not originary events of the turn but rather interconnected micro-turns that, in their own manner, perform the space in which they unfold. The unprioritised, non-exhaustive listing, although not random, tries to avoid passing judgment on whether they can be considered the origin of law's spatial turn (short

43 Mahmud, 2007
44 Kerruish, 2002
45 Bhandar, 2009
46 d'Souza, 2006: 16
47 d'Souza, 2006: 46

answer: no), and on whether there consequently exists a clearly established *direction* in the turn (there isn't). What it reveals instead is that *law's spatial turn is a series of turning spatial.* These are fractal instances of a larger turning that have bequeathed us with the following tools: a situatedness that exposes, a vulnerability that strengthens, a withdrawal that initiates new spaces of law, a dissimulation that affirms, a self-preserving ability of materiality that is pervasive, an embodiment that orients law where really needed, and a methodology that follows the question rather than the need for answer. All these set the stage for what I later call a *manifold understanding of the lawscape.* This space of law, this *lawscape,* is not devoid of contradictions, conflict and opposing positions. In this sense, it is fully *legal.* With this in mind, I would like now to proceed to a brief review of the current literature on law and space proper, and attempt a simple categorisation of the various connections. Unsurprisingly, what emerges from such an attempt is a fear of space on the part of law and legal scholarship more specifically. This throws even further into doubt the self-aggrandising sentence that opened this section, namely that the law has turned spatial, and respectively fleshes out the second caveat, namely that the space in which law is turning is not sufficiently spatial.

1.2 False turns

In turning spatial, the law has engendered a paradox: despite the prominent connection between law and geography, law's engagement with *space* is being increasingly despatialised.[48] The focus of this section is the problem of marginalisation of space in law, for which geographical terminology and references, however profuse, do not compensate.[49] This, I argue, can be attributed to a fear, on behalf of the law, of the peculiarities of space and its import. Fear is meant here as an anxiety that trammels law's understanding of itself and its usual textual boundaries.[50] This fear of space risks

---

48   If indeed it has ever been 'spatialised'. As Doreen Massey, 2005: 17, writes '[spatial conceptions] are unpromising associations which connotationally deprive space of its most challenging characteristics.'
49   Exceptions of course are both many and luminous, only a small part of which can be referred to in this chapter. It is, however, interesting but beyond the scope of this chapter to look at it from the perspective of different 'jurisdictions'. Thus, an imperialist Anglo-American scholarship has been found to marginalise at least to some extent other scholarships: Nicholas Blomley, 2008: 290, notes the relevance of this for critical geography 'that aspires to internationalism and solidarity, reflexivity, and the analysis of power'. One notes, for example, that Francophone or Italian spatial/legal scholarship systematically goes beyond this fear of abstraction, while at the same time dealing with the issues in hand. On this, see Soja, 2009; and indicatively, Brighenti, 2013; Cacciari, 1997; Farinelli, 2003; Nancy, 1999; Xifaras, 2004.
50   In other words, fear is always fear of fear. I have dealt with this in 'Fear in the Lawscape', in Philippopoulos-Mihalopoulos, 2007d. See also Massumi, 1993.

turning what for law can be a defining self-reconfiguration into a comparatively narrow disciplinary excursus.

Let me clarify from the outset that there is nothing reprehensible even if law's spatial turn is simply a cross-disciplinary adventure that experiments with geographical terminology and attempts to situate the law in its geographical context. Whether as terminological flirting or geographical input, it remains an indication that the law is leaving its ivory tower behind (and slowly moving towards its tower of Babel). This move is welcome in terms of at least the legal discursive closure and its persistent exclusion of any interdisciplinary contingency. But the same move will have to be seen with suspicion if it remains a token that merely mollifies instead of becoming the epistemological and ontological *revolt* that a spatial turn signals. To put it even more metaphorically, a spatial turn is indeed happening in the law, but the law bargains its turning in ways that move away from, rather than within spatiality. These are the questions that the present section attempts to ask: can the law 'turn'? And if so, why is it that law's spatial turn seems to be turning away from space?

I argue therefore that law's engagement with space could reach further than either terminology or specific geographical emplacement. Space adds itself as a singular parameter to the hitherto legal preoccupation with time, history and waiting, and forces law into dealing with a new kind of contingency: one that emerges from the peculiarly spatial characteristics of simultaneity, repetition, disorientation, materiality and exclusionary corporeal emplacement. Law's spatial turn presents two opportunities: first, to rethink law's spatiality, namely the novel unpredictability of space that has now flowed into law, at the same time both grounding and ostracising. And second, to reclaim the concept of spatial justice from a socially diffusing, geographically applied concept of regionalism, while also making a case for it from within the law – for it is indeed the case that the latter is generally and quite unjustifiably the grand manqué of spatial discourse, widely thought to be adequately represented by the political discourse.[51] While I focus on spatial justice in Chapters 5 and 6, the discussion on law and space, whether in the form of the lawscape (Chapter 2) or atmosphere (Chapters 3 and 4), lays the base for a spatial understanding of a legal justice.

Spatial semiotics is by now *de rigueur* in several discourses whose analytical depth has been enriched by references to mapping, scale, horizon, domain, field, space/place, boundary, crossing, topology, and so on. Law's text has found its context in an ambiguous terminological strip that often remains metaphorical and in this way it enables the law to carry on judging

---

51 This is the reason, I believe, for which potentially important books for law and space (such as Massey, 2005; or Soja, 1996 and 2010) do not make any reference to law but simply to a generalised, politically mediated normativity.

without exposing itself and its own disciplinary closure too much. Of course, metaphors are not inferior to whatever the thing behind the metaphor might be. Metaphors are often the only means of overcoming the problem of disciplinary boundaries, thereby mutually revealing the other side.[52] At the same time, however, metaphors can become too comfortable.[53] They start working against the objective of confluence, facilitating instead a distance between, in this case, law and space, by appeasing the discourse with small chunks of meaty allusions. Metaphors remain a part of the legal discourse, which is far too integrated to allow law to exceed its boundaries and connect with the radical opportunities of space.

It is my firm intention in this book to move beyond metaphors.[54] The main reason is this: the need to consider a material ontology on the level of the law which goes beyond (while not excluding) the textual comfort zone of the law. But formed in a non-metaphorical way, law's spatial turn can be disturbing on many fronts. It obviously disturbs a certain positivist concept of the law as immaterial, universal and abstract. It also disturbs a more accepted sociolegal understanding of the law as grounded, empirically proven and geographically situated. It finally disturbs a certain critical concept of the law as particular and embodied. While the first is not surprising, the last two do sound odd. These two kinds of legal literature are justly assumed to be better equipped to deal with a spatial, material influx. It is after all through sociolegal and critical legal thinking that spatiality has been introduced. Yet, as I argue below, something has been left out. No doubt, when in the early 1990s Nicholas Blomley presented us with the powerful challenge to put law and space together on a solid philosophical and sociolegal footing,[55] things had to focus in specific, immediate needs. This was followed by some equally powerful attempts to respond to the challenge.[56] But the majority of the literature following that, especially the literature emerging from legal thinkers (as opposed to geographers looking into law) seems to be more and more indifferent towards a theoretical understanding of space for law, and consequently falling into a few rather too comfortable patterns. I will schematically list three types of pattern, in full consciousness of the unjust violence of such a categorisation.

The first way of putting law and space together is by constructing space in a narrow, legalistic way as jurisdiction.[57] Jurisdiction (space) may change eventually (time) through juridical developments or disputes. But in this formulation, space remains fixed, unchanging and simply following its

---

52 Cooper, 1998
53 Warf, 2009; Massey, 1994
54 On the metaphorical structure of this sentence, see Gunnar Olson's talk on spatiality of metaphors, in Merriman *et al.*, 2012.
55 Blomley, 1994
56 E.g., Delaney, 1998; Cooper, 1998
57 Blacksell *et al.*, 1986

traditionally more alluring antipodes of time.[58] A surprising amount of literature is still characterised by what I would call 'the parochial turn', namely the turning towards a conveniently constructed canvas that confirms hypotheses, barricading itself behind a geographical emplacement and never considering the spatiotemporal *continuum*. This criticism has already been made on the level of geography, namely the global as opposed to the localised city or countryside,[59] and demonstrates how the question (although not exhausted by globalisation) has repercussions that reach beyond the purely theoretical.[60] Space must be thought of as the ruptured continuum between the global flow and the territorial structure,[61] the tangible and the distant, the particular and the universal: or, as Austin Sarat and Thomas Kearns put it, 'the immediate and the familiar juxtaposed to the distant, strange, and cosmopolitan ... origin and home, point of departure and place of return.'[62] Space as world, or 'the opening of space-time' as Jean-Luc Nancy puts it,[63] is neither just the global nor the local, but the continuous space of immanence and questioning in which the law finds itself and which the law constructs.[64] In avoiding seeing how the law is part of the spatiotemporal continuum, the law indulges a double fear: fear of the resistance generated in the world; and fear of itself, the law, in that any look in the mirror may reveal mismatches between appearance and self-understanding, means and mission, force and justice. These fears are opened up and dealt with when the world, rather than jurisdiction even in its globality, is considered to be the space of the law.

The second approach constructs space as a process – thus, seemingly the opposite of the first construction. Closely following geography's understanding of space as relational,[65] here space is constructed as fluid, dynamic, ever-changing, a veritable *accueil* of difference. Space is idealised

---

58 See however an attempt to form a flowing concept of jurisdiction by Richard Ford as 'simultaneously a material technology, a built environment and a discursive intervention', 2001. See also McVeigh and Pahuja, 2010, on creative use of jurisdiction, and also Dorset and McVeigh, 2012, on the limits of jurisdiction. On time and space, see Soja, 1990; Harvey, 2001; Pred, 2004.

59 A valid point itself, but not conclusive, as Massey effectively argues in her *World City*, 2007, a book about London and the world.

60 'On the one hand space and places are increasingly the product of global flows; on the other hand we work with a politics both official and unofficial that is framed by a territorial imaginational and formal structure.' Massey, 2007: 14

61 Pearson, 2008; see also Brighenti, 2006, for a countertheorisation of space.

62 Sarat and Kearns, 1995: 5. Sarat and Kearns refer to the concept and practice of everyday, but its distinctive spatial language and its import make it, at least for my purposes, tautological with the spatial.

63 Nancy, 2007: 73

64 Delaney, 2002. The concept of the world, and consequently of Nancy's *mondialisation* or Luhmann's *Weltgesellschaft*, is different from the usual terrain of globalization. See, e.g., Stäheli, 2003.

65 See Jones, 2010 for a sympathetic critique.

as a panacea for social injustice, casting anathema to time and history.⁶⁶ The law clings onto this new 'ideal' space and delivers itself from its normative obsession by allowing the spatial influx to operate as law's new clothes. While this is arguably preferable to a parochial turn, it is still not adequate. First, it idealises space in ways that space cannot sustain. To put it rather bluntly, only law can deliver law from its obsessions.⁶⁷ Second, it ignores the always-already spatial conception of the law,⁶⁸ the materiality of law and its inevitable emplacement in space.⁶⁹ Third, it marginalises the disorder, fragmentation and unpredictability that come with space, in favour of a limpid and linear spatial construction – for even as process, space and its modes of production are thought of as given to prognosis and essentially inert. In that sense, the potentially radical nature of space becomes institutionalised, co-opted as part of the institutional discourse, streamlined to serve the purpose of the system. Or even worse, space becomes law's ideality, an instrument for law's escape from itself. The consequence is perilous: the law simply carries on feeding its own sense of superiority, 'its importance, history, and its disciplinary identity',⁷⁰ further swallowing up the supposed factuality of space for purposes of its own imagined co-extensiveness with an imaginary social totality. Thus, law's all-inclusion has mastered even space. This is an odd over-indulgence to an idealised state of law's continuum that accepts no interruption, indeed no *rupture* and no *withdrawal* (more of which in this book). In many ways, this is the opposite of the first category.

Finally, the third category of space and law literature is characterised by the phenomenon of 'add space and stir'.⁷¹ This approach reduces space to yet 'another' social factor, 'another' perspective which does not offer anything more than at best a context and at worst a background. This is probably what Lefebvre wanted to avoid when he wrote 'space is not a thing among other things, nor a product among other products: rather, it subsumes things produced and encompasses their interrelationships in their coexistence and simultaneity – their (relative) order and/or (relative) disorder.'⁷² If the spatial turn exhausts itself in considerations of background, thus failing to function as the epistemological ground on which

---

66 Thus succumbing to what Lefebvre, 1991, has called the fetishization of space.
67 This is a poststructural interpretation of Luhmann's ideas. See Philippopoulos-Mihalopoulos, 2009
68 This connection is not a new thing for law but a coming-forth of what has already been there and whose origin can only retrospectively be located. In 2002, Blomley was writing about how he had been working on the subject for 'nearly two decades', 2002: 17.
69 Philippopoulos-Mihalopoulos, 2007a; see also Koskenniemi, 1989 on how law's universal claim can only claim validity if the particular resonates in it.
70 Blomley, 2002: 21
71 Ellem and Shields, 1999
72 Lefebvre, 1991: 73

such 'coexistence and simultaneity' can be demonstrated, then we can more accurately talk about geography rather than space. Geography, the imaging of the world, the *grapheme* (-graphy) of the earth ('geo-'), is a representation.[73] As such, it reveals but also conceals its reference, namely space itself. As David Delaney puts it, geography 'seems *to stand for* spatialities, places, landscapes, materiality, and the thick and sensuous domain of the visible.'[74] Geography indeed *stands for* all that, itself an epistemological avenue through which some of these things are sketched. And geography converses with the law – but what *stands for* law? Is the distance between law and talking about the law (e.g., in the form of legal theory) comparable to the distance between this 'experiencing' space, and talking about space? Geography as a discourse can only to some extent facilitate law's workings with space. With this I do not want to denigrate the epistemological power of geography and its potential connection to law on that level. But epistemology can be a treacherous thing and can ineradicably affect the ontology of the spatiolegal. Tayyab Mahmud has shown how the geography of modernity, embodied in its hordes of empire-cradling intrepid explorers, surveyors and adventurers, has played a foundational role in the emergence of the racialised 'Other', and kept on promoting similar patterns in international law, proving even more resistant than actual colonisation.[75] Epistemologies do not develop separately from ontologies, and both are equally treacherous and given to withdrawal. There is no reason to trust ontology more than epistemology, even if one could indeed distinguish between them in a way that would actually set them in a separate rather than co-emerging and co-evolving parallelism. We do however have a situated responsibility to dig underneath layers of epistemological constructions that naturalise ontology and present it in abstract and universal ways that incapacitate in advance resistance or questioning (this is how atmospherics function, as I show in Chapter 3). Whether my present, necessarily epistemologically-mediated access to ontology yields anything different, or simply adds to the layers of construction is something that can only be judged ex post facto. But residing in epistemologies of disciplinary boundaries is certainly not the way.

1.3 Abstractions beyond metaphors

It is worth repeating that metaphors are not condemnable as such. The problem begins when metaphors occlude what has led to them and what is

---

73   Gregory, 1993
74   Delaney, 2002: 67, added emphasis
75   Mahmud, 2007: 545, 'Colonial geographies survived the demise of formal colonialism.' Mahmud goes on to link it to current debates on globalisation and empire. See also Mahmud, 2010a.

to be done on an ontological and even transformative level. Metaphors have this odd ability because they tend to become too comfortable. This comfort is attributable to the fact that metaphors freeze what has lied before their emergence. The conditions that allow the emergence of a metaphor (a crisis, a problem, a new legal challenge) become fixed within the metaphorical discourse and allow for a distance with the material. Yet metaphors offer a tangible means of understanding what is often difficult to grasp due to its abstraction. In the case of law's spatial turn, metaphors allow the filtering of the abstraction, which otherwise might have been too difficult to register.

But what is there beyond metaphors when it comes to space? It is indeed doubtful that, at least at first instance, one would have much ground to stand on. In a way, space beyond metaphor is an abstraction that vies with the traditionally conceived abstraction of the law (as opposed to a situated, concrete law in space). What is space for law if not another abstraction that needs to be taken into account, like time, individual, causality, levels of harm, liability, constitution, human rights, human beings? The list is endless and potentially all-inclusive. But if David Cunningham is correct in his suggestion that it is only through another form of abstraction that capitalist abstractions can be fought,[76] then on a different level, the abstraction with which the law dominates the geographical discourse is threatened by this different abstraction, the abstraction of a space beyond metaphor and facile terminological dealings. Thus, even a radical reading of the law that accords special emphasis on the particular, fears space in its abstract, theoretical dimension because the law may then lose its newly founded and only with great difficulty conquered embeddedness.[77] And this is quite right: the fact that both law and space are seen in their concrete production is a result of recent sociolegal and critical scholarship,[78] certainly constituting a radical step in an important direction. This hard-won achievement is by no means beyond threat.

It is precisely for this reason that an adequate theorisation of the connection between law and space is needed. It is very important when it comes to the increasing production of law and space literature, not to underestimate the risks inherent in the act of bypassing a thorough discussion on the new abstractions that originate in this law and space turn (such as the turn itself, the plurality of emerging spatiolegal concepts, embodiment and corporeality, movement, affects and how they affect the

76   Cunningham, 2005 and 2008
77   Itself not entirely innocent – the example of 'European legal space' seems to be illustrative of precisely this kind of territorial, bureaucratic and essentially jurisdictional authority. See the ECHR judicial exploration of the concept in *Issa and Others v. Turkey*, Application 31821/96, Judgment 6 November 2004.
78   Indicatively, Blomley, 1994; Chea and Grosz, 1996; Harvey, 1996; Rose, 1993; Mitchell, 2003

spatiolegal, flatness, assemblages and so on). A discussion of this sort is a defence against capitalist and market-oriented, materiality-diminishing discourses, as well as a tool for appropriation of the same discourses that are regularly used *against* the importance of law's spatial turn. It brings forth Samir Amin's observations on capitalism's supposed territorial disembodiment, and addresses them on the same level.[79] Issues such as aspatial globalisation, non-embedded economic flows, climate change, technology and so on, rely on the fact that the spatial turn will never address them on the very same level of abstraction. But if it did, the following assertion would have been inevitable: that all these constructions are simply de-spatialising mythologies. It is no longer surprising that all the philosophical, sociological, financial and even urban discourses that in one way or other push for a despatialised globality of flows, and regardless of whether this is done intentionally or not, end up contributing to the impoverished impact of actual spatialisations. Only on their level of abstraction can despatialising mythologies be fought, and it is only by taming its fear of abstraction that law's spatial turn will be able to fight this fight.

Derek McCormack's work on abstraction in geographical thought addresses precisely this tendency for oversimplification of abstraction as 'a malign process of generalization and simplification through which the complexity of the world is reduced at the expense of the experience of those who live in the concrete reality of this world.'[80] At its base, this hides another oversimplification: that there is an opposition between the concrete and the abstract, the lived and the theorised. From the opposition, a prioritisation often emerges that tends to favour either the abstract, in what appears to be the critical strand of legal geography, or the concrete in what appears to be the sociolegal strand. What is more, and in full awareness of the risk of simplification, the abstract route is more heavily criticised because of its perceived distance and lack of connection to reality, the field, the ground, indeed space. But McCormack warns against this tendency:

> without remaining 'vigilant in critically revising' our own relation with abstraction and the terms upon which these relations are based, it may well be the case that geographers run the risk of missing opportunities for attending to the surprising ways in which abstraction can participate in how we think through and become involved in lived spacetimes.[81]

McCormack's warning to geographers can equally be heard (or not, as it might be the case) by legal scholars working on the intersection. If any-

---

79   Amin, 2011
80   McCormack, 2012: 715
81   McCormack, 2012: 716, references ommitted

thing, the intersection between law and space is even more abstractly burdened than the mere geographic. Law's traditionally abstract self-description is precisely the enduring condition towards which McCormack urges vigilance, and which brings about reduction and universalism. Such venomous abstractions tend to be so deep-seated that end up becoming naturalised, as I show later on in the context of atmosphere. It would thus not be far fetched to trace a parallel between Henry Lefebvre's understanding of capitalist space on the one hand, and law's abstraction on the other:[82] for Lefebvre, capitalist space is a concrete abstraction, namely space removed from reality yet constitutive of the very reality by virtue of its removal. The concrete abstraction of capitalist space is a *sine-qua-non* of today's reality, without which (that is, without its removal from reality) reality could no longer work. In the same way, law's universalising and reductive abstraction operates as another concrete abstraction because it remains constitutive of reality. At the same time, space as concrete abstraction can only be understood through a practice, a concretisation that is linked to the space of everyday.[83] This is the paradox: we *need* the law to be considered universal and abstract in order for us to carry on believing in it and its possibility of delivering justice. Law's beloved but estranged daughter, justice, has also taken up the traits of her father: blind and violent, she can only see herself acting from a distance of universal equality and grand abstract reductions that bind not only the entirety of her jurisdiction but also future decisions. At the same time, this abstraction can only be dealt with through an emplaced, embedded, material space of everyday. A way needs to be found, in which both the need for dissimulation and the need for grounding are met.

While I return to this below, it is important here to repeat that abstraction is necessary. Its necessity lies both in the above sense of law's necessary dissimulation, but also in its potential to relate to reality more intimately through withdrawal. Once again McCormack writes:

> we might question those critiques of abstraction that take it to task on the basis that it is a withdrawal from experience: this is precisely because something of lived experience is always partially withdrawn from us. Experience needs to be understood in relation to that which is always withdrawn from it – otherwise it would be so self-contained that change or becoming would not be possible.[84]

The concept of withdrawal is particularly relevant to the present book and indeed the conceptualisation of spatial justice I am attempting. It is

---

82   Lefebvre, 1991
83   Lefebvre, 1991. See also Butler, 2009.
84   McCormack, 2012: 721

perhaps relevant to mention here that abstraction is a form of withdrawal (and here seen as an ontological move) that generates the necessary (epistemological) rupture for concreteness to become more present. In the context of spatial justice, withdrawal operates as a shaking-up mechanism that challenges taken-for-granted constructions of what becomes at every point accepted as official, granted or even claimable. When withdrawing, the details of concrete spatial situations come to the fore without being occluded by historicised layering. The role of law in all this is decisive. Law's contact with space shakes up such taken-for-granted abstractions, in its turn reconsidering its abstraction on a different level: no longer as the universalising reduction but as a withdrawal that operates both as a necessary dissimulation and as a tool though which the concreteness of living is brought forth.

Here I take into consideration the paradoxical nature of space as both concrete and abstract, and the specific ramifications that such a paradox might have for the law. This is arguably the only way in which the law might be able to deal both with the concreteness of space without becoming sucked in by an unproductive narrow localism, and the abstraction of space without becoming sucked in a diluted and equally unproductive universalism. In the first case, the law ends up being unable to fulfil its role as an institutional guarantee of social expectations of what is lawful and what is unlawful; and in the latter case, the law becomes unable to bridge the distance between its own perceived universalising abstraction and the particularities of the cases which the law is called to resolve. Consequently, I am not making an argument for de-concretisation of space in law, for space as a universal abstraction, or for a return to grand philosophical understandings of space. On the contrary, I support a full embracement of law's emplaced concreteness, but in parallel to a continuous rethinking of the connection between law and space and its repercussions.[85] For I find that this fear of space affects the way in which the concretisation of law is played out. Precisely because the interest in constructing the theoretical foundations of such an enterprise has waned, we are left with a literature that keeps on reproducing spatial clichés without venturing into the radical territories so gallantly promised by the concept of space. Once again, the most disappointing case of lost opportunity is the concept of spatial justice. Hence, the argument here is for a reinstatement of the particular embeddedness of the law, in a parallel and continuous rethinking of the ways in which this happens, with a view to a fuller, more potent understanding of the connection between law and space. For this text fears something too: if the spatial turn carries on unfolding itself only on the concrete, and ignoring the abstract, philosophically examined side of either space or law, the discourse itself will be impoverished and debilitated. Thus, to David Delaney's caution 'any effort to effect a

---

85 See Steinberg, 2013, for a similar argument in the context of the sea.

dematerialization of law must be regarded with suspicion',[86] I would also add a caution against the fetishisation of legal materiality. By not armouring itself against the conceptual minefield that abstraction can be, and permitting itself an unanchored embeddedness, law's spatial turn risks being co-opted by conservative political and social thinking (just like other 'grand' ideas, such as sustainability, globalisation, identity and so on). By not facing its fear of abstraction, the space of the law allows whoever feels more at ease with it to manipulate its embeddedness, thereby converting it from a radical tool to a hegemonic presence.[87]

So, what does space bring into law? As said earlier, space for law is not (just) jurisdiction, ideality or geography. It may at times be, represent or indeed be represented by all that; but these leave out some truly 'awkward' and upsetting facets of space. Let me refer to Doreen Massey's description of space: a product of interrelations and embedded practices, a sphere of multiple possibilities, a ground of chance and undecidability, and as such always becoming, always open to the future.[88] This seeming openness is firmly conditioned:[89] multiple possibilities indicate lack of direction and possibly destination; continuous becoming means also instability and unpredictability; interrelations denote a difficulty in pinpointing causality, origin, actors. Interestingly, from an applied perspective, the law is not that much different from space. It is already contingent, with multiple possibilities, often unpredictable and with an internal understanding of causality. In that sense, space brings to law a different awareness of law's operations. As I show in the following chapter, space and law are a tautology that unfold their difference in various ways. Thinking that all law is spatial means that law is embodied movement and rest. This is quite the change for a law whose frequent self-description is abstraction, textuality and universality. Even critical legal or sociolegal scholars find it difficult to accept that the law is always embodied. What about the textual hierarchies, what about constitutionality, transnational law, legal history even? There are always ways in which law can appear non-embodied and beyond spatiality. But this is precisely the point: appearance of the law is a play of dissimulation. Accepting that law is embodied and spatial means that all bodies participate in its emergence. Once this is accepted, then one can look into the hierarchical differences and deal with them. The difference is that they will be dealt with on a material, already emplaced level, rather than the textual regime of dissimulation. So, on a very basic level, spatiality brings a new self-understanding of the law.

86 Delaney, 2002: 80
87 It is widely observed that the language of the material and the particular is no longer the prerogative of progressive politics but also of significant portions of conservative ideology. Whether this is simply rhetoric, it is simply irrelevant.
88 Massey, 2005
89 For Massey, 2005: 10–12, through its parallel development with politics.

Space also brings an awareness of (other) spaces, both within and significantly beyond the reach of the law, which, in turning spatial, the law is progressively taking into consideration. Gendered labyrinths, sexualised basements, refugee camps, spaces of vulnerable bodies, postcolonial reterritorialisations, leaking spaces, graffiti-written cities, transboundary pollution flows, and so on. These spaces have increasingly been claiming their own non-metaphorical legal space, that is to say, their own spatiolegal situatedness. At the same time, space embodies the violence of being lost, of being uncertain about one's direction, orientation, decision, judgment, crisis. In this space of contingency and unpredictability, the violence of drawing lines, of *horizein* (of delimiting the horizon, of judging) takes place, and for this reason is all the more violent. The act of judgment is the act of line-drawing that separates while bringing together.[90]

This violence is counteracted by a disorientating sense of freedom. Spatiality in law brings such issues as the possibility to amble aimlessly, to feel ecstatically lost in wanderlust,[91] a Lefebvrian right to the city,[92] or even the right to make use of a variable open space characterised by a 'polyvocality of directions'.[93] A spatial law has the potential of becoming a law whose boundaries are lifted for the flow of bodies to move independently of property lines. This does not mean that one has roamers through one's kitchen. But it does mean that what is still called public space is safeguarded against private homogenisation and moral panic. A spatial law could preserve public space as an uninterrupted line of flight in the middle of the city, as an opportunity for creative normativity where the law guarantees the conditions for the spontaneous movement of bodies and where one can, even briefly, fancy oneself a nomad, a trailing presence that belongs to the earth and not to the high street shops.[94]

It is clear that what space brings to law is not only concrete, geographically specific benefits. It is that too, but it is also immaterial (which is not meant as metaphorical but as the way in which the materiality of the act is understood, namely as an idea, as I show below). In other words, spatial judgments (and allow me to state axiomatically here until further discussion, that all judgments are spatial) are both physical things and ideas. Yet the two are not in some relation of cause-effect. Neither generates the other. Rather, they co-emerge. This is the infamous Spinozan *parallelism* in action, where counteruntuitively, body and mind do not connect causally yet develop together.[95] They can be understood as the same mode (namely

---

90   Ingold, 2007. Law as *nomos*, as Carl Schmitt, 2006, famously put it, which divides and allocates, partitions and governs, includes and excludes.
91   Solnit, 2000
92   Lefebvre, 1996
93   Deleuze and Guattari, 1988: 421
94   Bottomley, 2007a; Layard, 2010
95   Spinoza, 2000, Book III, E2p7

the changeable properties of a substance, the way it affects and is affected by others) of the one substance, yet expressed by different attributes (namely ways of perceiving the substance, namely either thinking or spatial extension). Yet, by virtue of their being modes of the same univocal substance (for Spinoza, this is god or nature),[96] they always develop in parallel and fold into each other. Moira Gatens puts it thus: 'the mind is constituted by the affirmation of the actual existence of the body, and reason is active and embodied precisely because it is the affirmation of a particular bodily existence.'[97] Parallelism is the way univocity operates. Here, rather than subscribing to the one, I multiply it and overlay it, imploding it into several *ones*, all of them claiming the vastness of continuum and the complexity of parallelism through ruptures. In other words, it will become clear as we progress through this book that I refer to several overall enclosures that construct fractal totalities in the manner of Spinozan univocal substance. While it is not my aim to develop an ethics out of the connection between law and space, I find that their interweaving takes place everytime within a continuous totality (lawscape, atmosphere), and that the way in which either of them emerges in this totality corresponds to the Spinozan parallelism between extension (physical things) and thinking (ideas of things). In practice, this means that, first, ontology (what we know) and epistemology (how we know) co-emerge (continuum) but find themselves in a parallel relation to each other (rupture), each one a different way of approaching the same one thing;[98] and second, more concretely, that law and space exhibit both attributes of extension (materiality) and ideas (immateriality) simultaneously. These two outcomes are interconnected. Thus, the position from which we approach the spatiolegal is to be found on the same surface as the spatiolegal itself. The 'how' and the 'what', although not the same, by virtue of the univocity of the totality whose modes they are, co-emerge and evolve in parallel to each other (this is the problem of continuum and rupture that runs through this text). Likewise, materiality and immateriality are not opposites but parallel attributes of the same surface. In terms of law and space parallelism in particular, it is important that both are understood in their respective parallelism. Thus, space must be considered in terms of its tangibility as *earth* (as well as moon, recently Mars, plus all the mapped and mappable

---

96   Spinoza, 2000, II, P7s 'The thinking substance and the extended substance are one and the same substance, which is now comprehended under this attribute, now under that. So also a mode of extension and the idea of that mode are one and the same thing, but expressed in two different ways.'

97   Gatens, 1996b: 57. Commentators on Spinoza have pointed out how this ontological parallelism is supplemented by an epistemological parallelism, which refers to the way physical things and ideas are represented or become knowable. See Deleuze, 1990; Della Rocca, 2008.

98   Deleuze, 1990

universe), and its intangibility as *world*. The latter is characterised by an indifferent universality: space withdraws from the human. Yet this withdrawal is mediated through such concepts as 'place', 'identity' or 'agency', just as time is mediated through future and past constructions. In comparison, however, time is gentle: time heals whereas moving in space is mere escapism;[99] time is only now, and all can be embraced (even illusionary) in the all-contained present moment, itself folding within its history and its desire; but space is always parallel, always elsewhere, always withdrawn. Time is stationary, space is moving.

Law is also characterised by the parallelism of materiality and immateriality, things and symbols, bodies and ideas.[100] What is more, law is both a field of knowledge, and the court, the contract, the prison. Law always dissimulates its materiality, because of its apparent incompatibility between that and universality. But bringing law and space together demands an awareness of parallelism, with all its dissimulations, visibilisations and invisibilisations. Arguably, this is the greatest import of spatiality into law: the revelation of the strategy of dissimulation in which both space and law indulge in order to maintain their conative abilities. Legal spatiality allows us to understand both the vicissitudes and the virtues of the necessity of dissimulation, and yet not necessarily judge them as either bad or good. Dissimulation is employed both as a strategic tool for societal control, and as individual survival strategy. From this net of dissimulation, responsibility emerges as one of situatedness and indistinction, as I discuss below in the context of the Anthropocene. This is a distinctly spatial understanding of responsibility, which takes place on the same surface as everything else.

So to return to the initial questions on law's spatial turn: the law *can* turn since the law is spatial through-and-through. Indeed, the law *has* always been turning, folding unto itself in order to create new lawscapes, new ways of dissimulation, new strategies of in/visibilisation. Turning has been happening on a micro-level in law, fractal instances of practical, applied and theoretical spatiality that alert the law to its surroundings. The various moments of spatial turning described in this chapter and spread temporally and geographically, are now coming together and enable the law to perform perhaps the most radical turn: law's turn towards its very own space, its own process of lawscaping. In turning spatial, law gathers force from the archipelago of micro-turnings, and affirms its disciplinary distinction while at the same time taking along its sweep a variety of other disciplines. Yet, only the law can turn *itself* spatial – no other discipline can 'force' this turn. This should answer in a pacifying way the question of disciplinary resilience for the law. Namely, how can the law open up to space (that shakes up and resemiologises law), without at the same time

99 Kaplan, 1996
100 E.g., Butler, 2013a

making itself implode, collapse under the weight of its spatiality? Law turning itself spatial means that the law emerges through the bodies of its constitution. Law's spatial turn is the process of emergence of law's always-already spatiality, its connection to space and its questioning qualities.[101] Thus, *the spatial turn* is not a process of invention but a bringing forth.[102] In this way, law is made aware of both its limits and, significantly, its limitations in a confident way:[103] law becomes confidently modest as it were. This is because, in order for the law to bring forth its spatiality, the law needs to suspend itself, go beyond and even against itself in order to invite, welcome and accommodate this destabilising presence. This, however, does not stop the law from being fearful of such an opening. The law is threatened by the expanses of contingency opened up by space, at the same time both more material and more abstract than the equivalent complexity originating in law's mingling with other 'guests', such as culture or social context in general. The law has to act through its limitations *and* despite its own limits, indeed despite its own fears of spatiality, in order to reap the fruit of what this new spatiality bears. This does not mean that the law is pioneer – far from that, law's temporality is by definition slower than most other institutions.[104] Politics, economics, sociology – all law's regular bedfellows – have turned earlier than law. In turning spatial, law confirms and exceeds the spatial revolt of other disciplines, including them in its own turning and resemiologising them, indeed assimilating them. What we can look forward to, therefore, is a spatial law from which, in a parallel, non-causal way, a spatial justice might at some point emerge.

---

101 This is how the spatial turn has been understood in some disciplines, such as religion or indeed history: Arias, 2009; Bergmann, 2007.
102 Derrida, 1999. See also Delaney's, 2002, distinction between 'law-in-space' and 'space-in-law'; and the space of law in Haldar, 1994, and Mulcahy, 2007.
103 On this, see, Sarat *et al.*, 2005, the introduction to the volume.
104 Luhmann, 2004

Chapter 2

Welcome to the lawscape

While you are reading this, you are in the lawscape. As you walked down the street, or earlier as you came into a building, or even earlier when you woke up at home: it is all lawscape. The question 'where is the law?' can only be answered with an ambiguous 'all over'.[1] Less metaphorical than it might sound, the law spreads on pavements, covers the walls of buildings, opens and closes windows, lets you dress in a certain way (and not any other), eat in a certain way, smell, touch or listen to certain things, touch other people in a certain way (and not any other), sleep in a certain space, move in a certain way, stay still in a certain way. In the lawscape, every surface, smell, colour, taste is regulated by some form of law, be this intellectual property, planning law, environmental law, health and safety regulations, and so on. Law regulates traffic, allows you to cross the road or not, allows you to drive your car, to go to the cinema, to enter the zoo, to stay in your own home. It allows you to switch on your TV, to access the internet or read a newspaper. Even the simplest acts are controlled to a greater or lesser extent by some legal agreement, limitation or prescribed direction, whether this is in the public or private space. A simple visit to the bathroom, this sacrosanct of private spaces, is regulated by legal provisions of water procurement, building regulations with regards to the material and placement of pipes, legal ownership of sewers and regulations on waste disposal, planning relation of the bathroom space to the rest of the home in the sense of where it is and what provisions have been made for emergencies, and the kind of paint and other materials used.[2] And even more intimately: the way in which one goes to the bathroom, or whether one can afford extra time or has to hurry up in view of a waiting family or co-workers, depends on a whole set of rules that come to characterise society, culture, the individual. And even further, the fact that there is still water to use in view of the world's water problems, that one has a bathroom to use which is not an open facility but affords a certain privacy, or indeed a bathroom at all in view of geopolitical or ecological disasters, such as wars, population displacements, forced migration, tsunamis, flooding or bush fires, depend on

1 Sarat, 1990
2 See Braverman, 2009

a legal framework of international, regional and bilateral agreements, domestic legislation, regulation of aid distribution, pollution regulations, use of banned materials, burden of proof on whether a substance is harmful, and so on. A series of regulations, contractual agreements, statutes and cases visit you in your bathroom, and the whole thing quickly becomes overcrowded if you also call insurance. This is law in the service of space.

Now think of an airport. The law makes its presence felt even before entering the airport building: you arrive by public transport or taxi, for which you have to pay, or your car, which you have to park for, again, a fee. You then walk in specific corridors, desperately following signs, screens, uniformed personnel. You are checked several times, you might be frisked or you may be asked to lay out your underwear on the security table. You have to hide behind your legal persona, your passport, your request for asylum, your visa, your reason for visiting. Your steps are determined by anti-terrorist regulations, aviation agreements, environmental treaties, as well as local prejudices, religious customs, cultural preferences that often end up normalised in the form of law. If you are about to depart, and in some cases even when you arrive, you are delivered to the comfort of duty free, brimming with triangular chocolate bars and overperfumed clients. The airport is still the lawscape, but represents a different side of it: in the bathroom, the law is taking the back seat faithful to the idea and practice of the 'private'. In the airport, on the other hand, law is visibilised in order to make you feel threatened, securitised, surveilled. You only feel even more 'free' when you emerge from the duty free area, where freedom is a small rectangular piece of plastic. This is space in the service of the law.

To put it simply, the lawscape is the way space and law unfold their tautology as difference. This takes place through an interplay of in/visibilisation. In what follows, I am entering deeper in the lawscape, and trying to understand its legality, corporeality, spatiality and temporality. I invite you to walk with my students in the lawscape, discovering its angles and angularities, its chiaroscuro. And I finally look into the proliferation of lawscapes across the surface of space and law and their temporal connections. But before that, I am opening a parenthesis in which, continuing from the previous chapter, I look more closely into the kind of spaces currently emerging that make the conceptualisation of the lawscape possible. These spaces can no longer be ignored by scholarship that deals with the way law is thought, because they alter the law radically.

2.1 Emerging spaces, emerging bodies, emerging law

Space can no longer be considered merely a measurable object.[3] The 'relational' turn in human geography that follows to some extent Henri

---

3   Amin and Thrift, 2002

Lefebvre's work,[4] has alerted us to the possibility of space being much more than something pre-existing and empty of meaning, a measurable entity, a container, or a background. Rather, space has been found to be both a product and a producer of relations.[5] Space is now seen as an active factor in constructing bodies, society, law.[6] At the same time, relationality has been shown to have its limits, not least because of its ultimate inability to displace the human from the centre,[7] but also because of its inability to accommodate withdrawals and ruptures. As a result of this, and as I have shown in Chapter 1, there has been a move towards a more conceptual understanding of space. One could say that the old attempt at discovering the 'essential' qualities of space has been revived, albeit in a distinctly 'fluid' way, where essence is replaced by a process of becoming, inexorably leading to an elevation of contingency to a determining (in its unpredictability) factor in understanding space. In what follows, I go through an array of spaces, laws and bodies, observing them in their co-emergence, and enabling them to yield the various elements for the formation of the lawscape that follows in the next section. Just to foreshadow, however, the main movements of this section, I begin with the common surface between the linguistic and the material in order to sketch the characteristics of the *manifold*, namely the space of bodies and body of space in which becoming takes place. Manifold as spatial and corporeal organisation emerges in various configurations, such as cyberspace, Australian aboriginal space, labyrinthine space, constantly folding and unfolding space. The manifold is a continuum of space and bodies, which regulates itself through its own folds and movements, while at the same time resisting uniformity or total relationality through its ruptures. Ruptures emerge in various guises on the continuum, enabling the emergence of singularity and material agency. In all this, the law is intimately implicated, both as *logos* and *nomos*, namely as state law of boundaries and nomadic law of passage. Law is also inscribed on the common surface between the linguistic and the material. This is the kind of space that gives rise to the lawscape, where space, bodies and law are folded together.

I would like to begin with the idea of space as process that is both discursive and material. For Jacques Derrida, space is always 'spac*ing*',[8] namely a process rather than a fact, becoming rather than being. *Spacing* is what language necessarily entails, namely the process of creating an interval, of opening up a space between the text and its producer.[9] The text of space is a 'citational performance' as Gillian Rose puts it, 'extraordinarily convoluted, multiply overlaid, paradoxical, pleated, folded, broken and,

4   Mainly Lefebvre, 1991
5   See the classic Gregory *et al.*, 1994; Crang and Thrift, 2000
6   Hubbard *et al.*, 2002
7   Clough, 2010
8   Derrida, 1972
9   See also Blomley's, 2002, term *splicing*.

perhaps, sometimes absent'.[10] Spacing is the rupture of presence, the discontinuity between utterance and understanding, being here and listening to you over there. In his *Language of Space*, Foucault writes: 'such is the power of language: that which is woven of space elicits space, gives itself through an original opening and removes space to take it back into language'.[11] Michel de Certeau has famously written about the production of urban space through walking: 'they are walkers, *Wandersmänner*, whose bodies follows the thicks and thins of an urban "text" they write without being able to read it ... The networks of these moving, intersecting writings compose a manifold story that has neither author nor spectator'.[12] Foucault and de Certeau's explicit link between language and space, although very different from each other, is comparable to Derridean spacing in that they all see the spatial sharing the same surface as the textual. This surface, however, is interrupted: spacing is always the process of producing discontinuity, difference, *différance*. In Derrida, rupture is the moment of deconstruction. In Foucault, the interruption is circular, continuous, reinscriptive. In de Certeau, rupture is the excess that emerges from the inability to read the traces of walking. Yet one keeps trying.

De Certeau's textual and material manifold fleshes out the concept's main trait, namely its ability to fold and fold again, accommodating new spatial necessities and rupturing challenges. Space as fold finds its apogee in Deleuze's treatment of Leibniz and the Baroque.[13] For Deleuze, the fold replaces the point as the minimum element. A fold is necessarily multiple, a manifold. This manifold space keeps on spreading by folding and unfolding itself like an origami, or as Manuel de Landa puts it, a spatial multiplicity, a field of rapidities and slowness defined internally, by and through itself and its inclusive immanence.[14] A fold is the way bodies inhabit space, folded with other bodies and co-constituting the manifold. Although folded unto themselves, bodies also fold into other bodies. The fold is the space of becoming but also that of becoming-other: this is the excess of the self that keeps bodies together while allowing bodies to pass and become other. This excess is constantly present whatever the folding may be, only to be passed on to further foldings, and foldings within foldings, interfolded in a plasmatic depth but always ready to resurface on that plane of contiguous, interdigitated space of folds.

10   Rose, 1999: 247
11   Foucault, 2007: 166. This is Foucault as a literary critic, which brings to mind Blanchot's, 1982, work on space and literature.
12   de Certeau, 1984: 93
13   Deleuze, 2006; see also Deleuze, 1993
14   De Landa, 2005, for whom the manifold is a spatially pronounced description of multiplicity.

The manifold manages to accommodate new necessities. Perhaps the most prominent example of such necessities would be the cyberspace. Bela Chatterjee's work on cybercities as a manifold of textual and material aspects of space,[15] understands even electronic space as material, crossed by processes of spacing in the form of software and code. Likewise, Kitchin and Dodge have worked extensively on the software as a material production which can only be understood from a spatial perspective. They write: 'software is not an immaterial, stable, and neutral product. Rather, it is a complex, multifaceted, mutable set of relations created through diverse sets of discursive, economic, and material practices.'[16] Software's complexity unfolds in the space of what the authors call *code/space*, namely what emerges 'when software and the spatiality of everyday life become mutually constituted, that is, produced through one another'.[17] The above can be read in terms of the lawscape. Just as code/space, the lawscape is a co-emerging manifold. As such, it is given to contingency ('the result of all this contingency is that software development has high failure rates'[18]) and for this reason is unable to guarantee that things will work in the planned way. The main difference, however, is that the lawscape is always emergent, as opposed to code/space that, at least according to the authors, may or may not emerge. The emergence of code/space leads to what Greenfield has called *everyware*.[19] Everyware is the spatial ubiquity of computational devices that saturate society to the point of ambient intelligence.[20] In other words, everyware is a material and immaterial continuum that supplements software space by revealing how intensive surveillance and securitisation is often an unavoidable reality. From code/space to everyware is a short leap, one that is arguably already happening. This is also the connection between lawscape and atmosphere, as I discuss in the next chapter.

Manifold as continuum can also be found in the thought of Deleuze and Guattari,[21] who call it a *plane of immanence* or *consistency*, namely 'the "holding together" of heterogeneous elements.'[22] Here, spatial contingency is reflected in the multiplicity of positionings characterised by an immanent spatiality and corporeality. In epistemological terms, the plane of immanence is the parameters of thinking that determine the problem in

15 Chatterjee, 2007
16 Kitchin and Dodge, 2011: 37
17 Kitchin and Dodge, 2011: 16. See also Wood, 2012 for the constitution of game spaces.
18 Kitchin and Dodge, 2011: 37
19 Greenfield, 2006
20 Kitchin and Dodge, 2011: 221. Ambient Intelligence is defined as a saturated interrelational network of objects that are sensitised to humans or other objects around them. This is closer to the concept of atmosphere, as developed in Chapters 3 and 4.
21 This has subsequently inspired strands of what Nigel Thrift, 2007, calls *non-representational theory*.
22 Deleuze and Guattari, 1988: 323; Deleuze and Guattari, 1994: 141

answer to which a concept is suggested.²³ Moira Gatens describes it as 'a plane of experimentation, a mapping of extensive relations and intensive capacities that are mobile and dynamic'.²⁴ Drawing also on Spinoza, Jane Bennett talks about

> an ontological field without any unequivocal demarcations between human, animal, vegetable or mineral. All forces and flows (materialities) are or can become lively, affective, and signaling ... portions congeal into bodies, but not in a way that makes any one type the privileged site of agency.²⁵

In that sense, the plane of immanence is the surface on which the concept of the lawscape emerges as an answer to the problem of the connection between law and space. Just like the plane of immanence, the spatiolegal manifold is both ontological and epistemological. At the same time, the symbolic/textual/immaterial is found in a relation of Spinozan parallelism with the spatial/material. It is important, therefore, to state the paradox that, on the one hand, neither space nor law are exclusively material or immaterial; yet either of them is, through and through, *both* material and immaterial. This classic strategy used by both law (which often dissimulates its materiality as immaterial) and space (which often dissimulates its immateriality as material) is carried through in the lawscape and affects the way space and law fold into each other.

From a more traditional legal perspective, space is linear, divisible, measurable. In a spatiolegal manifold, this is replaced by a labyrinth: 'a continuous labyrinth is not a line dissolving into independent points, as flowing sand might dissolve into grains, but resembles a sheet of paper divided into infinite folds or separated into bending movements, each one determined by the consistent or conspiring surrounding' as Deleuze puts it.²⁶ This new manifold legal space can be alluring, not because of any promise of certainty but because of the continuous uncertainty and ambiguity. The bodies that move in the labyrinth do not always want to get out. The labyrinth offers the pleasure of disorientation, the need for wanderlust, the desire to hide. At the same time, the labyrinth may fold within itself the *pain* of disorientation. Labyrinthine space can become oppressive. It all depends on what body we are talking about. Luce Irigaray's critique points

---

23 Deleuze and Guattari, 1994
24 Gatens, 1996a: 165. This is how I deal with the various spaces in this book: the city of London, the Triveneto plains, Robinson Crusoe's island are extensive and intensive planes of immanence, on which the question of the connection between law and space arises, and for which I suggest the concepts of the lawscape and atmosphere.
25 Bennett, 2010: 117
26 Deleuze, 1993: 18

to the way sexed bodies (women, but one could add queer and trans bodies, as well as nonhuman bodies that are captured in the labyrinth, such as animals, plants and inorganic matter) are never seen as embodying the knowledge of the labyrinth, that is, the ways out, in and through the labyrinth.[27] In her reading of Nietzsche labyrinths, Irigaray finds that Ariadne and her knowledge of the labyrinth which saved Theseus from the Minotaur, does not fit in existing conceptualisations of the spatiotemporal. For Irigaray, female subjectivity is always connected to that 'other of space', namely space as irrational, unstable and ever-changing, as opposed to Euclidean space as measured, divided and stable. Ariadne's spatiotemporal understanding, however, is neither solely the first, nor the second kind of space. She is simultaneously the way out and deeper in the labyrinth, both containment and excess, stability and impermanence. The space of the sexed subject for Irigaray is a manifold of irreducibility to essential forms, rather like the sea as opposed to the land. If land is the locus of the Nietzschean *übermensch*, the sea is the threatening excess of space that cannot be fully revealed. This is the threat posed by the unknown to the controlling, a point reiterated in another elemental study by Irigaray, this time on Heidegger's *forgetting* of the air.[28] This is where the sexed subject is usually deposited and imprisoned: an essentialised unknown as an irrational unknown. But there is another unknown, one that organises itself in inaccessible but no less relevant patterns. Irigaray refers to Lacan's description of the famous Bernini statue *The Ecstasy of Santa Teresa* (Figure 2.1) as yet another indication of the unknown in which women are thrown, especially when, as the sculpted saint abundantly embodies, they experience the jouissance of sexual pleasure.[29] Lacan's logocentric discourse cannot understand Teresa and for this reason silences all female jouissance as supplementary (to male) and inexpressible even for the women who experience it. Irigaray's critique is accurate and well-placed; however, there is something about the statue that is not statuesque. There is no doubt that Bernini wanted to *freeze* the unknown entrails of female jouissance and make them evident to the world. At the same time, however, when I (unknowingly following Lacan's encouragement to go to Rome to see a female orgasm …) first saw the sculpture in the Cornaro Chapel of the church of Santa Maria della Vittoria in Rome, I was overwhelmed, not so much by the obvious orgasmic expression, but by a blatantly overlooked quality of the statue: that the largest part is taken by the rising and falling, folding and refracting riot of the saint's robe. It is marble, but the marble is moving. It is captured, but the captive is breathing. In a strange way, it seems to embody what Irigaray writes elsewhere: 'our pleasure consists in

27 Irigaray, 1991
28 Irigaray, 1983
29 Irigaray, 1985a

Figure 2.1 The manifold – or Gian Lorenzo Bernini's *The Ecstasy of St Teresa*.
© Andreas Philippopoulos-Mihalopoulos.

moving, being moved, endlessly'.[30] This continuous folding seems to me to create a whole new space, entirely Baroque yet still excessive even for Baroque aesthetics: the space of the absolute immanence of the manifold.

In the manifold, bodies cannot be differentiated from space. Henri Lefebvre has been writing on how bodies produce space: 'each living body *is* space and *has* its space: it produces itself in space and it also produces that space'.[31] Bodies and space are not found in a relation of foreground/background but fully folded to the extent that, ontologically, bodies are space and vice versa. Add to this the Deleuzian fold of becoming-other, and what we have are bodies that are space but are also other bodies, themselves also space ... This is intimately connected to the fact that a body is not a neatly defined, contour-bound entity. The body is not defined by its outline. If they were a painting, bodies would be Venetian sprawls of colour

30 Irigaray, 1985a: 210; Irigaray, 1985b
31 Lefebvre, 1991: 170

without drawn boundaries, staging through their expansive leaking a radical withdrawal from the Florentine canon of humanist containment. Deleuze writes: 'the edge of the forest is a limit. Does this mean that the forest is defined by its outline? ... We can't even specify the precise moment at which there is no more forest'.[32] All bodies are leaking. By 'all bodies' I mean human and nonhuman. While previously, whenever included, nonhuman bodies have been either resource, context or the negative of the dialectics of humanity, here I follow the schools of thought largely identified as new materialisms, non-representational theory, speculative realism and object oriented ontologies,[33] themselves generally drawing from a Spinozan/Deleuzian understanding of the body. Thus, for Deleuze, 'a body can be anything: it can be an animal, a body of sounds, a mind or idea; it can be a linguistic corpus, a social body, a collectivity.'[34] All bodies are *assemblages*, namely aggregations of human and nonhuman bodies that are contingent upon the conditions of their emergence and which do not presuppose the centrality, and certainly not the exclusive presence, of the human. What is more, assemblages are both actual, namely space and matter, and virtual, namely potential but still real. Actual and virtual are not found in a dialectical opposition; nor does the actual determine the virtual.[35] Rather, there is no ontological distinction between the two, and if anything, the actual is determined through its folding with the virtual, both actual and virtual aspects of the continuum. This means that each assemblage has a spatiality and materiality that is both, schematically speaking, its own (actual), and part of the wider continuum with other assemblages (virtual). Assemblages are properly speaking *geophilosophical*, namely spatial and temporal, part of the great earth continuum yet individually different from each other.[36]

How do bodies differentiate themselves to each other? How do singularities emerge from the continuum? And what is the role of agency in the continuum? Let me start by saying that the spatiolegal continuum is not uniform but immanently different. As we shall see below, there is a multiplicity of lawscapes within the lawscape, each one a singularity. These singularities emerge through what I have referred to in the Introduction as *ruptures*. Ruptures can be Deleuzian folds, namely co-emerging assemblages, and therefore ontological differentiations. Or, as we shall see, they can be atmospheric distinctions, namely a differentiation between an interior and an exterior, and therefore epistemological differentiations.[37] They

---

32 Deleuze, *Cours Vincennes: Sur Spinoza*, 17.02.1981 in Lambert, 2013a: 74
33 See for example, Coole and Frost, 2010; Bennett, 2010; Thrift, 2007
34 Deleuze, 1988: 127
35 Deleuze, 1986
36 Bonta and Protevi, 2004
37 Luhmann, 2012

are also ontological necessities, such as the inherently ontological move of withdrawal that every body performs. Or indeed, they can be an effect of withdrawing strategies, as I show in Chapters 4 and 5. Whatever they are, they are part of the continuum, constituting it in difference. I have employed the term continuum in different contexts,[38] but it is consistently characterised by this one quality: it is always ruptured, indeed self-ruptured, through foldings and distinctions on its surface. Sometimes illusionary necessities and other times ontological facts, ruptures are the locus of singularity in the continuum. Ruptures constitute the continuum, to the point that one is left only with a continuity of ruptures.

Through ruptures, bodies differentiate themselves from other bodies. Rupture can be caused by a difference in movement. According to Spinoza, bodies differentiate themselves from each other on the basis of their differentiated velocities or pauses – it is all about how a body moves or rests.[39] A human body will be faster than a chair will be faster than the plant will be faster than the fossil, and all of them will be slower than a typhoon. Rupture enables a difference in movement that results in a repeated difference. Yet, precisely because the continuum is a series of ruptures, ruptures do not rupture the continuum as such: they are all inscribed within. So, the continuum is crossed by lines that produce meaning, while continuously being ruptured by them. Lisa Blackman in her seminal work on the body, has called this the problem of 'the one and the many', namely the ontological difficulty of being coherent yet multiple, in other words, self yet othering. The main challenge is how to avoid categorising either of these as inferior, namely how to escape the trap of mapping the continuum and the rupture 'onto differentiations made between the civilized and the primitive, the superior and the inferior, the simple and the complex, and the impulsive and the environmental':[40] indeed, how *not* to make rupture and continuum a dialectic of opposites, namely a question of positive presence and negative absence, but a co-emergence. This is not easy and the struggle remains in the core of this book.

These ruptured pockets of differentiation are the way in which social systems and institutions (just as any other body) emerge: what is the legal system if not a snapshot of the differentiated velocity of its body, itself an assemblage that includes humans, technology, organic and inorganic matter as well as other social systems, taking place on a space of continuity with other bodies? Deleuze and Guattari employ the term *agencements*

---

38   The lawscape is one of them, as I show below, but also atmosphere, open ecology, materiality continuum, surface and so on.
39   See the remarkable study of mobility negotiation, all about law but without mentioning it, by Jensen, 2010. See also Cresswell, 2006 and Urry, 2007 for two of the basic writings on movement.
40   Blackman, 2012: 59

(which has been translated as 'assemblages') which gives a clearer idea of the kind of role envisaged. As the editors of *New Materialsms* Diana Coole and Samantha Frost write, we have moved away from an era where

> agents are exclusively humans who possess the cognitive abilities, intentionality, and freedom to make autonomous decisions and the corollary presumption that humans have the right or ability to master nature. Instead, the human species is being relocated within a natural environment whose material forces themselves manifest certain agentic capacities and in which the domain of unintended or unanticipated effects is considerably broadened.[41]

This is what Jane Bennett in her book *Vibrant Matter* means when she refers to the vibrant self-organising nature of both organic and inorganic matter.[42] These new corporeal agencies are not effects of an agency-giving social system (such as the law or politics or the state) but emerging assemblages that, by virtue of including law or politics or other social institutions in their assembled bodies, they acquire agency.[43] The body does not have a pre-given agentic outline. Deleuze writes: 'the limit of something is the limit of its action and not the outline of its figure'.[44] Agency is an open emergence, each time differently positioned in relation to its constituent elements. Karen Barad puts it clearly:

> Agency is not held, it is not a property of persons or things; rather, agency is an enactment, a matter of possibilities for reconfiguring entanglements. So agency is not about choice in any liberal humanist sense; rather, it is about the possibilities and accountability entailed in reconfiguring material-discursive apparatuses of bodily production, including the boundary articulations and exclusions that are marked by those practices.[45]

Agentic, corporeal matter unfolds in space, taking up forms that are co-determined with this very space. Ben Woodard, referring to Sten Odenwald,[46] writes:

> Space…blurs the distinction between base materiality and pure formalism … Space cannot be taken as merely the stage of matter. It is

---

41   Coole and Frost, 2010: 10
42   Bennett, 2010
43   See also Lenco, 2012
44   Deleuze, *Cours Vincennes: Sur Spinoza*, 17.02.1981 in Lambert, 2013a: 74
45   Barad, 2012: 54
46   Odenwald, 2007

a quality of gravity that controls the motion of matter and is warped by matter in kind.'⁴⁷

So, in space and of space, matter and form, bodies and symbols emerge, subject to a gravitational force that brings them together.

At the same time, these bodies are kept apart. Connection in the form of assemblage and disconnection in the form of withdrawal correspond to the broader discussion on continuum and rupture. Again, these values are not dialectic. Continuum is made of ruptures and ruptures are continuous. There is no difference; there is folding. Likewise, assemblage and withdrawal are constituted of each other. Every body assembles itself in larger bodies while at the same time withdrawing from the ontology of the assemblage. Graham Harman in his work on Bruno Latour, writes:

> objects are not defined by their relations: instead they are what enters into relations in the first place. Objects enter relations but withdraw from them as well; objects are built of components but exceed those components. Things exist, not in relations but in a strange sort of vacuum from which they only partly emerge into relation.⁴⁸

I understand objects and bodies as synonymous yet coming from slightly different theoretical trajectories.⁴⁹ For my posthuman-focused purposes, a body retains the connection to the human but allows for an extended, posthuman understanding. On that basis, and replacing objects with bodies in the above quote, bodies are at the same time assemblages of other bodies, part of assemblages with other bodies, and withdrawn from every relation. Withdrawal is ontological: every body is a closed, autopoietic system that withdraws from full openness, connectivity or exteriority, and into a monadic singularity that is gathered around its autopoiesis, its self-perpetuation.⁵⁰ Its openness rests on its closure. Its connection with the exterior takes place only through the systemic interior. Assemblaging does not take place in an exterior but in a fractalised interior. Withdrawal is taken inside the body, takes place from within the body, and ultimately becomes self-withdrawal, as I show in Chapter 5. Assemblage rests on withdrawal. The world is what each body makes of it. But this is not relativism or subjectivism. The world withdraws as much as the body withdraws. 'Nothing "points" toward anything else or bleeds into anything else. Everything

---

47 Woodard, 2013: 12
48 Harman, 2009: 132
49 Bodies come from a Spinozan/Deleuzian understanding. Objects continue with this by adding Heideggerian/Latourian/Whitehead influences, generating what has been called Object Oriented Ontologies (OOO).
50 Philippopoulos-Mihalopoulos, 2009

withdraws into itself'.[51] Harman's work on withdrawal has affected my understanding of the term in this book, and has complemented my previous understanding as a corporeal movement in the context of spatial justice. When I first started working on withdrawal, the corporeality of the move was its main characteristic, followed closely by its ethical direction. What I have gained from Harman's work, as well as other theorists' working on new ontologies, is the foundationally ontological notion of withdrawal. This, in combination with a broadly autopoietic materiality framework that I have been developing in the context of law and agency,[52] yields a concept and praxis of withdrawal that is both ontological and strategic.

Agency is determined both by withdrawal and by assemblage. In a way, this is what Karen Barad refers to when she speaks about 'agentic separability,'[53] namely the rupture that constructs agency without denying the indistinction between this and that side of the rupture. Material agency's emergence from within and as an assemblage, is tantamount to gathering one's singularity around one's own materiality by means of withdrawal. Bodies are singular because they withdraw from the assemblage, and they can only be part of an assemblage because of their withdrawing. Bodies are self-organising and therefore withdrawing, while at the same time fully immersed and constituted by assemblages. Both assemblaging and withdrawing come from *conatus*: the Spinozan term for the striving to affirm and maintain oneself, and in systems theory/autopoietic terms, the *autopoiesis* of matter, namely the desire/innate programme of matter to maintain itself, offer resistance and simultaneously expand its extensive (namely, spatial) physical qualities. Deleuze writes:

> the conatus defines the *right* of the existing mode. All that I am determined *to do* in order to continue existing (destroy what doesn't agree with me, what harms me, preserve what is useful to me or suits me) *by means of* given affections (ideas of objects), *under* determinate affects (joy and sadness, love and hate ...) — all this is my natural right. This right is strictly identical with my power and is independent of any other ends, of any consideration of duties, since the *conatus* is the first foundation, the *primum movens*, the efficient and not the final cause.[54]

Deleuze adds something quite decisive in his reading of conatuss. Conatus is the 'natural right' of the body, that necessitates encounters with other bodies in order to create assemblages. At the same time conatus is excessive: it cannot be measured by practices of fitting in or appropriate

---

51 Harman, 2009: 113
52 Philippopoulos-Mihalopoulos, 2014
53 Barad, 2006: 175
54 Deleuze, 1988: 102

emplacement. It *can* be strategically engineered in order to encounter bodies that affect the conatus positively, and thus make a body stronger, but even that does not always work. Engineering is contingent and excessive, because the ontology of the body is also contingent and excessive. The conative excess is the reason for which every body withdraws. The vastness of the conative machine, always overflowing the body itself and affecting all others, renders a body immeasurable. This immeasurability is at the core of ontological withdrawal: no body ever presents itself fully ontologically. Even conatus exceeds itself. In other words, conative strive and withdrawal are one and the same thing.

Legal theoretical work from stem cells and co-operatives, to the role of glass partitions in the courtroom,[55] has demonstrated how a posthuman conception of matter and legal agency allows us to trace the process by which material conatus becomes engineered. Pottage's work on genetically modified crops for example,[56] has shown how the entelechy of the best-before date or the predetermined way of use, which are inscribed within the very material body of the crop, do not exhaust themselves in the vertical, hierarchical superpower of corporations (that too, of course) but also in the way the modified organism as such extends spatially, tending to occupy surrounding space, physically and legally. Worryingly, genetically modified crops and their in-built obsolescence go against the conative/autopoietic trait of matter. This ingrained move amounts to an ordered self-extinction that overwrites the conative power of self-perpetuation and that of withdrawal since matter becomes measurable and fitted in a spatio-temporal engineering process. GM crops exemplify a much larger risk that has befallen all contemporary bodies: to fall into the economy of *affordance*, which contains even the excess of conative drive. This is Reza Negarestani's warning: 'The aim of affordability is to make the discrepancy between the inherent desire for self-preservation and the inevitability of death consistent with the economic order of the organism'.[57] The difference between perpetuation and (self-) extinction is the unresolvable question of capitalist society, in which event, withdrawal from desire has limited effect and in which atmospherics manage to mop up any excess, as I show in Chapter 3.

Agency emerges in space and through an assemblage that generates space. This is what Marcus Doel means when he repeatedly announces in his seminal work *Poststructuralist Geographies* that 'space takes place'.[58] Space is encountered in its ruptured immanence as *a* space: a space of a specific assemblage. Space takes place and indeed spacing constantly through the

---

55 See Persaud, 2013; Bottomley, 2009; Mulcahy, 2010
56 Pottage, 2011
57 Negarestani, 2011: 196
58 Doel, 1999: 10

assemblage. Each body is an assemblage, and each body can be organised in assemblages with other bodies. Bodies are and produce space. Space is the repeated taking-place, again and again, simultaneously rather than additively, horizontally rather than vertically, repeatedly yet with every time as the first time, every time once, unpredictably, contingently, defined internally and by itself, not as part of a uniform globality or a transcendental exteriority but through its very immanence. Space itself is folded in the event and provides for what can be called the context – but, alas, this context is the event itself, the very *text* that surrounds and is surrounded by its *con*text. To quote Doel again, 'context is never closed in on itself ... context is untimely ... Were there to be a law of space and spacing, it would have to be a law of context – and a non-saturable context at that. This is why it is a *harsh* law of space.' And further on, 'such is the harsh law of spacing: everything is disadjusted and dis-located; everything is (s)played out according to the contingent folding, unfolding, and refolding'.[59]

If this is the law of space, in all its harsh disorientating reality, then how does it differ from the law of law? And can we think of law outside or without space (and vice versa)? Law without space or space without law is an epistemological simplification or even worse, a disciplinary violence. We have, however, successfully managed to keep them separate, thus ontologising the epistemological and, in a distinctly Marxian way, naturalise and normalise it. We have succumbed to law's dissimulation, as an abstract, universal, linguistic/symbolic, closed discipline whose spatiality is to be found, if anywhere, closed up in a court. Nathan Moore puts the paradox thus: 'at the core of the normative system is the surprising presumption that to be specifically embodied is already to depart from the norm'.[60] This is arguably a feat of Western law (itself a form of abstraction), which, in its colonising form, has managed to undo the connection between space and law in other jurisdictions too.[61] Take for example Christine Black's work on indigenous (Australian aboriginal and Maori) jurisprudence. Her proposition, itself emerging from the practice of indigenous jurisprudence, is that land is the law. Black's work is 'from within', between indigenous but also addressing the settlers' law, and always on the basis of the tautology between law and land. The tautology is also the source of responsibility towards the earth as a whole – not unlike the responsibility of the Anthropocene, more of which below. Black shows how indigenous jurisprudence is shaped in three concentric circles: cosmology, the law of relationship, and rights and responsibilities. The three are folded into one another, spanning from the local to the cosmological. The law of the land runs through them, and is indeed the line that connects all three. The first circle, cosmology, is always of the particular

---

59   Doel, 1999: 114, 123 and 131
60   Moore, 2013: 62
61   Black, 2011

indigenous group, inscribing the tribal cosmogony onto the territory. The second, the law of relationship, takes place between two non-oppositional moieties, and assumes the form of songlines or Dreaming Tracks across the land. Finally, responsibilities and rights are law in relation to the position that humans hold within the verticality and horizontality of the land. This is not always a harmonious relationship. There is very little idealisation in Black's insider's account. There is violence, misunderstanding, miscommunication, conflict. Yet, all this takes place within a wider cosmology of situated law:

> humanity here does not colonize cosmology. Rather, the world around humans moves from being a space subordinate to the human desires to one of a superior informant, of the human's need for survival – a survival based on the interpenetration of the knowledge found in the seen and unseen.[62]

The interplay between seen and unseen, the inclusion of the world as immanence, and the involvement of the collective in the individual are central in the way I conceptualise the connection between law and space. It is interesting how property, the most standard Western form of connection between law and space, is understood very differently in aboriginal jurisprudence. It is worth quoting from Bruce Chatwin's book *The Songlines*:

> Aboriginals, it was true, could not imagine territory as a block of land hemmed in by frontiers: but rather as an interlocking network of 'lines' or 'ways through'... Before the whites came, no one in Australia was landless, since everyone inherited, as his or her private property, a stretch of the Ancestor's song and the stretch of country over which the song passed. A man's verses were his title deeds to territory. He could lend them to others. He could borrow other verses in return. The one thing he couldn't do was sell or get rid of them. Supposing the Elders of a Carpet Snake clan decided it was time to sing their song cycle from beginning to end? Messages would be sent out, up and down the track, summoning songowners to assemble at the Big Place. One after the other, each 'owner' would then sing his stretch of the Ancestor's footprints.[63]

The indigenous understood space and law in the form of an affective, embodied and emplaced songline, individually materialised but part of the greater cosmological continuum of a song cycle. Drawn from a different context but immediately relevant is the following quote from Nathan Moore's work on law and architecture:

62   Black, 2011: 27
63   Chatwin, 1988, 56–7

> 'this space is a curious one: less a space contained within a boundary line, and much more a line that, by being infinitely divisible, seems to contain within it an infinite space, even when the line in question is of a very definite length or circumference. In short, the line implicates space, not the common sense space of everyday experience, but a space of virtuality, a space that rather than act as a background to be moved in or across, is itself in movement – a space that moves in place, without going anywhere.'[64]

This whole space, a songline that is infinitely divisible yet pulsating in place with its own movement, is moved from within by the indigenous tribal song. This could be seen as an application of Deleuze and Guattari's concept of *refrain*, namely an expression of rhythmic difference that marks a territory against other territories. The discussion in Deleuze and Guattari spans various types of refrains, from bird songs to Wagnerian *leitmotifs*, all of which have the same function: to mark territory, and at the same time to make this marking obvious to other refrain-singing bodies. Through a refrain, a territory acquires perhaps the most important of its characteristic: it becomes *expressive*, it assumes qualities that differentiate it from others. A refrain endows an assemblage with spatial range and temporal consistency: it allows an assemblage to carry on altering itself while at the same time, allowing it to ground itself by literally bringing it down to earth: 'the role of the refrain ... is territorial, a territorial assemblage ... it always carries earth with it; it has a land (sometimes a spiritual land) as its concomitant.'[65] The aboriginal song is the expressive medium of the territory, holding land, earth and country folded within, but without differentiating between its line and the law. The law is both land and its expression through the song. Stephen Muecke's work shows how the Australian aboriginal word for law is also the word for land, and is experienced as gut feeling for country.[66] Law and space are not different things but a spatial and embodied knowledge that fuses human and nonhuman interiorities and physicalities. Chatwin again:

> white men, he began, made the common mistake of assuming that, because the Aboriginals were wanderers, they could have no system of land tenure. This was nonsense. Aboriginals, it was true, could not imagine territory as a block of land hemmed in by frontiers: but rather as an interlocking network of 'lines' or 'ways through'.[67]

---

64   Moore, 2013: 59
65   Deleuze and Guattari, 1988: 312
66   Muecke, 2008; Shoemaker, 2004
67   Chatwin, 1988: 56

It is my intention to claim something of the above for the spatiolegal connection I am sketching here. Thus, by spatiality I mean an emplaced materiality that affects the way the law is embodied. As I have shown, law's materiality is not just courts and wigs but the way the law emplaces itself, its measures, commands and prohibitions that determine the distance and propinquity between bodies. Law determines these spaces but also defines which bodies are to be brought in affective contact with the law, and through the law with each other. At the same time, being part of the assemblage, law is affected by these bodies and connections to such an extent that the law has to move, thereby affecting its extensive qualities, its spatial emplacement. Law emerges in various assemblages without tying itself to any one in particular. Law is carried by and within the bodies. It does not exist somewhere out there – there is no out there. All is space; all is continuum. Bodies embody the law, carry the law with them in their moves and pauses, take the law with them when they withdraw. Elizabeth Grosz puts it eloquently: 'bodies speak, without necessarily talking, because they become coded with and as signs. They speak social codes. The become intextuated, narrativized; simultaneously social codes, laws, norms, and ideals become incarnated.[68] What is more, bodies bring out the law even when the law is not here (at least not immediately visible): regulations trammel our everyday life in ways that are more insidious than a simple top-down legal application. We are all law-makers and law-enforcers with and against each other.

Law, and specifically the law that Deleuze and Guattari have in mind and from which Doel takes his inspiration, is the Oedipal law, the law of the Father, law as *logos*, rationality and language. This law traditionally fixes space, turns it into points, tight measurements of distance and propinquity, normative geometries, lines of connection that do not allow any excess to surface. Space is terminally unfolded by law, spread like a canvas on which legal operations take place. This kind of law reduces space into its own saturable, controlled context while refusing to (admit that they) operate together in a folded becoming. As I have shown in Chapter 1, law is typically spacing itself away from space. It turns against its own turning, unadventurously returning to the banality of the locality, the incantation of the particular and the hasty concealment of a certain fear of space and its manifold, uncontrollable, unpredictable folding. Or, it recoils to the well-rehearsed fantasies of waiting, of the temporality whose finality brings redemption, or even the 'now' of revolution. Yet, law is not just that. There is another law that functions along space, folding and unfolding while eavesdropping on spatial operations. This law is not necessarily prohibitive, exclusionary or hierarchical. Its intimate connection with space means that

---

68   Grosz, 1995: 35

this law is perfectly aware of its spatiality, and understands its operations to be stemming from and returning to it. To law as *logos*, Deleuze and Guattari oppose *nomos:* 'the *nomos* came to designate the law, but that was originally because it was distribution ... one without division into shares, in a space without borders or enclosure.'[69] *Nomos* is the uncountable, incalculable law that distributes emplacements and *lines of flight* (namely, creative processes that push the limits of immanence) on smooth space. *Nomos* is 'wedded to a very particular type of multiplicity: nonmetric, acentred, rhizomatic multiplicities that occupy space without "counting" it and can "be explored only be legwork"'.[70] *Nomos* is also connected to what Deleuze has referred to as *jurisprudence*,[71] the space of flows of desire that is addressed not towards a lack, such as the always elusive Lacanian *petit objet a*, but towards an empirical, context-specific plenitude conditioned by the open potentiality of the virtual.[72]

For Deleuze (and Guattari), the *logos* of hierarchy and stricture is nothing but a repeated call to transcendence. To transcend is to offer the teasing connection to a prior Other (Levinas), to a more worthy 'quasi-concept' inside metaphysics (Derrida), to an instrumentally necessary idea that will guide reason (Kant). To transcend is to reduce one to one's own limitations by showing what one has not achieved: that there is always more one could have done but will never be able to. Law's commands are filled with impossible leaping feats that demand a moralising reaction. Law's disjointment from its own spatiality allows an overblown transcendentalism to emerge. 'The law is not enough; we are asking too much of the law; law is not the answer': typical transcendental admonitions. Not because they allow the law to stay within its limits and limitations (this can be a good thing, considering that a body has no outline), but because in essence they probe the law towards a continuous and self-reproducing inadequacy. To put it differently, the problem with transcendence is that it impedes the ethical (immanent, ontological, inscribed within one's capabilities) and

---

69   Deleuze and Guattari, 1988: 420. This differs radically from Schmitt's, 2006, *nomos* since here *nomos* does not distinguish between friend and enemy, but constitutes a passage.

70   Deleuze and Guattari, 1988: 409

71   'What interests me isn't the law or laws (the former being an empty notion, the latter uncritical notions), nor even law or rights, but jurisprudence. It's jurisprudence, ultimately, that creates law, and we mustn't go on leaving this to judges'. Deleuze, 1995: 169. See Lefebvre, 2008 for an analysis of the use of the term and Moore, 2007: 34, who defines jurisprudence as 'the mode of practising the law, where the law is engaged with anew in each and every situation'. Deleuze uses the term creatively, as a contextualised, positive and problem-specific thought located in the middle of concrete situations. Jurisprudence is a philosophical thought that amounts to praxis, and that facilitates the production of radical encounters between bodies and, consequently, the 'prolonging of singularities'.

72   For a comparison between Lacanian and Deleuzian desire, see Milovanovic, 2007; also Smith, 2007. On emergent law see Murray, 2006 and 2013.

encourages the moral (expansive, colonising, capitalist).[73] Immanence pushes from the inside, whereas transcendence pulls from the outside. Ethics is inscribed in law's immanence, whereas morality is the demand on the law to transcend itself, try more, be something other. The stentorious retort to such demands, however, must be one of immanence, namely a continuum that is both vast and closed upon itself, internally proliferating through its ruptures, but without waiting for salvation from the outside. This immanence can only be achieved through the conceptualisation of law *as* space, and space *as* law. At the same time, spatial immanence allows for a proliferation of law within: law is not just *logos* but also *nomos*. Despite first impressions, *logos* and *nomos* are not different laws. They are overlapping, interfolding aspects of the law. The Deleuzian/Guattarian preference for the nomic must be modified and contextualised in the space of immanent continuum. No doubt, law provides readily available avenues of thought and action, it binds expectations of how to move, and in that way binds thought and behaviour in narrow, blind corridors of striation. But at the same time, law makes its own walls collapse, betrays expectations, reveals smoothness where only pillars used to be. Even in its quality to be disobeyed, the law opens up smooth spaces of new distances and proximities. This is an intensely spatial process – nothing metaphorical here – grounded on material, shifting space. Law is both *logic* and *nomic*, both prohibitor and enabler. The space of law is simultaneously *striated* and full of prescriptions; and *smooth* and open with possibilities.[74]

This is nowhere more obvious than in what I have elsewhere referred to as law's repetition,[75] namely the way law produces difference (and potentially justice) through its repetitive, even sometimes mechanistic, application. What Nathan Moore finds problematic, namely the self-referential nature of the law, I consider law's only way to remain immanent. In his influential reading of Deleuze, Moore finds that the law produces

> an uncritical and unthinking mode of legalistic being that never departs from the vicious circle of interminable self-reference. On the other hand, jurisprudence is the mode of working through, and acts as the event or abstract machine of the legal assemblage. It is jurisprudence that demands a thinking through, while the law and laws require mere application.[76]

---

73   As mentioned, the distinction comes from Luhmann, 2004, himself following Weber, and Spinoza, 2000.
74   See Davies, 2008 for an in-depth discussion of what can be put as vertical and horizontal law.
75   Philippopoulos-Mihalopoulos, 2011b
76   Moore, 2007: 43

The distinction between law and jurisprudence is of the Roman legal tradition – because here and in Deleuze's distinction, jurisprudence is used in its continental sense of legal cases rather than legal theory – and one that seems to lose its relevance in Common Law systems. But is this really the point? The distinction is structured as uncritical application versus applied thinking of the law. This kind of distinction can be applied to both Roman Law and Common Law traditions, and in every case will be found wanting. This is not an apology for the law. It is precisely the opposite: it is an argument for the non-clear-cut nature of the law, its expansion in thinking application as well as uncritical generality. The law is not just legalistic. This would have been too simple (to deal with, understand, counteract, oppose to, revolt against). Rather, the law repeats itself, every time producing a difference inscribed in its self-referential immanence. Within the folds of its repetition, the law might render itself just, might turn itself into justice, as I show below in Chapter 5 when I talk about the connection between the lawscape and spatial justice. The law is both legalistic *and* just.

While this is not the kind of book that attempts to answer the question of what the law is, it is important to mention this: my referring to the law as simply 'the law' rather than this law or that law or the law there or the law then, I am emphatically not attempting to essentialise the law. Rather, my objective is diametrically opposite: I am aiming at *banalising* the law. By mentioning the law without qualifying it and, as we shall see below, in situations with which the law seemingly has little to do, the law becomes as ubiquitous yet as imperceptible as the air one breathes. Naturally, air quality variations interfere with breathing. Likewise, the law puts on its appearances with different intensities at different times and spaces. Its main characteristic, however, is the way in which the law is folded with space at all times, whether as *logos* or *nomos*. In that sense, talking about 'the law' denudes the law from a potential impression of functional priority it might have over other social institutions. By allowing the law to emerge everywhere, and by pointing to such emergence, this text contributes to a generalised attempt that aims at making the law leave its conceptual closure and start encountering its own multiplicity. This closure is often accompanied by a certain sense of superiority, of überfunction, of a peculiar ability to provide 'the' solution (one needs only briefly to speak to a bunch of second-year law students in order to get a sense of law's own self-assurance and confidence in how relevant and necessary it is). Indeed, law's decision-making abilities (via court judgments, arbitration decisions, statutes, legal titles and so on) are definitive enough, absolute enough and irreversible enough to convince anyone of such relative omnipotence. But what happens when the law turns out to be not just statutes and judgments, in short not just text, but also symbol, and then building, human body, animal, land, movement, pause: what happens when the law is nothing more than just one part of an assemblage with other bodies? What happens to law? I would like to suggest that 'we' of a critical sociolegal persuasion are fortunate to observe but also to be active parts of a process of productive banalisation of the law, and its slow fusion with other forms of disciplinary thinking and acting.

Where do 'we' then stand after this? Let me briefly sum up: space as manifold is a whirlwind of bodies generating and becoming space. These bodies are physical human and nonhuman, and even 'ordinary human bodies', such as the human body or the legal corpus turn out to be assemblages of human and nonhuman bodies. Each of these bodies emerges with its own contextual agency, which I have described as material agency, itself always spatial. Materiality, however, is co-extensive with immateriality. In that sense, both law and space are simultaneously and in parallel material and immaterial, body and idea. Law is statutes and judgments, but also movement, senses, distances, all of which keeps on repeating in the manifold. Law is immanent to space, and space is immanent to law. The manifold that includes space and law keeps on unfolding in materiality and immateriality, labyrinthine, continuous, actual or virtual, everywhere and everyware, including all the emerging bodies that produce the many folds of the manifold. But these bodies differentiate themselves from other bodies: they move faster or slower, they withdraw from the continuum, they pause and let the world whirl around them. These are the ruptures on the skin of the manifold. The ruptures do not break the continuum but constitute it. Each body is a singularity, forever withdrawing, yet forever claiming its material agency, its singularity in space through its conatus. This is the ontology of bodies. And this is the epistemology: we need to rupture the continuum, create pockets of difference, regardless of whether they already exist ontologically or not, and regardless of whether they are illusions, necessities, or necessary illusions. Smaller ruptures make bigger ruptures. Bodies make assemblages make bodies. Yet with rupture comes the responsibility of situatedness, as I show below. This is both ontological and epistemological. Let me therefore briefly explain my ruptured epistemology, before moving to the first large rupture of this book, namely the lawscape.

2.2 Posthuman epistemology

A posthuman epistemology is a situated epistemology. It assumes the responsibility of its corresponding ontology by taking into consideration the particular epistemological situatedness, both in terms of discipline and in terms of ontological responsibility. This is the epistemology that this book follows. I return to the posthuman more specifically in the next section of this chapter, in the context of the posthuman lawscape. It is important, however, to note the characteristics of its epistemological parallel here. First, posthuman epistemology is a radically interdisciplinary epistemology. Rosi Braidotti puts it in terms of both a challenge to humanities and an opportunity to reach out to new constitutive horizons.[77] Cary Wolfe puts it more specifically as the necessity for any discourse or critical procedure to take account of the constitutive (*and* constitutively paradoxical) nature of its own distinctions, forms and procedures'.[78] Why is this any different from a critical interdisciplinarity? For the simple reason that posthumanism decentres the centre of observational ability. The epistemologist who has an overview of the field of observation is the only one responsible for her choices, and her observations are contingent, on the one hand, upon the observations of other epistemologists, in their turn rupturing and undercutting the observational field; and, on the other, the grand ontological continuum that includes the epistemologist while at the same withdrawing from her. This is what Niklas Luhmann refers to when he talks about the blind spot of observation.[79] This is not merely a thing that eludes observation. Rather, it is *the thing itself, the body as a whole, that eludes observation.* Every body withdraws not just from observation but also from itself. We do not know the limits of a body, whether seen as a positive understanding of Spinozan power, or as a Luhmannian understanding of ignorance of the myriads of blind spots produced with every movement. Posthumanism requires both epistemological humility *and* ontological ambition. Posthumanism is intimately connected to what I have referred to in Chapter 1 as *ontological vulnerability*, namely the condition of finding oneself in the middle of the continuum, exposed in a space where skin does not separate, walls do not protect, seas do not isolate. There is nothing soothing about the continuum: it is precipitous, antigravitational, whirling, and on that precarious surface, the epistemologist attempts to construct ruptures that yield folds of distance and safety. No matter how well protected humanity feels behind its technological armour, it remains exposed to the flow that surrounds it, be this in the form of natural disasters, autopoietically-produced technological risks, human-caused accidents or hybrid forms of life between the natural and the artificial. 'Coexistence is in our face: it *is* our face. We are made of nonhuman and nonsentient and nonliving entitites. It's not a cozy situation: it's a spooky, uncanny situation'.[80] This is the reason for which the space of vulnerability exposes the futility of looking for a centre in the way life, society or anything else is structured.

77 Braidotti, 2013: 143ff.
78 Wolfe, 2009: 122
79 Luhmann, 2002: 86, 'the operation of observing, therefore, includes the exclusion of the unobservable, including, moreover, the unobservable par excellence, observation itself, the observer-in-operation.'
80 Morton, 2013: 130

Second, posthuman epistemology is directly related to the new position of responsibility in which the human has found herself. It derives directly from the Anthropocene, namely the geological era that is determined by the presence of a single species on earth: the human. The term was coined by Crutzen and Stoermer in 2000,[81] and has been the source of a cross-disciplinary debate on the repercussions of both the name and the idea of a separate geological era. One of the clearest affirmations arising from it is that 'with the arrival of the Anthropocene, this division [between human/nonhuman or nature/culture] is de-ontologized; as such, the separation appears instead as an epistemological product mistakenly presumed as a given fact of being.'[82] This advances the posthuman epistemology in radical ways: one is forced to consider again the connection between ontology and epistemology.[83] The Anthropocene augurs the de-individualisation of the human in favour of a human collectivisation as a geological agent that affects the Earth.[84] This paradoxically means that even epistemology is now ontologised through sheer geological imprint and co-relation between human presence as cause and sign of a new ontology of indistinction. Yet, although of the same continuum, epistemology and ontology remain *parallel* to each other, evolving separately yet similarly, one incubating the other while being affected by it.

This gives rise to what I would call the *responsibility of indistinction*, which refers to the ontological indistinguishability between bodies. This is what Karen Barad says: 'we (but not only 'we humans') are always already responsible to the others with whom or which we are entangled, not through conscious intent but through the various ontological entanglements that materiality entails'.[85] We are all responsible because we are all situated in a continuum of indistinction. Yet, human responsibility is heightened *because* of the indistinction.[86] The counterargument, that assemblages are just another apparatus for neoliberal practices of responsibility deferral and biopolitical control, has of course purchase.[87] There is no reason why indeed assemblages are not and cannot be used in that way, if the context allows them. Assemblages are inherently amoral. Yet, the position that a body (individual or collective) assumes with regards to the assemblage is significant, and can be mobilised politically against the body partaking of an assemblage. Hannah Arendt writes on how the responsibility emerging as a result of often arbitrary involvement in a collectivity is

81 Crutzen and Stoermer, 2000
82 Turpin, 2013
83 See Brasier, 2011
84 Morton, 2010
85 Barad, 2006: 393
86 See Lorimer, 2012 on a similar take from an environmental protection point of view.
87 See Boltanski and Chiapello, 2007; Joseph, 2013; Mirowski and Nik-Khah, 2007

inalienable.[88] Such presence can be turned both into an ethical tool and a tool of political oppression by association. Merleau-Ponty writes that a body is 'an attitude *directed* towards a certain existing or possible task. And indeed its spatiality is not ... a *spatiality of position* but a *spatiality of situation*'.[89] While indeed the assemblage 'makes us do things' that we cannot even control, to detach oneself from the assemblage in terms of demoting oneself to the position of mere pawn given to the whims of the big bad assemblage wolf, is both dishonest and rehashing of old ontological distinctions. Assemblages are bodies, and each body, individually or collectively, has some power to move the assemblage in specific directions. This is emphatically not a neoliberal discourse. It is not about one's individual, or even collective, freedom to decide one's own fate. If there is freedom in it, it is a Spinozan freedom, namely the *necessity* of self-actualisation of each body, rather than an expression of free will. Since every body is part of the continuum, freedom is the actualisation of one's situatedness within that continuum, as mediated by that continuum. This is a distinctly *collective* freedom[90] that operates as the enabling bondage of the responsibility of being situated with regards to other bodies. Freedom is a responsibility, and the responsibility of freedom is to unveil the illusion of freedom (in the sense of free will) and allow this to inform our knowledge of the causes for which bodies move the way they do. No body decides its own fate, not because there is a teleology that determines it, but because a body is part of a continuum.[91]

The second qualification of indistinction is that flatness is not equally distributed among bodies. There are stronger bodies in the assemblage. There are prejudices that count more than formal law. There is formal law that is in the service of corporations. Some bodies weigh more, whether they are a corporation, a billionaire, religion or tsunami. This means that not every body is equally able to act and react. This, however, also means that the old categories between 'the powerful' and 'the weak' are just that: old categories in the service of existing structures. When we eternally recycle these categories (and others, such as private/public, us/them, north/south, formal/informal), we remain with the problem. Of course there are inequalities. But it is much more complex than this. 'The easy luxury of guilt'[92] is as easy to feel as it is to attribute. To retain distinctions of this sort and to keep on thinking of the equation as the ones who decide as opposed to the ones whose fate is decided, is an abstraction in the service

---

88 See Arendt, 1987 who cautions against a legal understanding of collective responsibility, and Gatens and Lloyd, 1999 expanding on this and contextualizing it in terms of Spinoza.
89 Merleau-Ponty, 1995: 100, added emphasis.
90 Braidotti, 2013, refers to it as posthuman ethics.
91 Spinoza, 2000: Part IV
92 Williams, 2006: 127

of the distinction itself. All bodies are part of an assemblage, and while one is always determined by the structures, one is also able to move this assemblage, including the structures, in some way. The fact that some bodies manage to move assemblages more efficiently than others does not necessarily happen because they are more powerful, but because the rest of the assemblage allows them to move in such a way. We can go even further: some of the assemblages *desire* to be moved towards, say consumerism, environmental degradation, political apathy. Once again, this is not a mechanism of guilt attribution. On the contrary, it works as an emancipatory mechanism that resists existing categories that imprison in their determination, while at the same time resisting the idea that each body is capable of heroically changing the world, or even one's own everyday world. One is not necessarily able to change anything just because one is an individual or even a multitude.[93] But one can organise one's own body (itself an assemblage, namely a collectivity) in relation to the rest of the assemblage, in its turn in relation to the world. Each assemblage is a lawscape, and the lawscape keeps moving.

Thus, a body retains the possibility of affecting the assemblage in its own way. Assemblage-thinking allows one to reposition oneself in relation to the continuum, both in terms of freedom and responsibility. At the same time, an assemblage is also a temporal event, bearing the inscription of its past on its bodies. The situatedness of responsibility is a spatiotemporal position: 'our bodies retain the traces of past modification; and those modifications in turn reflect the effects of modifications of other bodies, in chains of causal determination going back to the distant past'.[94] The difficulty of situating oneself responsibly in relation to a past that determines present and future is not easy to deal with.[95] The answer lies in the possibility of redescribing things in a way that neither absolves a body from the responsibility of situating itself, nor inebriates this body with the illusion of control of the assemblage, or indeed the whole world. The question now becomes how to take advantage of the human omnipresence and not be fooled by the superficial impression that to be *everywhere* equates to being *central* to everything.[96] It is clear that responsibility now becomes fully spatialised: it is the responsibility of situating one's body within an assemblage of indistinguishability. The spatiality of responsibility is first of all *extended*, that is both material as well as not restricted to the immediate or the local. Doreen Massey puts this as 'the Russian Dolls issue of care and responsibility: we always begin with the proximate, home, and then move outwards. But care diminishes as we move out'.[97] We might think we try

93   *Contra* Hardt and Negri, 2001
94   Gatens and Lloyd, 1999: 81
95   Amin, 2008 on 'ethics of the situation'.
96   Chandler, 2013
97   Massey, 2004: 9

to be 'responsible' about our presence in a locality, even a locality that 'thinks globally' as the motto goes. But 'the problem goes beyond how to dispose of human-sized things, like the stuff that gets flushed down a toilet. What should we do about substances on whose inside we find ourselves?'[98] This is why the spatiality of responsibility is *geological, ecological, and future-tending*. Clare Colebrook writes:

> the positing of the anthropocene era relies on looking at our own world and imagining it as it will be when it has become the past ... We can see, now, from changes in the earth's composition that there will be a discernible strata that – in a manner akin to our dating of the earth's already layered geological epochs – will be readable.[99]

Our geological imprint situates the responsibility not just with regard to other humans, not even with regard to future generations of humans, but with regard to the globe which might outlive us. The Anthropocene demands a *geophilosophical* situatedness of thought in relation to the earth. Reza Negarestani writes:

> geophilosophy is a philosophy that grasps thought in relation to earth and territory ... it is a philosophy that, perhaps unconsciously, grasps thought in relation to two traumas, one precipitated by the accretion of the earth and the other ensued by the determination of the territory. While the former trauma lies in the consolidation of the earth as a planetary ark for terrestrial life against the cosmic backdrop, the latter is brought about by a combined geographic and demographic determination of a territory against the exteriority of the terrestrial plane and fluxes of populations of all kinds.[100]

These two traumas, distinctly anthropocentric and a direct effect of the Anthropocene, are specifically legal. They both require an intervention that guarantees limits, while at the same time allocating responsibility. This is an anthropocentric legality that chops up the surface of the earth in territorial modes that include other (human and nonhuman) populations as resources (rather than allowing for multiple territories as it happens with animal populations); and then delegates the whole earth to the status of resource ('ark') for the human future. A geophilosophical position, and indeed a responsibility of indistinction, attempts to mend these traumas through the only way possible. Colebrook writes:

---

98   Morton, 2013: 140
99   Colebrook, 2014: 26
100 Negarestani, 'Triebkrieg', 5, unpublished manuscript, cited in Woodard, 2013: 14

the anthropocene thought experiment also alters the modality of geological reading, not just to refer to the past as it is for us, but also to our present as it will be without us. We imagine a viewing or reading in the absence of viewers or readers, and we do this through images in the present that extinguish the dominance of the present.[101]

In other words, the responsibility in the era of the Anthropocene is one of *withdrawal*: withdrawing from the present in order to read it as nonhuman future; withdrawing from centrality while retaining omnipresence; withdrawing from ourselves in order to read our traces. Withdrawal is a spatiotemporal move, fully embodied (rather than metaphorical), fully active (rather than passive), brimming with situated understanding (rather than retreating haphazardly).

2.3 The lawscape

The exact origins of the term 'lawscape' are not easy to establish, though it appears to have been independently applied within academic work across continents. The first example I am aware of took place in 1987 at a mock trial by and against lawyers in Atlanta, Georgia, where the term was merely referred to as self-explanatory. It is implicit that it denotes the always-already proliferation of legal norms in which especially a new lawyer inevitably lands. The lawscape was almost used as a criticism by lawyers and against lawyers. There is a strong ironic and even comic element in this mock trial, which otherwise followed strict court procedure, and produced a strong consensus for conviction for such crimes as 'billing practices, trial process abuses, and the rise of a competitive "what's in it for me" mentality'.[102] In Australia, Nicole Graham's 2003 PhD thesis entitled *Lawscape: Paradigm and Place in Australian Property Law*, and published in 2011 as a monograph, dealt with the ramifications of the spatiolegal in the Australian country.[103] I first used the term in 2007,[104] having 'encountered' it while cycling in Copenhagen and feeling surrounded (while at the same time embodying) a net of regulations that were infravisible and that applied to my peculiar posthuman formation of me and the bike. Nicole Graham shared with me that she 'encountered' the term while driving in the Australian bush and noticing the dephysicalisation and intense legalisation of the landscape. Her use of the term is different from mine since Graham focuses on the rural and is closer to issues of property rights than mine. These three provenances (and it may well be the case that other

---

101 Colebrook, 2014: 30
102 Goodwin Dunleavy, 1994: 259, of the original 1987 court transcript.
103 Graham, 2011
104 Philippopoulos-Mihalopoulos, 2007b, 2007c and 2007d

scholars have utilised the term in different ways at different times) are a remarkable indication, not only of the plurality and proliferation of lawscapes, itself an indication of the ubiquity of the spatiolegal; but also of the way in which the lawscape can be utilised as a strategic tool to comment on and hopefully change the very lawscape within which it emerges.

When I first suggested the term *lawscape*, my focus was the connection between law and the city, and especially the variety of urban lawscapes across jurisdictions.[105] Since then, the term has developed, both through my writing and in the way it has been received by other scholars either critically or in applied ways,[106] and has returned to me in a proliferated, sharper form. The lawscape I am interested in here refers to the spatiolegal in general, without differentiating between the urban and the non-urban. In what follows, I briefly ruminate on what law without space and space without law might be, moving on to define what kind of law and what kind of space constitute the lawscape, and then to define the lawscape itself.

I define the lawscape as *the way the tautology between law and space unfolds as difference*. On some level, the neologism risks making the individual use of the terms redundant. For what is space without law or law without space? The previous discussion on emerging spaces has offered a manifold answer to this. Anything less than that would be artificial. Space without law can only be this fetish of absolute smoothness, the absurdity of a holy city of justice (which, theologically refers to society as a whole), perpetually floating in a post-conflict space where everything is light and forgiveness. Likewise, a law without space is a law without materiality, this other fetish (this time of legal thinking) that considers law to be a universal: What? Thing? Breath? Divine will? Act of violence? Both law without space and space without law are fantastic beasts that operate at best as horizon and at worst as cheap rhetoric. As I have shown, law is always spatially grounded, embodied, materially present. Law as an abstract universal, free from the constraints of matter and bodies and space is one of the illusions that law itself (and some strands of legal theory) insist on maintaining. Law as control is by necessity material (meaning spatiotemporal and corporeal), for it is only through its very own emplaced body that the law can exert its force. Law comes nowhere but from within the controlled, their bodies of appearance and the corridors on which they move, as Foucault's work and subsequently "governmentality and postcolonial theory have taught us.[107] This is more than just biopolitical control, since it addresses the material nature of the law itself: only from within matter can law control. Thus, to posit a law without space is tantamount to positing, say, a universal human

---

105 Philippopoulos-Mihalopoulos, 2007c
106 Bottomley and Moore, 2008; Carr, 2013; Young, 2013; Jeyaraj, 2013; Pruit, 2014; Arvidsson, 2014
107 Bhabha, 2005; Dean, 2010

right that applies to everyone without the need for contextualised application. The latter is not 'just' the context. On the contrary, as I have discussed above, it is the supreme need to close in and eavesdrop on the particular body's specific circumstances. Even in the theoretical, indeed *horizon*tal, possibility of a just space, law's withdrawal would be a material one, its movement traced on the skin of the world, its back turned to the spatial deification. In its turn, a just space has captured time itself, engraved it right *here* in its Edenic intramuros. There is no other way: Fainstein's 'just city' is an Augustinian theological concept and cannot accommodate anything that falls sort of divinity.[108] And the risk of course is that this can always be co-opted on behalf of cheap demagogy.

Let me first look at what law is in the lawscape. This will take some analysis. To start with, law in the lawscape is not just the standard, written law. It is that too, and this includes both its actual form as the law that, say, directs bodies in a prison, and its potential, latent form that might emerge at any point of a social interaction. In this context, David Delaney refers to the nomospheric juridification that readily converts rules of social behaviour into illegal actions.[109] Future law is already part of the legal system, according to Niklas Luhmann.[110] The law is the main operation of the topology of the autopoietic legal system. Law is whatever is included in the legal topology (space). There is no law until there is law (time). (Autopoietic) law is a spatiotemporal emergence that cannot be predicted with certainty, since it is always given to contingency. Everything is potentially law if the legal system understands it as law. Circularly, the legal system is defined as the sum of laws at any point in time. It is not just state law or corporation law or social norms or even natural law – it is the *virtuality* of the law that is potentially law at any point.

While this is important, the law I am interested in goes further than that. It is not just future or potential law that is included, but law as life of the bodies that bear the law in them. Law as state command and law as embodied command are not that different. Margaret Davies has put it memorably in her formulation of flat law:

> flat law, as I have called it, names and prioritises the multiple locations, manifestations, and interpretive possibilities of law. Flat law is state law understood as a national and cultural artefact which cannot in the end (or even in the beginning) be separated from the social spheres of which it is such an intrinsic part. It is state law expressed, constructed and transmitted by relationships between people.[111]

---

108 Fainstein, 2010
109 Delaney, 2010
110 Luhmann, 2004; also see Philippopoulos-Mihalopoulos, 2009
111 Davies, 2008: 289

State law spreads on the bodies of its emergence and determines relationships between people, as Davies writes; but it also spreads between people and things, between bodies material and immaterial, that demand a different human positionality. The law of the lawscape is state law (and by this I mean, national, transnational, international, supranational law) but also the law of space that brings bodies into encounter with other bodies, which in its origin or beginning might be state law but by the time it is incorporated, its origin as state law cedes priority to the emergence of a specifically situated law. The law in the lawscape is co-determined with the space between bodies (as in a social interaction or a multinational treaty or the slave trade); the space that is produced and is occupied by bodies; the movement of bodies; the desire of bodies; and the withdrawal of bodies for another law. At the same time, the law in the lawscape is born in the distance between these bodies: each lawscape is its own law and its own space, singular yet also part of the greater lawscaping continuum. Each lawscape withdraws from the continuum of other lawscapes, just as each body withdraws from the continuum with other bodies: this is why the lawscape is a rupture. At the same time, the lawscape is a continuum unto itself: it has no a priori limits; it moves along the bodies of its emergence; it continues for as long as its bodies continue to lawscape.

Law in the lawscape is what Spinoza calls *rules for living* and what has been variously understood as the normativity of the everyday, legal cultures, the production of norms that are not 'strictly legal' yet contribute to the production of the law, and so on.[112] Spinoza's rules for living were ways in which one could achieve the best from living in society (such as to get along with other bodies, to follow accepted rules, to seek sensual pleasures and so on) and had a prescriptive character.[113] Yet Spinoza's teleology is always mediated by the univocity of his philosophical schema. In that sense, I understand Spinozan rules for living as the law that determines the way one body encounters another on the continuum, and how the effects (on these or other bodies) of an encounter are being dealt with by the continuum. Although law production is a prescribed process, the law itself is not. It changes according to its context, which is the space and bodies that carry it. Robert Cover puts it thus: 'Law is a force, like gravity, through which our worlds exercise an influence upon one another, a force that affects the

---

112 See for example Sarat and Kearns, 1995; Lowe, 2000; Sacco, 1991; Nelken, 2012
113 Spinoza, 2009; see also LeBuffe, 2007 on the normative power of the rules, and Kisner, 2011 on how these rules are propelled by conative considerations. I have decided not to focus on the rather more conservative writings of Spinoza on laws to be found in *A Theologico-Political Treatise* because there a sovereign is always assumed to be present and responsible. What I am trying to construct here is a law without a centralised sovereign but of acentral emergence which, at the same time, is not natural law since it is based on each body's incorporation and acting out of the law.

course of these worlds through normative space'.[114] Yet, the law that I am interested in is not merely operating 'through normative space' but *is* normative space and normative body and normative movement. This is the law in Agamben's study of monastic life, the 'vital precepts' that refer 'to the rule insofar as … it can coincide, not only with the observance of individual precepts, but with the monk's entire life'.[115] What is significant is not so much that the monk obeys the law but that the law becomes co-extensive with life. The result is what Agamben calls 'a zone of undecidability with respect to life. A norm that does not refer to single acts and events, but to the entire existence of an individual, to his *forma vivendi*, is no longer easily recognizable as a law'.[116] Law becomes the body of the monk to the point that neither life nor law can claim independence from each other. Although not all bodies are monastic, the lawscape makes bodies co-extensive with the law, to the point of law's (at least partial) invisibilisation: as I show in the following section, the invisibilisation of the law in the lawscape takes place because of the law's corporeal nestling in the bodies of its emergence. And the law variously appears as freedom, choice, preference, desire.

To this, I would like to add another consideration, rather obvious by now in view of the earlier posthuman references: nonhuman bodies as well as inanimate bodies and objects are included in the concept of the body. Just as a body, an object is already functionalised, normalised, never independent of its normative position in the world. Every body/object determines the functions and normalisation processes around it – it generates its own *zones* and as Timothy Morton writes, it emits its own spacetime in which its surrounding space becomes captive, fixed firmly to the spot.[117] *Every body lawscapes.* The law in the lawscape emanates from every body, without one discernible origin. In that sense, human, natural, artificial bodies come together in determining and being determined by the law. For this reason, I would talk about the law as an expansive *institutional affect* that permeates the formal and the informal, the abstract and the material. This expansive and diffused form of law is not necessarily a good thing. Its flipside is that it might easily act in compliance with the current surveillance culture. A diffused law takes few risks and delegates

---

114 Cover, 1986: 9–10
115 Agamben, 2013: 25
116 Agamben, 2013: 25
117 Morton, 2013: 144, while disembedding us from the world. Morton's object zones are not direct experiences but set of shifting emissions in which we are left fuddling for how to behave, 'how to dispose myself relative to the zone … I can feel the irreducible dissonance between my idea and the zone'. Objects, as we shall see in Chapter 5, withdraw from each other (and from humans) in ways that render both zoning and withdrawal ontological categories, thus affecting the way spatial justice is being practised.

conflict resolution to what it considers to be higher levels of judgment-making – indeed, to go back to Spinoza, a sort of guardian authority that pursues efficiently the individual interests of its subjects:[118] we are happy to be in a 'nanny state', we are not concerned about the problem of political apathy. The more diffused law is, the easier it is to engineer it in ways that, on the one hand might cover specific needs, such as issues of belonging, constructions of home and community, as well as emplacement; and on the other, encourage legal subjects to recede from actively questioning the law (complacency or reassurance). This is how a lawscape becomes atmosphere, as I show in the following chapter: by diffusing the law in the various bodies and spaces to such an extent that the lawscape itself withdraws. Even so, things can on occasion overflow, exceed themselves and embark upon a flight of radical self-redefinition. In such cases, the already 'contagious' (in the sense of epidemic imitating[119]) nature of the legal affect doubles up and becomes rapid, horizontal and fiery, engendering such eruptions as demonstrations, revolts, revolutions, coups. In all these cases, the law does not leave the stage. It is merely supplemented by a different direction, a higher velocity and a possibility for reorientation.

Space in the lawscape, on the other hand, is the continuum of material and immaterial bodies that includes humans, nonhumans, linguistic bodies, disciplinary bodies, buildings, objects, animals, vegetables, minerals, and so on. This continuum is ruptured by the singularity of each body, expressed through its differentiated movement and its ontological withdrawal. This manifold spatiality is a manifestation of what I have elsewhere called, based on Guattari's *Three Ecologies*,[120] the 'open ecology',[121] namely the assemblage of the ecological territories of the natural, the human and the artificial. These 'territories' are found in a relation of interfolding, or as Andrea Brighenti puts it, 'a series of territories, which can be thought of as superimposed ... or mutually exclusive ... or even crisscrossed and overlapping'.[122] Hinterlands, the globe, outer space, hybrid technohumans, prosthetics, technologically manipulated meteorological phenomena, and so on are all 'territories' grounded in space. Whatever their position, they emerge as an ecological continuum, or what Vicky Bell following Whitehead calls 'the creative moment of concerns between elements in relation to each other',[123] whereby concern is the mode in which the elements attend to each other and through which they achieve their self-attainment. Space in the lawscape is emergent, manifold, contin-

---

118 Which for Spinoza, 2000, is the State.
119 Tarde, 1903
120 Guattari, 2000
121 Philippopoulos-Mihalopoulos, 2011a
122 Brighenti, 2006: 80
123 Bell, 2012: 112

gent, withdrawing, labyrinthine, given to disorientation, lack of direction and easy predetermination. It poses a challenge to the law but also to geography. It requires different strategies than the ones relying on measurement. It also goes beyond relationality, which although important, does not exhaust spatiality. As I have already mentioned, bodies are never ontologically fully present, and therefore never fully relating. There is always a part of a body that withdraws. Bodies are space, and in its turn space always withdraws from being fully present. This is not a phenomenological observation of the type 'we cannot fully perceive space'. It does, however, have its epistemological parallel: one can never *know* space fully. The fact that the forest has no outline, as Deleuze says, means that it is never fully present to any body, whether this is an ambler, a fox or a satellite. We do not know what a body can do.[124] Spinoza's maxim, made explicit by Deleuze,[125] means not only that the capabilities of a body are infinite (in the manner of Nietzsche), *but also that corporeal space is characterised by a constitutive ignorance (epistemologically speaking) and an immanent withdrawal (ontologically speaking)*. This means, not only that one cannot fully know space but also that space can never be fully present.

The lawscape operates as a continuous manifold in which law and space emerge interfolded. Being a manifold, the lawscape does not constitute a new unity. Rather than positing an origin of a fusion between law and space, it builds on an existing *becoming* (becoming lawscape entails that space becomes law becomes space *ad infinitum*). It does not assume the role of a synthesis since it does not presuppose a dualism between space and law. On the contrary, it assumes that the two have always shared the same ontological surface and even the same epistemological lines of flight.[126] In this sense, it is an always-already emergence of ontological and epistemological parallelism. There is no causal link between space and law on the one hand, and the lawscape on the other, since there is no distance that

---

124 Spinoza, 2000, Book III, 3s
125 Deleuze, 1990
126 David Cunningham, 2008, revisits Lefebvre's Rome as the locus of the production of abstraction as a direct result of the city's legal impositions. Rome rather than Athens is the first to do this, not so much because of a comparatively larger geographical expansion but, I suggest, because Rome achieved the invisibilisation of its law by abstracting it to the point of material tautology with the city itself. The Athenian spatial mythology of law and politics was but a preparation, a journey of discovery towards the Roman invisibilisation of the law. Rome acknowledges the hitherto inseparability of law and the city – the Greek *polis* – and delivers the coup that eventually operationalises this inseparability and expands the *polis* to *urbi et orbi*, to city and the world, namely to the space of an empire. Thus, from being co-originating, the two now become tautological. However, this tautology unfolds as difference. Ruptures, conflict, and excess are squeezed between the fractious surfaces and render the lawscape 'febrile', in the platonic sense, 2000, of necessary conflict within the city.

needs to be bridged, no logical step that needs to be taken. Likewise, there can be no causal attribution in the lawscape – say, law comes from space, or space comes from law. Space and law emerge as lawscape, in their excessive tautology, their parallel difference.

Let me unpack this. On the epistemological level, the link between space and law is tangible: the one operates as a means of a better understanding the other, or at least certain aspects of it. Thus, law's obsession with naming, categorising, organising and 'tidying up' is revealed in the fact that space, despite its elusiveness, somehow 'works': whether through property regimes, state boundaries, zoning etc., space allows bodies to move in a certain way (while no doubt impeding them too). Conversely, spatial acentricity helps visualise law's materiality, especially its relation to violence in the sense of its force of perception/application,[127] its attempt to control power struggles, and its role in the process of capital production and consumption.[128] Seen through a spatial lens, law's presence is magnified to a deafening extent: planning restrictions, environmental regulations, zoning, social control, borders between private, public and restricted access areas, pavements, roads, traffic lights, metro barriers, flow of people, headscarves in schools, hoods in shopping malls, power architecture and landscaping, are just a few of the lawscaping moments. Space renders law's presence concentrated and overt, in close contact with the production, consumption and disposal processes.[129] Space is the great testing ground for law, its loudspeaker and its gaming table. And, in turn, the law is space's measure, the (in)flexible, (un)reliable metallic ruler that makes its presence felt through inches and centimetres of propinquity and distance, determining identity and difference. Law is the regulator of spaces between places, connecting and severing bodies.

The facetious question, therefore, whether the law determines space or space determines the law can only be answered circularly.[130] It is indeed the case that law is put in the service of spatial order with a view to producing a better society (and one only needs to think of Plato's normative *polis*, Descartes's 'enlightened' space, or More's *Utopia*), just as spatial conditions beget different legal responses (such as Rousseau's envisaged role for the law and the police against what he thought of as inherently 'scheming' and 'depraved' urban nature[131]). However, this is somewhat misleading: the point is not so much that different spatial needs (even differently perceived) create different legal reactions and vice versa. The latter is only to be expected, and has already been captured by what the editors of the *Legal*

---

127 Delaney, 2002: 79
128 Lefebvre, 1991
129 Bauman, 2003: 106
130 For the same question but formulated between space and society, see Harvey, 2000
131 Rousseau, 1960: 58

Geographies Reader call the 'irreducible interpenetration' between law and geography.[132] What is important here, and has until now remained largely uncommented, is the ontological interweaving of desire and response to desire of the bodies involved: the circularity between law and space in the form of an always-already desire to be conditioned by the other. The desire for the other to intervene is immanent to the body of the other. In that sense, the desire comes even before the need to invite. Neither merely a tautology nor a simple disciplinary coincidence, lawscape is the continuous *becoming* between space and law, emerging from their parallel connection of invitation by the one (the law/space) to be conditioned by the other (space/the law). The lawscape is a continuous manifold where law and space are fused in an embrace of escaping distance.

The embrace unfolds as difference. The tautology is excessive, and the continuum is ruptured: the crux of the lawscape does not lie in its reconfiguration of legal and spatial epistemologies, or the tautology between law and space; rather, it lies in its constitutive ontology of in/visibilisation. From noun to verb, lawscape is always a process. *The lawscape is the interplay of in/visibilisation between law and space.*[133] The act of *lawscaping* is the immanent process of the lawscape that in/visibilises space and law depending on the conditions. In the lawscape, law and space are in a constant process of negotiation of in/visibilisation. More internally, and perhaps more accurately, law invisibilises its spatiality and space invisibilises its legality, since law is spatial through and through, and space is legal through and through. The internal fold repeats itself fractally ad infinitum. What I am interested in here is that law and space have always been co-emerging, yet in a process of mutual in/visibilisation. Not entirely though: in/visibilisation happens in degrees and through various tools, depending on the needs of the lawscape at any point. *Neither law nor space disappear fully from the lawscape.* This, as I show in the following two chapters, is the atmospheric ambition of the lawscape. But until that point, that is until and if ever the lawscape becomes atmospheric, the lawscape regulates its chiaroscuro of in/visibilities according to its needs. These needs are not transcendental but immanent: the bodies in the lawscape determine what and how it gets in/visibilised. The lawscape is not here to maintain some sort of public order. The process is not a top-down imposition but a negotiation that rests on explicit and implicit desires. Thus, law can be visibilised when disciplinary conditions need to be enforced more stringently; or, space becomes more visible when consumerist or leisure practices need to be encouraged. There is a teleology behind this: first, in/visibilisation is a necessary ontological feature of both law and space,

---

132 Blomley *et al.*, 2001: xvi. However, they subsequently equate the connection to 'an identity', which is markedly different from what is suggested here.
133 Philippopoulos-Mihalopoulos and FitzGerald, 2008

since it allows them both to carry on with their self-perpetuating myths. Law's myth of universal applicability is not fazed by concepts of legal pluralism, since behind it all, there is always law – just a different kind of law. This of course becomes strongly qualified by understanding law as an always-already lawscape. Regardless, the law *needs* its self-perpetuating myth for questions of identity, disciplinary distinction and so on; likewise with space. Space is mythologised as *accueil* of difference in the form of a new land of opportunity and supposed freedom; as multipolar, which conceptualises 'private' or 'place' as free from legal intervention; as exoticised and sexualised; as the breeding grounds of communitarian nostalgia, or indeed as the healing locus of escapism. The list can go on. The point is that these are self-perpetuating myths on both sides which, in order to carry on having some purchase, must engage in the process of in/visibilisation.

The second purpose of the in/visibilisation is the perpetuation of control over bodies and spaces. Following Foucault, one would say that the lawscape is part of a political technology of the body that invisibilises the law, and is subsequently shaped as (spatialised forms of) disciplinary institutions within the sphere of public morality.[134] This is not, however, the ultimate purpose of the lawscape. Control is a by-product of the *conative* trait of the lawscape. The lawscape is not a tool in the service of the elite but a diffused ontology of desire that cannot be controlled centrally. This is not to contest the incontestable, namely that bodies are controlled by stronger bodies, in the form of disciplinary institutions, corporations, political elites and so on. But it is important to remember that there is no one sovereign power deciding on the state of in/visibilisation. Mariana Valverde puts it accurately when she writes that 'governance techniques do not necessarily have a built-in or default politics'.[135]

The lawscape is not a *landscape*, namely an ocularcentric human intervention that frames what is visible.[136] The lawscape is an emergence. Panopticon's central point has been replaced by a multiplicity of bodies, controlling each other, limiting each other's ability to move or pause. Chris Butler, quoting Lefebvre, writes:

> the body is broken down into various locations, with prescribed uses and normative values attached to them. In fact, there is a great similarity between the ways that 'space is ... carved up' and how 'the body is cut into pieces'.[137]

But who cuts the body of the lawscape if not the body itself? This is what Foucault means when he notes:

---

134 Foucault, 2003a
135 Valverde, 2011: 279
136 Deriu and Kamvasinou, 2012, as well as the whole journal issue.
137 Butler, 2013b, quoting Lefebvre, 1991: 355

the emergence, or rather the invention, of a new mechanism of power possessed of highly specific procedural techniques ... which is also, I believe, absolutely incompatible with the relations of sovereignty ... It presupposes a tightly knit grid of material coercions rather than the physical existence of a sovereign.[138]

This grid is no other than the embodied and spatialised lawscape. The difference is that the lawscape as a totality, through its constituent bodies and their spatiotemporal movement or pause, has replaced the sovereign and regulates its constituent bodies, as well as its own lawscaped and lawscaping body. This is an ontology of both invention *and* emergence, building on the desire of bodies for precisely such a grid. In that sense the lawscape shares with Arjun Appadurai's 'scapes' their emergent, arbitrary nature that accommodates a variety of confluences, but differs from them since lawscaping does not take place on the level of individual human agency,[139] and at the same time is fully spatialised.[140]

Thus, in/visibilisation is not just a game on the side but the lawscape's only mechanism of ontological continuation: its ultimate purpose. This is all that the lawscape is. This is why the lawscape makes things much harder for discourses of resistance, revolution, social change and even protest, than the usual schema adopted by classical critical theory of capitalism versus protesters, or powerful elite versus victims. The lawscape is mobilised by all its bodies and only by its bodies (nothing outside). While some bodies may be more powerful than others and consequently have a greater 'pulling' power, they do not necessarily need to fall back on this. The main issue with the lawscape is that its constituent bodies themselves *want* to lawscape (one wants to invisibilise the law when walking about in a shopping mall, wants to forget money when paying by direct debit or credit cards, wants to invisibilise the human landscaping hand when creating an English garden), and their desire *is* ontologically the lawscape. This is much more banal than a resistance strategy. It is simply a carrying-on in the form of narrating, name giving, street delineating, landscaping, train boarding, people watching – all of which are primarily practices of regulating the everyday and only incidentally can be seen as resistance, nomadism or conflict.[141] The illusionary distinction between law and space that is offered by in/visibilisation is a necessary condition for simply being,

---

138 Foucault, 2003b: 35–6
139 For Appadurai, 1996: 33, scapes are 'the building blocks of the imaginary worlds ... associated with the individual and with agency'.
140 This is also Mahmud's, 2007, criticism of Appadurai.
141 See also Valverde, 2012. This does not exclude the opposite, namely that resistance is the primary, visible reason for the quotidian. But this would presuppose and engender a legal conflict: see the example of endless renaming in Panjim in Fernandes, 2007. See also Brighenti, 2006, where the above is described in terms of territorialisation.

if this is the right word. No one wants to be reminded all the time that one lives in a CCTV-infested dystopia. Thus, a fully legal lawscape is the place where K arrives in Kafka's *The Castle*, full of procedural labyrinths, representational nooks and crannies, overseen by a towering sovereign in whom all originates and to whom all ends. Now, there is no doubt that most spaces are fully legal spaces, part of the spatiolegal continuum. But they are cleverer than this. They have learnt to dissimulate their legality, even to make it hyperpresent. *The Castle* becomes *The Truman Show*. One *desires* invisibilisation. Or, as Andrea Brighenti puts it, 'the invisible is what determines us'.[142] A lawscape whose law is constantly visibilised is a panopticon in the perfect, unprioritised Benthamite fragility of the perennially visible *and* unverifiable: a visible spatiality *and* a visible law allow no space from which either of them is to be *seen*.[143] Their simultaneous visibility and unverifiability straighten lines up, drain depths out, and bring the horizon in as inhaled claustrophobia. This is utopia at its most resplendent, renaissance and perspectival: life saturated with control: it is atmosphere.

The ontology of the lawscape is non-anthropocentric. The interplay of in/visibilisation is a geological necessity that resides within while exceeding the human. The preservation of the lawscape addresses much bigger issues than political regimes, justice and everyday survival – yet at the same time nothing is bigger than these. Political oppression, capitalist constructs of desire, allures of Western comfort are bodies from which such lawscaping practices as climate change, cancer, radiation, nuclear accidents, natural disasters, global poverty, infant mortality (the list is endless and growing) emerge and with which they constitute the lawscape. This does not mean that the lawscape opposes these, nor however that it encourages them. *The lawscape lawscapes, and all these issues are part of the conatus of the lawscape.* It does not mean that things cannot improve or indeed deteriorate. But the lawscape does not have a predetermined direction or moral value attached to it. It is, in its own way, both a tool and a body. As a body, it follows its autopoiesis, its conative effort.

As a tool, however, the lawscape can be and is used strategically. Bodies negotiate the degrees of in/visibilisation. As mentioned, the process runs on a scale of in/visibilisation that changes constantly depending on the conative needs of the lawscape (that is to say, the bodies that constitute it, collectively and as an emergence). Brighenti writes, 'power is not only exercised in seeing without being seen, but also in seeing the invisible through specific procedures for visibilising it'.[144] The quote reveals the complexity of lawscaping. The invisible can be visibilised *as* invisible, namely as special access privileges or ineffable sense of luxury or document leaks. In

---

142 Brighenti, 2010: 67
143 See Brighenti's, 2010, understanding of the connection between seeing and being seen.
144 Brighenti, 2010: 34

practice, it often remains at least partly invisible (not many read WikiLeaks in their entirety) but its invisibility is visibilised. Permutations of in/visibilisation are standard lawscaping practices: from signs on the streets to uniformed (or not) bodies strategically positioned, to makeshift corridors, to street barriers, to open doors in shops, to phone tracking to rituals and habits, these are variable valves in the in/visibilisation mechanism that shift according to the needs of the lawscape. In the process, there are positions of hypervisibility or infravisibility. Invisibilisation is not always the privilege of a power position, in the form of Gramscian hegemony that dominates from within or Durkheimian society that remains unseen by the individual in society (akin to Timothy Morton's hyperobjects[145]) or classic Foucaultian power that withdraws in order to discipline. Invisibilisation is also a tool used to keep certain bodies in the margins. Thus, the lawscape visibilises some social movements, awarding them with spatial prominence. Often, however, it also facilitates their simultaneous invisibilisation so that they get caught by the law whatever they do ('resistance against being overvisibilised'[146]). Brighenti again:

> when persons move, or are pushed, above the upper threshold of correct visibility, they enter a zone of supravisibility or super-visibility in which any action undertaken, being overly visible, becomes so enormous that it paralyses the person performing it. This is a paradoxical double bind whereby a person is prohibited from doing what s/he is simultaneously required to do by the set of social constraints to which s/he is subject.[147]

Rosi Braidotti writes about the female bodies that are positioned

> at the intersection of some formidable locations of power: visibility and media representations produced a consumeristic approach to images in a dissonant or internally differentiated manner. Female embodied subjects in process today include interchangeably the highly groomed body of Princess Diana (like Marilyn Monroe before her) *and* the highly disposable bodies of women, men and children in war torn lands.'[148]

Lawscaped in the paradoxical position of both hypervisible and faceless, bodies of refugees, victims of abuse, pensioners, children, are ping-ponged between the psychotic and the neurotic in a lawscape that moves and tilts and stops constantly.

---

145 Morton, 2013
146 Brighenti, 2010: 66
147 Brighenti, 2010: 47
148 Braidotti, 2002: 17

The above notwithstanding, in/visibilisation remains a tool with which bodies can reorient the lawscape and negotiate a different regime of in/visibility. There is no guarantee that things will turn out well – the lawscape is 'evental',[149] namely profoundly contingent and slippery when it comes to prediction. But its political potential is fully embodied and spatialised in the *here* of the process. Bodies are always assemblages, that is to say both part of an assemblage and constituted of further assemblages. This means that bodies are always collective. Depending on their position in the lawscape, namely their ability to move and to mobilise, bodies have powers to negotiate in/visibilisation, to use or diffuse the law in their advantage, to spatialise or to despatialise according to their needs. The political potential of the lawscape does not promise solutions but negotiating positions, if in/visibilisation is used as a strategic tool. From this to spatial justice the leap is not unimaginable. Indeed, it would seem that this is the only way in which spatial justice might emerge, as I discuss in Chapter 5 and 6. However, a desire for spatial justice has to fight another desire, often more powerful: the desire of the lawscape to become atmosphere and to fix bodies in positions of illusionary certainty. There is more on this in Chapter 3.

A final clarification: despite its etymological connotation, the lawscape is not about the visual. As said, the in/visibility of the lawscape is an ontological condition for the survival of the lawscape. There is no human looking here. At most, the lawscape is an autopoietic observer, observing itself observing.[150] It is obliged to draw the distinction between in/visibilisation, indeed to draw *any* distinction,[151] in order to carry on. One part showing more, the other part a bit deeper in the dark. But no one here to look at it, no eye to landscape the lawscape, no 'I' to orchestrate it centrally. This is a distinctly posthuman understanding of the lawscape. Once again, it is not just a question of epistemological choice, but of ontology. In a moving passage, Clare Colebrook writes:

> The very eye that has opened up a world to the human species, has also allowed the human species to fold the world around its own, increasingly myopic, point of view. Today, we might start to question the appropriate point of view from which we might observe and evaluate the human viewing eye: from our own greater will to survive,

---

149 Brighenti talks about the field of visibility as relational, strategic and eventual. He connects it to the abilities of subjects to manipulate visibilities, which in our vocabulary would be bodies as assemblages. Brighenti, 2010: 39, writes: 'visibility is a rippling, anadyomenic phenomenon'.
150 'observing systems in the double sense of the English -ing form. We ourselves may be observing systems observing observing systems', Luhmann, 1992: 1420
151 'observing means making a distinction and indicating one side (and not the other side) of the distinction', Luhmann, 2002: 85; see Spencer Brown, 1969

or would it not be better to start to look at the world and ourselves without assuming our unquestioned right to life?[152]

The human eye becomes just one of the apparatuses of in/visibility in an expansive lawscape, decentred and fighting for its conative perpetuation just as any other apparatus. There is of course a way of restoring the eye but as 'a machine. This machine would not be a computer, for a genuine machine does not have a central organizing program but is put to work through connections.'[153] This eye has no way of organising a unified vision of the lawscape, but only a relational, situated and radically reoriented in/visibility. One of the possible lawscapes emerging out of this machine is a lawscape after the extinction of the human. Something that has been routinely imagined in the twentieth century, a century ravaged by two World Wars, and has moved to the invisibilised side in the era of climate change and nuclear disasters.

2.4 Posthuman, immanent, fractal: one lawscape

Three qualities characterise the lawscape, as the title of the section indicates. These qualities vary in degree according to the conditions of the specific lawscape but generally characterise the lawscape across its emanations. First, the *posthuman* lawscape. The lawscape lies beyond such distinctions as human/natural/artificial.[154] However, to speak about a *posthuman* lawscape specifically, merits a deeper discussion, especially since the concept of the posthuman is not without its problems in the way it can be co-opted by neoliberalism as an 'anything-goes' assertion.[155] The context is what has been broadly called the 'Anti-Copernican revolution', namely the objection against Kantian logic that reinstated the human in the centre. Alfred Whitehead is the first theorist of the connection between the human and the nonhuman, namely the first theory that dealt specifically with the Copernican move of decentring the Earth from the solar system.[156] Whitehead has famously written that minerals and stones prehend the world, humans included, in just the same way as humans do through their mental abilities.[157] This is not an anthropomorphisation of the nonhuman, but a levelling of the playing field, where the various bodies (human and nonhuman) relate to each other. From that moment onwards, several theories

---

152 Colebrook, 2014: 24
153 Colebrook, 2014: 18
154 Wolfe, 2009
155 Rose, 1996
156 The Anti-Copernican revolution has been pioneered by Speculative Realism scholars such as Harman, 2011; see generally Bryant *et al.* 2011
157 Whitehead, 1978

emerged (such as Deleuze and Guattari's work, Michel Serres's theories of affect and senses,[158] Rosi Braidotti's posthuman,[159] Catherine Malabou's work on plasticity,[160] or the recent philosophical currents of object oriented ontologies) that refined these relations, and on which I am drawing. Thus, I am arguing for an understanding of *a decentred human/nonhuman surface that has destabilised the standard legal and political understanding of the human as a unified agent that acts independently*. The posthuman is the assemblage between human and nonhuman bodies that constitute the human as it actually is rather than idealistically stands. Thus, the (post-)human is determined as much by thought and awareness, as she is by corporeality, technology, animality, vegetality, inorganicity, all of which more often than not completely bypass thought and awareness. Rosi Braidotti puts it thus: 'not all of us can say, with any degree of certainty, that we have always been human, or that we are only that'.[161] The wet dream of the human as the white, male, heterosexual, Northern hemisphere-resident with a garden and one and a half kids has been found hollow by many an anti-humanist theories. Niklas Luhmann's infamous adage that humans are excluded from society, was meant as an urge to abandon the effigy and start thinking of the environment, considered traditionally to be 'outside' the human, as the main locus of human activity. In that sense, the lawscape is a 'mental, a natural and a cultural ecology',[162] elaborating on Guattari's triad of body (and its connection to mind as a thinking multiplicity), nature (in its sense of *earth*) and the movement of the social (in the sense of assemblages operating on the continuum). To this, an understanding of law as movement is added,[163] whose posthumanity is ascertained by its moving to new assemblages with bodies thus far excluded from its remit.[164] The movement is continuous and always in new foldings. To quote Deleuze and Guattari, 'we make no distinction between man and nature: the human essence of nature and the natural essence of man become one within nature in the form of production of industry.'[165] The warped dialectics of the phrase is dizzying: human essence of nature and natural essence of the human are first folded into the natural, only to be thrown into the form of the human. One has the impression that the sentence could carry on unfolding without ever stopping anywhere for good. Rather than a bad infinity, this is the operation of the fold, and with this the practice of the posthuman lawscape. Donna

---

158 Serres, 2008
159 Braidotti, 2013
160 E.g., Malabou, 2008
161 Braidotti, 2013: 1
162 Guattari, 2000: 20
163 Barr, 2010
164 See Foucault, 2003c, and indicatively Ruddick, 2004; Sharpe, 2007 and 2010; Braidotti, 2006. See also Herzogenrath, 2008 and 2009
165 Deleuze and Guattari, 1983: 4

Haraway has famously declared that 'the boundary between human and animal is thoroughly breached' and cyborgs, oncomice and coyotes are posthumanist dimensions of more traditional feminist bodies that transcend the natural/cultural, organic/mechanical, physical/non-physical divides.[166] Katherine Hayles's digital subjectivity is built on a discontinued and inherently unpredictable conception of the human.[167] Renisa Mawani's work on insects is remarkable, not least because it brings to the fore the way the posthuman turns against itself by using what Mawani, paraphrasing Haraway, calls 'companions of war':

> while the olfactory senses of honey bees are being trained to locate dangerous chemicals, including those found in landmines, snails and cockroaches are becoming 'animal/machine hybrids,' experiments aimed at harnessing their natural sensors and energies in pursuit of micro-surveillance.[168]

In *Philosophy of Vegetal Life,* Michael Marder argues that vegetal finitude (rather than consciousness or lack of) is the basis for a holistic ethical treatment that includes vegetal life and 'the vegetal being in us'.[169] This does not mean that vegetal value is anthropocentric, but rather that there is a continuum between human and vegetal. Rosi Braidotti's vitalist philosophy and her treatment of the posthuman are grounded on a space of encounters that are decidedly within the continuum of indistinction between human and nonhuman.[170] Closer to the law, the question on whether trees should have court standing has been long and validly rehearsed in legal thinking.[171]

The posthuman, therefore, is understood here as a bridging mechanism between what was traditionally considered human and what we now see that the human is: an extension of the human that is not one-directional (from human to nonhuman) but multidirectional, not given to control or easy manipulation, but rather exposed to the technologies that support yet also determine the human, thus making the human prone to misleading desires, uncontrollable affective relations and disruptive technological insecurities, all of which come together to construct the grand ontological vulnerabililty of the posthuman. In that sense, the posthuman lawscape is not just about animality or the inorganic as affecting forces (that too), but most importantly about how legality can no longer be thought of as a one-directional decision-making action, but an assemblage-emerging, control-eschewing,

---

166 Haraway, 2004: 32
167 Hayles, 2005
168 Mawani, 2014
169 Marder, 2013: 182
170 Braidotti, 2009 and 2013
171 Stone, 1974

predictability-resisting, self-rupturing yet continuing series of encounters between bodies with varying levels of strength.

Second, the *immanent* lawscape. I have been referring to the lawscape as a continuum. This denotes, first, the continuous flow and rest of bodies on the legal/spatial manifold. Everything is, actually or virtually, affected by everything else and everything continuously *becomes* other. This does not mean that everything is the same. On the contrary, the lawscape is an aggregation of singularities that remain different while becoming-other. This is both a spatial and a temporal element of the lawscape that takes place on the basis of repetition, which is the topic of the following section. But it is worth clarifying here that singularity is not an idealised form of identity. Even in its mundane everydayness, the lawscape operates through a particular (to each assemblage) in/visibilisation that generates difference: not all strategies work in every occasion, as the trip to the lawsape shows below. Different assemblages require different in/visibilisations. In that sense, singularity is not a question of 'identity', freedom and self-attainment, but merely singular positioning in the lawscape, partly through desire and partly through the conditions, which can be either beneficial or harmful for the assemblage. These conditions are always underlined by the ontological quality of withdrawal that characterises every body. The second thing that emanates from the immanence of the lawscape is the fact that there is no outside to the lawscape. The lawscape is concrete, territorially determined and jurisdictionally definable; and simultaneously, following Timothy Morton, a hyperobject, namely so large and immanently expandable that is never fully present.[172] The absence of outside is a result of both plenitude and rupture inside. I have dealt with the issue of the outside extensively elsewhere, and it is a conscious decision not to engage with this here.[173] Suffice it to say, however that no outside means just as easily no inside; further, that no outside means that everything inside is merely a fold of the outside;[174] and, finally, that no outside means that the only way to open the inside is by *being butchered open* from within.[175] I do not consider any of the above contradictory or even different. They all yield this: there is no transcendence outside the lawscape, nothing to wait for or escape to. This is of course both enabling ('act now') and asphyxiating ('nowhere to go'). The lawscape conditions absolutely, encompassing a body totally, that is materially and immaterially, and capturing, without however necessarily controlling fully, the body's potential becoming. There is margin for move-

---

172 Morton, 2013, more of which below.
173 Philippopoulos-Mihalopoulos, 2007a, where the outside is considered an absence inside; Philippopoulos-Mihalopoulos, 2013a, where there is only a radical outside; Philippopoulos-Mihalopoulos, 2013b, where the world has no outside.
174 Deleuze, 2006; Grosz, 1995
175 Negarestani, 2008; see also Chapter 5 below.

ment, negotiation, reorientation, political mobilisation. But they are always inscribed within the lawscape.

A brief digression before I reach the third quality of the lawscape. In her employment of the concept of the lawscape, Alison Young suggests that other lawscapes exist, such as the 'uncommissioned city', that operate as alternatives to what Young refers to as the 'legislative city'.[176] These other lawscapes emerge from encounters with street art (Young's main focus) which take place in a surprising, unpredictable way. The point is not that one might or might not see these art pieces in the street. Rather, following Jane Bennett,[177] Young suggests that the point is to *encounter* them and be properly *enchanted* by them, that is to say, to have one's way of thinking and behaving altered, however momentarily or insignificantly. It is important to note that an encounter is not necessarily a positive experience. There is no moral judgment here, not even an aesthetic judgment. In that sense, an encounter of the above sort is a rupture in the lawscaping continuum, epistemologically desired, even necessary in order for the difference between lawscapes to emerge. Young engages the political potential of such an encounter, which is also in a way what happens in the encounters made in the last section of this chapter, in the context of the walk in the lawscape.

Young refers to China Mieville's fiction and specifically to *The City and the City*,[178] a book that has become very significant to me albeit for slightly different reasons than those of Young's. In the book, two cities, Beszel and Ul-Qoma, intersect concretely and abstractly, having their respective lawscapes interfolded. Same physical area, same expanse, but two different cities. Yet there is no superimposition. Mieville is careful to avoid any sense of palimpsestic lawscaping. Each lawscape withdraws while sliding next to the other lawscape, a Beszel area next to an Ul-Qoma area, a Beszel street crossing an Ul-Qoma street, a Beszel building standing next to an Ul-Qoma building. Another important element of the two lawscapes is that the citizens of each city learn how to *unsee* each other. They walk on a street that is *cross-hatched*, namely where parts of it belong to Beszel and others to Ul-Qoma, and where citizens of either city necessarily walk next to those of the other. Yet, from an early age, they learn how to *unsee* each other, including structures, buildings, parks, cars, even car accidents that involve parties of both cities, or children who do not know and might inadvertently *breach*, that is look across the other city. This *unseeing* is based on a subtle code that refers to dress, architectural style, technological era and so on. The citizens of each city grow up with this code and allow it to

---

176 Young, 2013
177 Bennett, 2001 on enchanted materialism.
178 Mieville, 2011

become part of themselves, to the point of not even realising that they adhere to it.[179]

The novel's main find of the two cities unseeing each other is an exceptionally relevant tool for understanding a great deal of current lawscaping, from geopolitics to mere neighbouring disputes. Lawscapes often slide next to each other and generate each other through folded withdrawal. Reorienting the lawscape in relation to other lawscapes is part of the political potential of the lawscape. This will become again relevant in the final chapters, where spatial justice is discussed as the conflict over specific spatiotemporal positionings. What is immediately relevant here is the reciprocal invisibilisation and interfolding of the lawscapes. In/visibilisation, just as between Beszel and Ul-Qoma, is the *conditio sine qua non* for the emergence of the lawscape. While, however, this shadowy interplay works on an epistemological, experiential level in the novel as felt by the various lawscapers that populate it, it also operates on an ontological level. Operating on an ontological level means that the very nature of both law and space is altered when taking part in the process. Law is no longer the same when it in/visibilises its own spatiality, nor is space the same when it has to play down its own legal striation. Both become other. They become phases of hybridity and can no longer claim purity. They have left their essences behind. They have entered that ontological manifold of the lawscape which alters not just the way they are described by others, but their self-descriptions. They become beyond recognition, in that they can no longer be recognised independently of each other, but only as folds of the same surface, co-rhizomatics of the lawscape. At the same time, it is not enough that they might be recognised as law or as space in the way they are described. Their recognition is not the whole story. And their self-recognition is always incomplete. They, just like us, have become wilful victims of the incessant interplay of dissimulation. They are happy to *unsee*.

The other important issue that arises from Mieville's novel is the way lawscapes proliferate. In my first sketching of the term, when I was writing the introduction to the volume *Law and the City*, I referred to the various lawscapes of the cities that the contributors had traced legally. Manhattan,[180] Athens,[181] Mexico City,[182] Dar es Salaam[183] were just a few of

---

179 Strictly speaking, therefore, these lawscapes have become atmospherics, that is non-negotiable fixed lawscapes that can only return to the manoeuvring space of the lawscape through a movement of withdrawal (see below, following chapter). In the novel, the withdrawal takes place through *Breach*, namely the policing force of both lawscapes in which the protagonist is forced in order to investigate a case. This withdrawal (itself not a 'good' withdrawal but a forced and politically suspect one) gives both distance and immersion, allowing him to attempt a certain reorientation of the lawscape.
180 Goodrich, 2007
181 Chryssostalis, 2007
182 Azuela, 2007
183 McAuslan, 2007

the lawscapes clustered around a new formulation of law and space. Each one was very different but all shared the fundamental ontology of a space that is striated by law, and a law that is inhabited by space. With this we have arrived to the third quality of the lawscape, the *fractal* lawscape. The proliferation of the lawscapes across the earth is a process of fractalisation of the continuum. Each one is different from each other, yet all remain lawscaping instances ('recursive iterations'[184]) that reproduce the constituent interplay of in/visibilisation, lawscape's main refrain. Fractalisation means that all lawscapes are ruptures of the continuum that produce difference, vying on their own level for their perpetuation, whether this is a global urban identity, a local village pride of place, or indeed a local religious community: in her study of the *eruv*, namely the self-initiated delineation of an Orthodox Jewish community in London, Davina Cooper has shown the way in which fractalisation operates. Some viewed eruv as a way of freeing up movement within the demarcation in view of the Sabbath limitations; others viewed it as a problematic practice that put into question basic principles of identity, cosmopolitanism, commonality of space, difference and neutrality.[185] This spatially concrete, religiously limited lawscape practised the same in/visibilisation as other lawscapes, seeing the poles and wire that would delimit the space as either a legal presence that severed space, or a spatial implement that liberated from the law. Cooper shows how this lawscape operates fractally on the same level as the nation-state with its own spatial and legal delineations. In this sense, the lawscaping continuum is 'cuttable into many parts without losing coherence'.[186] This is not an issue of geographical scale in the traditional sense of nested hierarchies,[187] but a horizontal rupturing. It is scale the way Neil Brenner describes it, not 'a single nested scalar hierarchy, an absolute pyramid of neatly interlocking scales, but ... a mosaic of unevenly superimposed and densely interlayered scalar geometries.'[188]

Paradoxically, this proliferation constitutes the immanence of the lawscape. And we come full circle: there is only one lawscape which is ruptured in order to produce different densities, layers or velocities of lawscaping practices (thus, other lawscapes that slide next to each other, all of which together constitute the grand plane of lawscaping immanence); and second, there is no possibility of transcending the lawscape. One cannot go from Beszel to Ul-Qoma, and even if one were able to do so (indeed it does happen in the novel), one never leaves the lawscape behind. One merely

---

184 Chettiparamb, 2013: 682
185 Cooper, 1996; see also Cooper, 1998 for further territorial/symbolic lawscaping.
186 Morton, 2013: 47
187 For a critical view, see Delaney and Leitner, 1997 and the whole issue there; Collinge, 2005; Marston *et al.*, 2005
188 Brenner, 2001: 606

replaces it with another (in Mieville's case, and without giving away the plot, the lawscape of Beszel *and* Ul-Qoma). There is nothing that is not spatially and legally determined. This is Mariana Valverde's point when she resuscitates the concept and practice of 'legal nonconforming use' from the modernist planning law discourse, itself in the heart of lawscaping processes.[189] 'Nothing can be – even by omission – outside the law. Illegality, alegality, extra- or infralegality are forms that, either by opposition or subversion, necessarily take a legal system as a reference and are as such constituted by it', writes Guillermo López in the context of law and architecture.[190] There are no virginal spaces, other worlds, in-betweens, just utopias, past ludic paradises. Everything is in the process of being lawscaped, but not in some sort of mega-conspiracy theory where everything is controlled by a panoptical big brother. The lawscape does not need this: we are all indefatigable lawscapers, plodding in between our needs for security, comfort, ideological positions, conflict or whatever it is that takes our fancy, as has been abundantly shown by people like Nikolas Rose, Davina Cooper and before them Foucault.[191] The lawscape moves itself through the bodies of its emergence and takes the shape of the world.

The above discussion sets the lawscape apart in relation to other fusions of the legal and the spatial, such as Sloterdijk's *nomotop* or Delaney's *nomosphere*.[192] It shares with these important concepts a desire to ground the law but also to bring forth the normative nature of space. There are, however, several differences between them, such as the question of immanence, and the importance of the mechanics of in/visibilisation, all of which linked to the fact that the lawscape is an ontological category. Another difference between the lawscape and these other fusions is that the latter are characterised by a compartmentalisation of the nonhuman in relation to a spatially determined human community. Sloterdijk talks about 'the "tensegral" nature of human association in the nomotopic field',[193] Delaney about 'cultural-material environs … and the practical performative engagements [with them]'.[194] In both cases, the human remains a central figure of perception, performance or action. Sloterdijk's series of human 'islands', one of which is the nomotop, moves in the direction of a material emplacement of normativity through the use of architectural structures such as

189 Valverde, 2011: 291
190 López, 2014: 7
191 Rose, 1989 and 1996; Cooper, 1998; Foucault, 1980
192 Sloterdijk, 2006; Delaney, 2010
193 Sloterdijk, 2006: 1 'a normative architecture that exhibits a sufficiently supra-personal, imposing, and torsion-resistant character to be regarded by its users as valid law, as an apparatus of obligatory principles, and as a coercive normative reality'. The nomotop is situated between lawscape and atmosphere, since it lacks the possibility of negotiation (atmosphere) but does not engage in in/visibilisation (lawscape).
194 Delaney, 2010: 25

tensegrities (the thin but necessary structures that support air buildings). Even so, its post-phenomenological structure retains the centrality of a human, anthropocentric and anthropomorphic subject. On the contrary, what is proposed here, largely following Deleuze and Guattari on their 'alloplastic stratum',[195] namely the level of creative construction of signs not limited to humans, is a decentring of whatever residue of centrality might remain in the configuration of the connection between the human and the 'environmental'. This entails a radical opening of both understandings of space and law towards an unmediated wilderness, conditioned by the posthuman as the nonhuman extension of the human. At the same time, however, not all bodies are equal. Human, animal, vegetable, inorganic, semantic, discursive bodies are characterised by different viscosities, different concentration of power in relation to other bodies and consequently different affective abilities. Thus, in the lawscape, human affective abilities are regularly stronger than animal affective abilities. Still all are subject to the affective abilities of meteorological, geographical, or complex social phenomena that construct specific corridors of movement.

2.5 The repeated time of the lawscape

The lawscape spreads over the horizon. The horizon as both past and future is engulfed in the here/now of the lawscape. The time of the lawscape is one of constantly renegotiated now. Every here/now is the carrier of its own temporal horizons.[196] Future and past horizons mark the limits of the possible, but these limits are as fluid as the movement within, towards and away from the horizon. Amidst this horizontal stretch, lawscaping occurs. Each lawscaping is an event, a change in the in/visibility between space and law. However, each lawscape has its own temporality in which the in/visibilisation happens. In the lawscape, time is understood as change, that is a renegotiation of in/visibility. But since the lawscape is both a fractal and a continuum, at any point there are two temporalities: that of the specific lawscape and that of the lawscape as continuum. While simultaneous, they remain two different temporalities. Lawscape is real, in the Deleuzian sense of both *actual* and *virtual*,[197] or the Luhmannian sense of marked and unmarked, or actual and contingent.[198] Actual and virtual are real in that they are folded into each other as a univocal reality that determines a body's movement and rest. This means that the lawscape occupies both sides of the distinction between what is presently actual, marked by selection and therefore understood as present; and what is

---

195 Deleuze and Guattari, 1988: 314–17
196 Luhmann, 1995: 312
197 Deleuze, 1989; see also above, Chapter 2.
198 Luhmann, 1998

Figure 2.2 The lawscape – or a footbridge in Brisbane, Australia.
© Andreas Philippopoulos-Mihalopoulos.

potentially actual, *actualisable*, markable, contingently included (or excluded), virtual.[199] These temporalities are shared across the fractal lawscape and the continuous lawscape. The fractal lawscape, that is the individual lawscaping event, is actual (but folded in the virtual). The univocal lawscape, that is the lawscape as continuum, is virtual (but actualisable only in the actual). In that sense, the horizon of the lawscape is included in the lawscape as claims stemming from within, or as contingency, unpredictability, uncertainty – whatever it is – is always part of the game of in/visibilisation. Luhmann writes: 'whenever anything determinate occurs, something else also happens, so that no single operation can ever gain control over its circumstances'.[200]

Yet if the lawscape is to be a change in the in/visibilisation of space and law, it needs to be repeated. Change needs to find its correlative across the lawscape manifold, in another lawscape or in the lawscape at large. It is only when the correlative change emerges in its different context, that the

---

199 Deleuze and Parnet, 2007, especially 'The Actual and the Virtual'.
200 Luhmann, 1995:42

change is registered as irreversible in the lawscape. Both presents are here/now, but it is only through the second present, the second event that brings change, that the present emerges as irreversible, as already past. In other words, when change is within the (first) lawscape, any lawscaping is still reversible, still oscillating between past and future, still ontologically undetermined, still epistemologically imperceptible. It is only when this change is streamlined with the time of the lawscape continuum by being repeated (second lawscape) that change becomes irreversible. It is the correlation of the two presents, and indeed the exposure of identity to difference, that renders lawscaping irreversible. These separate yet contiguous presents are the condition for the irreversibility of the change of in/visibilisation (that is, for change *qua* change). It does not follow of course that the effect of change cannot be changed. But this would be another change and another, future present.

In other words, lawscaping is an act of repetition of the here/now and the way the past passes through it, that aims at capturing the future. This is the function of the law, which, according to Niklas Luhmann,[201] stabilises societal expectations over time: what is unlawful now (based on past experiences) is probably going to be unlawful in the future too. The lawscape follows the same function. The future becomes part of the movement of the bodies that embody the lawscape: the same bodies that keep on parking here because it is lawful, keep on shoplifting there because they can get away with it, keep on breathing the polluted air of industries because to pollute is lawful, keep on dying because the causalities cannot be proven. What this means is that the time of the lawscape is an always present repetition, a revisiting of the same which, however, every time produces difference. No parking, shoplifting, breath, death are ever the same.

It is worth dwelling on the repetition of the lawscape. On a spatial level, the lawscape is repeated across the globe, with different in/visibilisations depending on the conditions. Let me take an example: in Australian cities, the law is visibilised constantly through obsessive signage, road directions, arrows, penalties, full citations of laws and so on. Things do not abate in the countryside either. Nicole Graham talks about the dephysicalisation of Australian country and the constant legal marking (here, visibilisation of the law and invisibilisation of space) with material and immaterial boundaries.[202] Intense juridification here spreads across the lawscape, still in the service of (post)colonial purposes.[203] Or think of concepts and practices of preparedness, emerging from anti-terrorist measures, that have taken the global lawscape by force, twisting in/visibilisation strategies in a contagious way. At the forefront of these, UK and the US have led the way for other lawscaping

---

201 Luhmann, 2004
202 Graham, 2011
203 See for example Mohr, 2003

practices to follow.²⁰⁴ It is not certain that the contagious spread will take place. But it is only once the contagion begins, with the repetition of the lawscaping event, that the lawscape becomes consolidated. These lawscapes are different to each other but follow the same interplay of in/visibilisation. The strategy varies from lawscape to lawscape – there is no one teleological lawscaping. And yet, there might be something teleological albeit well hidden: this is what Deleuze, following Nietzsche, calls *the power of the false* and what I have previously referred to as *dissimulation*. Deleuze uses the false in the context of cinema.²⁰⁵ The use of an umediated, direct image of time in cinema takes advantage of the production of 'falsity' and allows a proliferation of temporal narratives to co-emerge. The power of the false liberates from linearity and historical fixity. In the same context, William Connolly talks about the false as 'that which falls below clean recollection because it was not consolidated enough at its inception'.²⁰⁶ The false 'helps propel new turns of becoming at strategic moments'. In order to do this, however, I argue that the false needs to dissimulate itself as *true* (the virtual as actual, the unmarked as marked), and this happens only through repetition across the lawscape. Connolly uses an example from a Proustian character who begins a sentence that pulls in at least two directions, yet (quoting Proust) 'immediately, like a conductor whose orchestra had blundered, checked the phrase which he had started and with infinite ingenuity made the end of his sentence follow coherently from the word which had in fact uttered by mistake'.²⁰⁷ Everything needs to land back to the continuity of the continuum. For this reason, dissimulation is at work in the lawscape, in the Marxian sense of inversion that contains (and reduces) contradictions by generating a distance between essence and appearance. The lawscape, by now a total event of repetition, dissimulates itself as factual, solid, the *true* experience of present. This is neither bad nor good. Law is both oppressive and enabling; likewise with space. What is important is that the chronology of the lawscape becomes reinvented (dissimulated) in every lawscape, depending on the needs in hand.

More on dissimulation later in this text, but one thing needs emphasising here: that the repetition of the lawscape over time produces difference rather than identity. This is the meaning of the two presents mentioned above: the first present, the one within the lawscape, is still reversible, changeable. It is only once the correlate present emerges elsewhere, in a different yet continuous lawscape, that the present becomes irreversible. Namely, it is only through the emergence of difference that repetition operates in time. Repetition cannot be an act of repeating the same. Identity

---

204   Amin, 2012
205   Deleuze, 1989
206   Connolly, 2011: 117
207   Proust, 1983: 247–8

is still reversible, its actualisation unmarked: *identity is irrelevant except through its repetition as difference*. It is only then that lawscaping becomes effective. Until then, it is mere tremor, imperceptible movement that cannot spread to the horizon, nor can it extend in time. But once it is repeated in another lawscape and its difference emerges that splits identity, lawscaping becomes effective, extended in time and space, ready to carry on repeating itself in a frenetic closure of self-perpetuation: only then is it ready to become *atmosphere*, as I show in the following chapter.

Before this, however, allow me to dwell on it through a brief excursus on Søren Kierkegaard's *Repetition*,[208] which contains the three Kierkegaardian traits of repetition: repetition is not recollection; repetition is spatial; and repetition can only take place when one does not mediate in order to reproduce it. The narrator (Constantine Constantius – one of Kierkegaard's sobriquets) moves from Copenhagen to Berlin in order to repeat, in our language, the way the urban landscape had been lawscaped by his presence there. Repetition is always spatial, always *here*, and always actual (it has 'the blissful security of the moment'[209]). Constantius/Kierkegaard urges that it must replace the historicality of *recollection*. This is where repetition differs from recollection: according to Kierkegaard, repetition is a 'forward' recollection in actuality that opens up to one's becoming. Repetition prioritises living over understanding, thus moving away from what Kierkegaard dismissed as the erroneous backwards historicality of recollection, and reinstating the immediacy and actuality of *here*.[210]

Back in Berlin however, Constantius discovered that 'everything was exactly the same, the same jokes, the same courtesies, the same patrons, the place was exactly the same – in short, uniform in its sameness'.[211] His endeavour to reinsert himself in the previous lawscape ends up in perfect tedium: 'the only thing that repeated itself was that no repetition was possible'.[212] So, on one hand, there could be no repetition, and on the other, everything was the same! The problem with Constantius's approach was that he *tried* for repetition rather than abandoning himself to the here of the present living (which is already replete with repetition). In his pursuit of pleasure, Constantius acts like Don Juan (another favourite Kierkegaardian figure[213]) who over-invests in the here but only as another recollection in his list of conquests; he tries too hard and misses actuality. At the end, Constantius does what he should have

---

208 I have specifically chosen not to focus on Deleuze's *Difference and Repetition* here, although deeply inpsired by it, because I find that Kierkegaardian repetition remains closer to the idea of continuum I am putting forth here. See my dealings with it in Philippopoulos-Mihalopoulos, 2011b.
209 Kierkegaard, 2009: 3
210 Nymann Eriksen, 2000
211 Kierkegaard, 2009: 38
212 Kierkegaard, 2009: 38
213 Kierkegaard, 1949

done from the outset: he withdraws from the pursuit. 'Repetition is too transcendent for me. I can circumnavigate myself, but I cannot get beyond myself. I cannot find this Archimedean point'.[214] At the same time, however, and here is the peak of the Kierkegaardian enigma, this Archimedean point is nowhere to be found except within: 'that which is repeated has been, otherwise it could not be repeated; but precisely this, that it has been, makes repetition something new'.[215] The Archimedean point is folded in the immanence of the continuum that withdraws ontologically. In that sense, *repetition is immanent yet folded in the fissures of withdrawal.* For Kierkegaard, this withdrawal refers to the resignation from the possibility of a Hegelian *Aufhebung*, the final solution which, according to Kierkegaard is not a movement but a 'commotion', a mere 'mediation'.[216] Repetition is 'the new category that must be discovered',[217] a movement that moves away from both recollection *and* mediation, and opens itself with abandon to the future. And although repetition is the absence of endeavour, at the same time is inviting of a *rupture*, a transfiguration that makes repetition every time different: for as Constantius reminds the reader by expressly addressing her through the materiality of a copy of a *carte-visite* included in the pages of the book, 'repetition is transcendence.'[218] But this transcendence is contained as a rupture within the continuum. When Kierkegaard contrasts the above with a grounded 'repetition is actuality',[219] he posits a transcendence that can only be performed and embodied in an actual materiality, indeed an immanent and continuous spatiality, ruptured momentarily in order to allow repetition as difference to emerge.

Let me assemble the above elements in their relation to the lawscape. First, repetition in the lawscape is not recollection. This is important: in the lawscape, the past is inscribed on the earth, on the surface of the buildings, in the distance between bodies. Its temporality is horizontal, fully spatialised and materialised, open not only to the lawscape's jurisdiction but, most importantly, to space as world, through its repetition in other lawscapes. The past is not mere recollection. It constantly changes, depending on in/visibilisations. Space might disappear and only a thirst for repossession might emerge, expressed in legal claims. The lawscape then becomes overly legal, matrix-like, replete with statutes and judgments and devoid of the materiality of space (which might fit all parties, might lend itself to a conflict-diffusion or might even describe the claim in very different terms). Whatever it is (and again, I pass no judgment here), repetition in the lawscape is not an ethical

---

214 Kierkegaard, 2009: 50
215 Kierkegaard, 2009: 19
216 Kierkegaard, 2009
217 Kierkegaard, 2009: 18
218 Kierkegaard, 2009: 50
219 Kierkegaard, 2009: 4

event, it is not a guarantee for justice to emerge, and it bears no necessary connection to historical 'truth'. It is merely the way it is repeated across the surface of the lawscape, fully contextual, fully given to power struggles.

Second, repetition is immanent to the lawscape. The various foldings within the lawscape that form the fractal lawscapes in which repetition emerges, do not annul the oneness of the lawscape; rather, they confirm it through the very repetition of the lawscaping events. Immanence is full of internal splits, ruptures, events of distancing, pausing, revolting. On this surface of continuous rupture, repetition occurs. This is more than simple globalisation, whose tendencies extend in space and across rupturing boundaries. This is also more than legal recognition, which demands that the other recognises you in order to exist legally (as citizen, living human being, sovereign state, and so on). The immanence of the lawscape contains all that but goes beyond recognition and globalisation since it builds on difference rather than identity.

Finally, repetition cannot be mediated. Although a great effort goes into in/visibilisation and the eventual conversion of the lawscape into engineered atmospherics of closure (as I show in the following chapter), repetition cannot be orchestrated fully. Contingency creeps in and makes it actually impossible and virtually improbable to predict always and with reliable accuracy how repetition is going to emerge. Different spatio-temporal conditions across the continuum are more in position to predict the repetition of identity ('everything was exactly the same') than difference. If mediated, repetition across lawscapes tends to ignore the local conditions and impose itself as legal transplant, colonisation, mapping according to specific perspective – in short, identity.

The discussion on time fleshes out one important element of the lawscape, namely its internal folding as continuum and rupture. As already said, the lawscape is one continuum rippled with foldings (other lawscapes) that rupture while extending the continuum. The continuum operates through its folding (and unfolding too, since unfolding is the continuation of folding[220]). Within the folds, lawscapes as difference emerge and repeat other lawscaping events. There is no central brain in this, but merely a rhizomatic, horizontal spreading of lawscaping. But rupture can be made to be more than a mere continuing folding. In different legal and political contexts, rupture is instrumentalised as distance, separation, otherness. This might not be as sinister as it first appears. Rupture is needed as a surviving technique: boundaries are needed in order to exclude, exclusion is needed in order to include. When I talk about the lawscape as one immanent continuum, I do not equate it with a political utopia of full inclusion. I am offering an ontology of lawscaping, namely a ruptured tautology of law and space that operates both as a tool of security and oppression, openness and

---

220 Deleuze, 2006

asphyxiation, encouragement and suppression. It does not have a direction. Lawscaping moves along the bodies of its emergence, and in turn these bodies are directable things, prone to engineering and subliminal persuasion, which also emerge from within the lawscape (and its bodies, ad infinitum). *This* is the meaning of the continuous manifold: no central point of direction, no easy panopticon, but immanent movements that often eschew perception. At the same time, lawscape as continuum denotes the end of transcendence in the sense of the outside. There is no outside, either übercontrolling or utopian. 'There is another world, but it is in this one' as Paul Eluard is thought to have said.[221] Any way out is inscribed within. One has to dig through layers of present pasts, and orchestrate one's own lawscape from within existing lawscapes.

2.6 Walking the lawscape

Let us take a walk in the only space we can, namely the lawscape. We will follow the students of my *Law of the Environment* class in their walk in London. This is a walk each student takes individually and then reports back to the class. The purpose of the walk is for the students to have an encounter with the lawscape and their own lawscaping practices, expanding their awareness of the way in which they employ their body, the law and space in the lawscape, so that they eventually manage to *reorient* the lawscape. It is also part of an *atmospheric* experience which employs senses (and not just vision) and dwells in their use in order to understand how the lawscape transcends the visual and engages all other senses. Finally, it is simply a fun exercise, which, however, ends up with a affect akin to an existential 'asphyxia' for most of the students: for the main purpose of the exercise is to discover whether there is an outside to the lawscape.

In terms of context, it is perhaps important to mention that one of my quests pedagogically but also conceptually, is to bring space in some sort of encounter with my class. I use the space of the classroom by inviting my students to occupy it differently from other classes, but also in the context of specific activities where the classroom's lawscape comes forth, with laws and spatialities differently visibilised depending on the exercise. I consider it very important to 'flesh out' the materiality of the law we are studying in the law school by allowing my students to build their own awareness of the ways in which their body lawscapes. Above all, I want to open up a space in which my students would feel their body moving along, between and even

---

221 The authorship is disputed, with W.B. Yeats a strong alternative. John Llewelyn, 2009: 307, gives up his bibliographical search, accepting that the world of the quote fits well enough with that of another quote by Eluard: 'Everything I love, everything I think and feel inclines me toward a particular philosophy of immanence, according to which surreality would be contained in reality itself, and would be neither superior nor exterior to it'.

against lines of law, constructing thus for themselves a different awareness of identity.

Here are a few words on the course, part of which is the exercise that follows. The course is called *Law of the Environment*, deliberately distinguishing itself from a generic Environmental Law course, and examining the connection between the human, the natural and the legal through a regularly performative, generally theoretical and always interdisciplinary way. The class (about 30 students each year) discusses in equal measure concepts like Sustainable Development, the Precautionary Principle or atmospheric pollution (all more or less typical environmental legal issues); deconstruction, radical ethics, ecofeminism, phenomenology, autopoiesis; and ideas taken from legal theory, philosophy of science, geography, literature, music theory, art theory, economics, biology. In the process, such standard distinctions as the one between natural and human/artificial/technological are questioned, alongside the ability of the law to deal with such collapsed distinctions. Initially the students are uncertain about how to go about the class, but they quickly take to it in earnest, fully immersing themselves in a performativity that involves their body, the space of the classroom and, in some ways, the world at large, and start pushing their limits and those of the class as a whole. They are fundamentally encouraged to pursue their own interest in composing a reflective log midway and a final project that emphatically does not have to be textual, on a topic of their own choosing in consultation with me. There is no lecture as such but we are all required to read a text, find information or complete a small project in advance of every week's meeting. The discussion is student-led but teacher-channelled, taking risks every single week by flirting with unpredictability originating in boredom, tiredness, lack of preparation or simply the wrong mood.

For the particular lawscape walk session, I prepare my students by asking them to read some literature on law and space, as well as parts of Deleuze and Guattari's *A Thousand Plateaus*. Admittedly, the latter is a demanding text, but by that time in the course (halfway through the second semester) the students are familiar with at least the struggle to master different terminologies. We have a discussion on the texts and analyse some of the issues in relation to environmental law. Once this is in place, I present them with the text shown on page 96:

I ask the students to take this text with them while taking the walk, and also trace their movement on a map if feasible and does not interfere with a spontaneous flow of movement. During the week, they upload their maps and whatever else their walk has generated (photos, audio, logs) and we discuss them in class.

The thinking behind the exercise is to make one aware of the lawscape through the medium of the moving body. This brings forth the materiality of the law both in its spatiality and its corporeality. Senses are promoted

> *Walking observing thinking acting breaking flowing going against listening taking notes mapping smelling touching do not touch forgetting remembering overhearing fantasising running escaping getting lost feeling lost fearing deviating diverting avoiding throwing yourself in manipulating never stopping moving being moved*

Walking the lawscape

Take a long walk, about 45 minutes or so, preferably alone. Try to map where you are going but without worrying too much about accuracy. Do not only take the main streets or the streets you already know. Take side streets, get tangled in places you do not know, feel a little lost (but always keep safe, use your instinct and reason to keep out of trouble, and carry with you a map so you do not *actually* get lost).

While walking

Keep in mind three things: **yourself, your movement, your surroundings**. Try to analyse the connection between all three:

1. Does your body move differently in different situations?
2. Do you think of different things depending on where you are?
3. Do you feel constrained in your movement by anything internal (you) or external (the city)?

Think also of the way you use your **senses** while walking:

1. Do you smell?
2. Do you touch?
3. Do you ever look upwards?
4. Do you notice what other people say to each other or on their phones?
5. Can you eat/smoke/chew gum and walk? Does it bother you?

The assemblage

Now you have an assemblage of your body/the city /the law. See whether all of it moves as one, whether you map the city you are creating, and/or whether the city determines your movement, and to what extent the law determines you and the city.

See how this assemblage deals with:

1. *The law:* where is the law in the city? Do you see it determining where you can walk and where not? How you walk? When you walk and when you stop? What distance do you keep from things/people? What senses can you use?
2. *The body as hybrid:* can you see any hybrids anywhere? Human/animal/technological hybrids? Are you a hybrid? Do you use your sense in an 'animal' way – e.g., does smell stop you from doing things?
3. *Space:* what is space? What is urban space? Can you differentiate space from body – the city space you are walking and creating from your body? Is your body part of the space? Is law spatial?

and their awareness enhanced, in an attempt to enable the bodies to connect more broadly with the lawscape, engage in their own lawscaping, and also feel whether the law inhibits or encourages certain senses. Part of the process is also to discourage the usual prioritisation of the visual and allow other senses (mostly hearing and smell, and to a lesser extent touch and taste) to claim engagement with the law. While this mainly forms the basis for a discussion on atmospheres (below, Chapter 4), the underlying aim is to engage with law and space in the lawscape by discovering that law is spatial, corporeal and more broadly material, and that this spatiality makes the law and space variously more or less visible. The connection between the body that moves, the space in which the body moves and the way that the law determines the body's movement is an assemblage.[222] The idea behind the use of the term is to make sure that the given boundaries between the various elements (human body, technological extensions, animal qualities, the body of law with its commands, space with its particular multiple organisation) are questioned, while at the same time allowing the students to feel as if they are no longer the self-contained, fully determined human body but part of an assemblage, which, however keeps on withdrawing. As Yolanda, a Spanish student realises, her moving body is part of the assemblage with the city: 'and this is the way others can see you, you are just another element in the street they are walking, in the form of yet another obstacle in their way.' An assemblage lawscapes while moving among other assemblages (in that sense, every other body, their own mapping of the city that includes other bodies, and their own lawscaping). At the same time, Yolanda realises that her own body, in the standard way of seeing one's body as independent and contoured, is withdrawing from being fully present to her. Am I Yolanda or am I an obstacle?

One of the first things that students comment on is the physical boundaries that determine movement. This is not merely the distinction between the private and the public. No doubt this is relevant especially in a place like London which is characterised by a plethora of privately accessed gardens in full public view, constant construction site detours or signs that designate 'private streets' in what seems a perfectly open public area. Physical boundaries do not necessarily consist of fence-like structures. They are to be found in the distinction between buildings and open spaces (one cannot easily cross buildings unless one practises *parcours* and even then only in specific ways), different functions of pavements and streets, traffic lights and other forms of signalling that channel movement. Clementine a French student, remarks on how the city is divided into the congestion charge area and the one which does not attract the charge – a legal

---

222 A multiplicity that contains other multiplicities, 'human, social, and technical machines, organized molar machines; molecular machines with their particles of becoming-inhuman'. Deleuze and Guattari, 1988: 36

measure against pollution and traffic that controls movement while relying on economics, and which makes for a specific lawscape of movement in terms of human and vehicle traffic. Maria took a picture of a sign placed on the inside of a window of a ground-floor private home bringing potential callers' attention to the fact that 'this house is a no cold calling zone', and then explaining that they neither buy nor sell or indeed 'enter into discussion' at the door. Maria talked about how the lawscape is determined by zoning, both by the state and by individuals, whether there is a sign on the window or not. Indeed, signs are often unnecessary since a lot of zoning relies on the 'personal space' idea. This is what Timothy Morton talks about when he mentions that the objects zone space and attract bodies in their own zoning – for our purposes, their own lawscaping abilities.[223] There was a long discussion on underground train practices, as well as platforms, and some students tried to push this further by testing the limits of the lawscape: they stared at people longer than 'permitted', they were conspicuously sniffing people, they started talking to them in unorthodox situations, posed for rather silly selfies in the middle of the commute. At most these provoked a frosty stare. All students, however, were aware of the fact that these limits are brittle, depending on the circumstances, the distance might not be so great between a mere frosty look of disapproval of the social and cultural insensitivity, and legal action against harassment. The lawscape is everywhere.

Understanding movement as lawscaping is found to be not only geographically but also culturally determined. Elisabetta, an Italian student, comments on the fact that she cannot walk in the middle of the street as she would in an Italian city. Likewise, Lena and Armand, both Swedish, note that they do not feel obliged to wait for the green light in order to cross the road as they would do in Sweden. Violetta, a Russian student, keeps on crossing the street where she is not supposed to, in order to see whether any reaction is caused, to no avail. Jessica, a Canadian student, observes how Londoners do not have a 'joggers' culture of running on the left side of the pavement while leaving the right to the walkers. One does not notice these things until one notices them. The lawscape is noticed when noticed. The lawscaping ontology changes through the lawscaping bodies. This is not as self-evident as it sounds. It takes a shake-up that removes one from the usual routes (both in terms of actual routes as well as the ruts of habit),[224] a previous legal trauma or satisfaction, a certain suspicion that makes one's fitting with the rest of the assemblage angular and uncomfortable. In such cases, the presence of the all-permeating lawscaping and its mode of in/visibilisation is rendered visible. This is what Anne Bottomley and Nathan Moore call 'a diagrammatic reading', that

---

223 Morton, 2013
224 Bottomley, 2004

reveals not the visible in the sense of the visual, but the perceptive. Drawing on Deleuze and Guattari, they write:

> '*percepts* ... work because they link together (carry through them) our ability (our need) to use sensory material as an aspect of thought. This is a very different notion of embodiment – in which it is not our eye which sees, but rather our flesh which feels, and in which we move through, rather than simply "in", spatial relations. Here, we are not concerned with a real-out-there and whether we can or cannot see it – but rather with finding means by which to render visible the spatial (and temporal) patterns within which we encounter, are encountered by, other bodies, other forms of embodiment.'[225]

Through this sensorial ambling, which is not a visual *flâneurie* but a full corporeal immersion in which the body itself is co-producing its space of immersion, all students realise very quickly that the law is everywhere, that is, it determines all the steps one does and does not take, without however necessarily making its presence obvious. The students realised very quickly the interplay of in/visibilisation that takes place in cities, but they also conceded that what was needed in order to become aware of the lawscape was a spasm in the way one moves, a push (even by a teacher who wants his students to go out of the classroom and into the wild lawscaping urban forest) out of one's normal lines of movement, and then fits of spatiolegal presence are brought centre stage.

Some students note the way their body moves in different spaces. This was connected to social class aesthetics or conditions of light/darkness. Both elements are connected to issues of safety, and have been often expressed in terms of gender. Women lawscaped differently to men, but not necessarily in the expected way. Thus, the women in the class as a rule seem to be more daring and adventurous in terms of the time and place of their walks. At the same time, however, they were more aware of issues of safety. Walking in an affluent area was noted generally to be slower and less purposeful-looking than walking in a poorer area. Likewise, one takes one's time when moving in a well-lit area at night (yes, Shabana actually took her walk at 10 pm), whereas the rhythm becomes more hurried when in a dark, not well-lit area. In conditions of perceived safety (and this might have nothing to do with actual safety), the assemblage body/space/law spreads itself more loosely and assumes a movement that enables observation of such things like smells, surfaces, clothes that other people wear, and so on. This does not mean that when movement is faster and more purposeful-looking one does not take these things into consideration. However, one is

---

225 Moore and Bottomley, 2012; see also Bottomley and Moore, 2008.

not *aware* of noticing them. In conditions of perceived lack of safety, one takes all these into account but for the instrumentality of keeping safe. When safety is not an issue, one's awareness of observation is heightened. When safety is an issue, one becomes animal as Deleuze and Guattari would have it: one uses one's senses instinctively and moves about in a way that is animalistic. One does not of course start barking or miaowing. There is no representation of animal, no mimicry. Still, the process is real:

> Becomings-animal are neither dreams nor phantasies. They are perfectly real. But which reality is at issue here? For if becoming-animal does not consist in playing animal or imitating an animal, it is clear that the human being does not really become an animal any more than the animal 'really' becomes something else. *Becoming produces nothing other than itself.*'[226]

In becoming-animal, one becomes the process of transformation itself. Becoming-animal enables simultaneous states of becoming to emerge so that one is both animal and human. One becomes, properly speaking, posthuman. When Gabriel is lost in a park, he finds that he uses his hearing more than his vision, listening out for flapping and rustling, 'at times feeling like an animal; like a predator tracking my prey.' The body as posthuman, namely an assemblage of the animal/human within and without, is an impermanent becoming whose form and spatial presence keeps on changing.[227] The posthuman body is one with a mobile phone, a cyclist, a woman walking her dog, a parking attendant 'with a mysterious metallic stick' as Lilli writes, or even the students themselves: Amanee consults her electronic map while walking, whereas Lilli is suddenly aware of her 'false teeth and some artificial liquid in [her] knees.'

The students usually associate a perceived lack of safety with an absence of law, namely an absence of signs that denote law as boundary against their body. Thus, the presence of a policeman has a calming effect on Shabana. Legal visibilisation is seen as an enabling spatial factor of enjoyment, contributing to a lawscape set up 'for your comfort and safety' to use the habitual parlance. For other students, however, the experience is different. Armand's steps brought him to the area of the US embassy in London, a heavily policed, barricaded area in an otherwise solidly bourgeois square. Armand felt threatened by the presence of the law. The armed police appear threatening to Armand making him immediately conscious of the way his body appears (posture, clothes, unshaven). The law is moving sides – from being the thing that barricades one's body against another's to the

---

226 Deleuze and Guattari 1988: 238
227 This is what Latour, 1993, has called the hybrid.

thing that repels one's body, not only from a specific space but more importantly from a specific behaviour. The law thematises the whole square; it overcodes it by making it a space of heightened self-defence and rendering every body a potentially suspicious presence. The lawscape here is fully given to the visibilisation of the law, and the consequent invisibilisation of space. No longer an open public square in the middle of London, space becomes a playing field for the law. Armand decides to dissimulate, to become absorbed in the urban fabric by taking out of his pocket the piece of paper with the walk suggestions and pretend it is a map. In that way, he hopes to appear as an unsuspected tourist that could not possibly pose any threat to the law. The presence of the law forces the body to see itself as a potentially criminal body and to hide behind a perceived innocent stereotype. The lawscape in this case becomes an angular, ill-fitting jigsaw of bodies made of elbows and spaces full of potential falls. The city is no longer mapped by the freely ambling body but is forcefully squeezed into a line that precludes normal movement.

A similar legal visibilisation is noticed by another student, Eleanor, who is intent on using her mobile phone camera to take photographs of her walk and of the junctures where law and space affect her body. She finds, however, that she is stopped from doing so by various other bodies, all looming high on their legal pedestal: first the body of some sort of odd moral attitude in the form of a school administrator stopping her from taking pictures of a sign in a school playground (although technically not illegal, her action was seen as inimical to a the particular administrator's embodiment of the law, one that is built upon the fear of children being photographed); and then a body of national security, in the form of an officious guard stopping her from photographing a government building (despite the fact that, as she discovered later, the building is visible on Google images[228]). Things turned for her and the lawscape became a more pleasant manifold altogether, when she encountered another body of the law, former London mayor Ken Livingstone who was canvassing on the streets for the forthcoming elections. He assured her that she had every right to take a picture of that particular building. Eleanor felt much happier in the space created between her and Ken, with the lawscape changing into something infinitely more accommodating, less legal, more material, human, connecting. Needless to say, she has had her photograph taken with Ken.

Jessica had a comparable experience of spatial prohibition when taking the walk, which for her, however, translated to her daily jog. She remarks on the fact that she is not allowed to run on the grass: cyclists and runners

---

228   www.google.co.uk/search?q=lunar+house+croydon&um=1&hl=en&rls=com.microsoft:
en-US&tbs=isch:1&ei=viGCTaGTPJS0hAeouujABA&sa=N&start=40&ndsp=20&tbm=isch

are supposed to use a designated lane. Cyril was provoked by the obsessive signage to avoid stepping on the green in the area around the House of Parliament. It became clear to him that the point of the signs were not for the protection of the green so much as for the protection of the distance between the people and the seat of parliament. This was a visibilisation of a certain kind of legal signage and of spatial distance, in order for the real legal reason to become invisibilised. The green space was put forth as the reason for cordoning it off, invisibilising thus the fact that the barriers were not meant to appear exclusionary, threatening or barring. It is a parliamentary democracy after all. Yet Cyril steps on it. His foot, hovering over the grass, is captured by the photograph he shared in class. Gabriel had a similar realisation in his encounter with a fence along a canal walk. He is suddenly confronted with the surprising fact that the law's presence at that particular space is 'utterly pointless'. Gabriel describes his realisation in terms of an encounter. Deleuze's analysis of an encounter is telling:

> something in the world forces us to think. This something is an object not of recognition but of a fundamental *encounter*... its primary characteristic is that it can only be sensed ... it moves the soul, 'perplexes' it – in other words forces it to pose a problem: as though the object of encounter, the sign, were the bearer of a problem – as though it were a problem.[229]

Gabriel's realisation that there is a fence along the canal walk determining specific and limited points of entry/exit to the walk is one such encounter that brings about not just confusion and critical positing of a problem but significantly a material shift in the movement of his assemblage in the lawscape. He becomes aware that the spatial presence of the law inhibits him from *simply walking* aimlessly in the city. He feels constrained to have a purpose, 'to achieve an objective'. The lawscape makes the *flâneurs* of today feel guilty, potentially threatening, undesired. Consider Eleanor's desire to use her camera: this is a perfectly understandable part of contemporary *flâneurie*, wanting to make full use of a technological/human assemblage and to immerse themselves – yet the lawscape stopped any posthuman extension of the body from crossing the boundaries erected in the name of security.

Not being able to walk about aimlessly is a lawscaping triumph. But the greatest triumph is that one gets used to it: all students notice the CCTV cameras and the generalised surveillance structure that controls one's movement, but only because they have had a shake-up of the way they normally walk, a 'mission' to walk about in order to become aware of the

---

[229] Deleuze, 2004c: 176

lawscape. Under normal circumstances, the lawscape becomes atmospheric, without a renegotiating margin, fully given and fully desired. Just like breathing or putting one foot in front of the other in order to walk, after a while one stops noticing the lawscape. Only when one decides to walk aimlessly, that is to take a walk with a teleology detached from destination and immanent to the walk itself, does one realise the law's avuncular presence. And then, most people would recognise the feeling that floods Lena, another student taking the walk and in the event failing to walk without a destination that would inform her stride: Lena was lost. This is not because she got actually lost but because, in her attempt and eventual failure to walk aimlessly, she felt lost, unable to be registered, parasitical. She tried to come out of the cocoon readily offered by the lawscape, but she failed to take the risk (for this is what it is about) of letting the hand of the law go and negotiate a differentiated space of in/visibility. Instead she sat in a café and allowed its calm atmosphere to soothe her (only vaguely dissimulating the fact that she was still very much in the lawscape). David on the other hand, managed to perform this walk without a destination and indulge the different mannerisms demanded of someone who walks on relatively unfamiliar ground: David felt more careful and at the same time an 'outsider' to the 'laws of the land' as he called them. He navigated himself through areas he was less familiar with and his observation was that he could notice more, even though integrate less.

This walk essentially asks the students to become nomads, to move across space, to encounter the lawscape in its in/visibilisation. It is also asks them, albeit implicitly, to *reorient* the lawscape and create a different striation and a different in/visibilisation that accommodates their needs. The lawscape is characterised by a relatively ample margin for manoeuvring, at least in relation to atmospheres. The lawscaping bodies, especially after an encounter of the above sort, can and do reorient the lawscape as we have seen. This means: modify the given in/visibilisation. The success rates vary, but in the process, one understands a few things, which I have touched upon in the chapter so far: perhaps most importantly, the continuous lawscape. Namely, that wherever one moves and whatever one does, one cannot escape the lawscape. This is not just because the law is everywhere in space, but rather that the law is carried in the various bodies, themselves part of space. The law is not abstract and outside one's body, but very much dwelling within and among one's molecules. As we move, we lawscape. As we stand, we lawscape. This 'we' refers to a posthuman lawscaping, where student bodies mix with professorial superegos, technological mappings, caffeine intakes, and so on. Either obeying signs or defying them, one moves in the lawscape. In the first case, one follows the existing lawscape, and this becomes clear very early on. What is harder to understand is the way in which even in defiance of existing lawscapes, one generates other lawscapes. Resistance is part of the manifold. There is no space outside from which to critique, resist, defy, revolt, withdraw. Wherever one moves, one carries along the lawscaping body – it is the precondition of lawscaping. But one can reorient this lawscape.

As I said earlier, this does not mean that all lawscapes are the same. Far from that, and this is the second thing that becomes obvious in the exercise: each lawscape moves differently because it depends on the way the various assemblages (among the bodies that constitute them) are formed each time. One follows (and constructs) different lawscaping commands when one is standing in front of the US embassy or when sitting in a café. They are both lawscapes but with different visibilisations. When Lena sat down, she lawscaped an invisibilisation of the law. There is no law in a café, right? Wrong. From the moment you walk in, the law engulfs you but as invisibility: entering the contractual agreement to pay for your coffee and sit down to drink it is not something that needs to be visibilised. On the contrary, it needs to remain relatively invisibilised so that the spatial comfort of the café is the only visible thing. In another exercise my students and I undertook, we attempted to map the Occupy St Paul's camp during the period (2011) when the movement was going strong through its spreading repetitions across the globe, and tens of tents were pitched in front of St Paul's Cathedral, in the middle of the financial district of the City of London. The idea was not to map only the physical structures but the flows of in/visibilisation in that particular lawscape. Thus, we had to trace boredom, cold, excitement, conflict, debate, ideologies, the police, the local government, the business people, the passers-by, god, priests, the university teachers giving lectures at the Tent University, the toilets, the information tent, and so on. The maps were becoming more and more complex but one thing emerged with absolute certainty and perhaps surprise: what seemed like a way out of the lawscape in the sense of resistance, had generated its own fully regulated lawscape with signs on what to do and not to do everywhere, corridors for bodies to follow in order to walk between the tents, functionally differentiated specially designated areas that recalled zoning, as well as zones of working with or against existing parallel lawscapes of the City of London, of the religious institution, of the tourists, of the drunk homeless (who were explicitly excluded from the movement) and so on. Time was also lawscaped: depending on whether one was going to be away from one's tent for an evening, a while longer or not at all despite the tent appearing unoccupied, different coloured pegs were attached to the tents so that others would know whether they could or not use that tent for the evening or longer or not at all. There are huge differences between a mainstream London lawscape and the Occupy movement, not least issues of property, ideological positioning, use of violence (and legitimacy thereof or not) and so on. Some things, however, are common: that every participating body lawscapes as it carries on in/visibilising, and either generates confluence or conflict. Every body affects and is affected by other bodies and in so doing, it generates the lawscape. It is not so much where the lawscape comes from, from the state or the commune or the occupiers. It is the way the lawscape is embodied and borne with and through the body's movement that makes it powerful.

The tilting surface of the lawscape means that stronger bodies determine the in/visibilisation of the lawscape in ways that the other bodies cannot easily or readily challenge or sometimes even perceive. This is the zone of indistinction between lawscape and atmosphere, where atmosphere is engineered as a totality. Invisibilisation of the law serves the lawscape-turning-atmosphere: one is not even aware of the discomfort, the pain even, behind the façade of comfort. Consumerist temples of easy, cheap 'impulse' buys often hide sweatshops, child labour and environmental disasters. Conversely, being caught in the revolutionary spirit of a conflict over regime changes (and revolution is ostensibly a move away from the lawscape, indeed it finds itself in direct conflict with the existing lawscape) is a triumph of body assertion, position and ideological superiority, but only impressionistically does it operate 'outside the law'. What emerges from the above is a surface that cannot be judged a priori and from a distance, but only through immersion in its very lawscaping operations. We are all complicit with anonymous materials as Reza Negarestani writes[230] – even when we think we are our own masters and have full control over our ideological, geographical, temporal positions.

The third thing (and arguably the ultimate hope) that one realises through the exercise is the way identity is formed in the lawscape. No longer a simple given structure of I am and I am called, a recognition by oneself and by others, nor a dialectic structure of ego/alter, namely I am ego because I am not alter. Beyond these structures, Deleuze and Guattari have suggested a new form of understanding identity, one directly drawing from Spinoza, that they call *haecceity*, namely the understanding that a body 'consists entirely of relations of movement and rest between molecules or particles, capacities to affect and be affected'.[231] Haecceity understands identity as difference: as a posthuman collectivity that does not focus on the individual but on the assemblage with other bodies in terms of speed or pause. The temporality of movement is paramount. For Spinoza, as we have already seen, bodies differentiate from one another through their speed and rest, namely how fast or how slow they move in relation to one another. Movement is the basis of singularity of each body. The fast movement of a storm differentiates it from the human bodies caught in its eye, themselves faster than the houses or plants also caught there. In that sense, and this is where I distance myself from Deleuze and Guattari, movement is always withdrawal, to the extent that all bodies withdraw (from other bodies when

---

230 Negarestani, 2008
231 Deleuze and Guattari, 1988: 262

moving) and withdrawal dons a body with its ontological singularity. Singularities, however, are also organised in assemblages: 'the street enters into composition [assemblage] with the horse, just as the dying rat enters into composition with the air, and the beast and the full moon enter into composition with each other'.[232] This does not mean that singularity is lost in an assemblage. A body always withdraws, even in assemblages (in any case, a body is always in various assemblages simultaneously). The air can never present itself fully: it does not have a contour, and it does not have limits. This does not mean that it is infinite, but that its limits always withdraw. It is the same with every body. *Haecceity should be understood as the manifold paradox between, on the one hand, an ever present yet ever withdrawing singularity; and, on the other, an assemblage between one's body and other bodies, including the space in which one moves, and the body of law that determines the movement.* This also includes the space within law that allows the uncertainty and directionless of spatiality to emerge. In other words, haecceity includes the potential individual and collective reorientation of the lawscape from within the movements and rest of the individual and collective body. Reorientation does not necessarily require grand gestures. It can also come in minor gestures of the everyday. As Deleuze and Guattari remind us, 'taking a walk is a haecceity'. This is indeed the guiding idea behind this exercise (and perhaps my whole class curriculum): to enable law students to create for themselves a haecceity of creative *withdrawal from* as well as creative *pleasure in* the atmosphere of the given. Through corporeal, spatial, immersive learning techniques such as walking, one rediscovers and reinscribes one's reality, repositions one's body in the middle of the lawscape, observes the law in both its controlling and enabling, becomes aware of the inevitability of materiality, and eventually becomes responsible for one's own position, one's own lawscaping. One learns how to at least try and 'free the diagonal'.[233]

232 Deleuze and Guattari, 1988: 262
233 Deleuze and Guattari, 1988: 295

Chapter 3

From lawscape to atmosphere
Affects, bodies, air

You walk into a room that smells of roses. The walls are painted a stimulating combination of red and yellow. The first notes of Beethoven's *Für Elise* are piped into the air. You sit on a comfortable chair, you switch on your iPad and get ready to check your Facebook account. There is even a dart board, should you feel like playing. You feel well, at ease, energetic. You perceive the surrounding atmosphere as pleasant, familiar, protective. You take a sip from your Coke, if that is your thing, and settle in.

You are still in the lawscape. You have never left it. But there is something about the almost total, and successful, invisibilisation of the law in the above – unless of course you have been reading this book from the start, in which case you would already be conditioned to question such an apparent absence of law. In the room, legal invisibilisation is served on a bed of comfort, corporeal emplacement and consumerist desire. This is the lawscape's wet dream: to be able to withdraw *as* a lawscape. Of course, if one looks for the law, one finds the law. But the point of the above room is that one *does not even want* to find the law. One has become complicit with dissimulation amply offered by technological, economic, legal, anthropomorphised matter. This dissimulation feeds our ways of living and ends up determining the human while drawing around us a circle of 'belonging'. Things are good as they are, cosy and all-inclusive, safe and sheltered from the outside world. A moment of Baudelairian *luxe, calme et volupte* that can go on, why not for ever? We and our desires are captured by it. We do not want to leave.

The above extreme and all-inclusive, safe and perfectly sheltered lawscape is what I call *atmosphere*. Atmosphere is the lawscape that has managed to reach its 'perfect' dissimulation as a non-lawscape. Atmosphere is the withdrawal of the lawscape from the very bodies of its emergence. It is what remains when the lawscape departs, that is to say, when the interplay between in/visibilisation has been replaced by one self-perpetuating, all-containing elemental bubble of air and water and earth, one grand dissimulation. It is the fantasy of the *arcana imperii*,[1] the famous shaded

---

1   See Canetti, 1984

space of invisibility to which authority withdraws and from which it commands the world. Atmosphere is the face of Veronese's *Glory of Venice*, always in the shade, always half-seen from down below, basking in the eternally serene chiaroscuro of the powerful.[2] The bodies captured in an atmosphere are there only for one purpose: the perpetuation of this atmosphere for just a little longer, as long as possible, or as long as 'needed'. This means that 'need' is constructed within an atmosphere, autopoietically eating its own tail. I discover my needs when I am faced with the 'product' that meets them. Or, my needs emerge in the particular context of the conflict that I want to perpetuate, or indeed curtailed by the particular atmosphere that only allows some needs to emerge. I can no longer distinguish between engineered needs and, what?, 'real' needs? Where do I find these? They may have been responsible for the fact that I entered the atmosphere, that I wanted to enter, if an inviting atmosphere; but atmospheric captivity goes beyond individual or even collective needs. They might still be here but are fundamentally recast in order to reflect the needs of the atmosphere. So our needs are converted into one foundational need: the need *of* the atmosphere to carry on existing, to perpetuate its *conatus*, its will to carry on. This is autopoietic atmospherics at work, namely an atmosphere that dissimulates the fact that has been created, and the only thing that counts is its perpetuation.[3] Bodies are placed in the service of the conative quality of the atmosphere.

Conflict can also be 'comfortable' in an atmospheric sense. An atmosphere of tension, confrontation and violence, whether physical or not, is as captivating as a confluent, traditionally 'easy' atmosphere. It will become clear in the chapter that atmosphere is not merely a comfortable lawscape in the sense of the above room, but also a space of conflict brimming with ideological charge, aggression, confrontation. It can be a political rally, a revolt against an oppressive regime, a moment of collectivity characterised by honest revolutionary spirit; it can also be a prison or a refugee camp or any other atmosphere of biopolitical manipulation. The same atmospheric capture applies there too. The difference lies in the way dissimulation affects bodies in every case: in a prison, law's ubiquity takes up the form of matter (walls, bars, yards, inside) that suppresses the law as *nomos*, namely the law that allows one to move across a smooth territory.[4] The law here is in the service of the atmosphere of suppression. In an urban revolt, on the other hand, the law is in the service of organising the particular spatiality and temporality of the revolt *in opposition to* the ancient regime that the revolt is fighting against. Thus, here a visibilisation of *nomos* and invisibilisation of the *logic* nature of any revolt is at work. Finally, in a department store, the law is

2 See also Goodrich, 2014 on visualisations of the sovereign.
3 See Philippopoulos-Mihalopoulos, 2013a for material considerations in autopoiesis.
4 See above, Chapter 2 for the distinction.

in the service of the atmosphere of excessive desire to consume, since what becomes dissimulated is law as *logos*, namely the oppressive nature of the law. And so on. The consumerist, non-conflictual paradigm additionally occupies this paradoxical position of being both a standard object of critique *and* the 'natural' environment from which the critique is often launched (think of us Western academics/artists/activists critiquing consumerism). The latter is not meant to be a critique of double standards but a call for an understanding of responsibility from within the spatial and temporal context in which a body is situated, as I show further in this book. Indeed, the point is not whether atmosphere is good or bad, and whether it serves benign or malign purposes. Atmosphere is not a moral judgment but a cut that tends towards all-inclusion and that renders inert any wilful resistance.

This chapter progressively becomes elemental, dealing with sounds, scents and airs. The first field of enquiry is the interstitial area between sensory and emotional occurrences, namely sensory experiences that are traditionally thought to be a causal result of external stimuli, and emotional experiences that generally allude to something internal. Instead, I am arguing, first, that both emotions and senses originate as much in the body as in its environment, and subsequently that there is no constructive difference between internal and external origin, namely senses and emotions. The distinction between a self-contained individual and an environment has collapsed, and this is not only because of the forced mendacity of either side of the distinction. I employ the term *affect* as the sensorial, emotional and symbolic flow circulating among bodies, and I suggest the concept of *atmosphere* as an attempt at understanding affective occurrences as collective, spatial and elemental. At the same time, however, I am not interested in mere atmospherics (if there is such a thing) but the *engineered* atmospherics of the lawscape.

Thus, your earlier entrance to the rose-smelling room was a piece of engineered atmosphere. The lawscape converts itself into an atmosphere by allowing certain affects to come forth while suppressing others. It typically entails a suppression of the legal elements of a lawscape and the preponderance of the material, spatial, corporeal affects of desire. Engineering atmospheres might well be used positively in an attempt to reduce crime,[5] or as a mechanism of withdrawal itself, emergent yet directed and potentially leading to spatial justice, as I show in the Chapter 4 in the context of transhumant shepherds. Or it might be used as a tool for political or economic strategies that guarantee and anticipate specific affective responses. Thus, in the following section I look briefly into how senses, emotions and information come together in order to produce institutional affects.

Let us, then, go a little deeper.

5 Borch, 2008

3.1 Affects: senses, emotions, symbols

Sensory stimulation in the lawscape is a beguiling, confounding, excessive affair. David Howes calls this 'hyperaesthesia', namely the sensory overstimulation characteristic of capitalist societies.[6] All senses are summoned to participate in one's experience of the lawscape: visual displays are in some ways the most innocuous stimuli since one has learnt to expect them, although even this is no longer the case with product placement and subliminal messages that catch us unawares. Supermarkets are perhaps the most typical example of consumer sensory stimulation, with the piped smell of baked bread and studied product shelving a thing of intense marketing scrutiny. One must not forget the relevance of textures of products that respond to the touch in inviting ways, or music meant to encourage either a lingering and luxurious or indeed a self-confident and energetic shopping experience. The new design credo is no longer beauty in appearance but 'beauty in interaction', as Howes reminds us.[7] The body is now firmly emplaced in a society of commodities, with chunks of its being claimed by various sensory stimuli from all angles. The body throws itself with abandon to a hyperaesthetic hedonism.

Hyperaesthesia does not necessarily need shop-windows and arcades.[8] The other characteristic element of modernity is crowds.[9] Crowds have traditionally been controlled in various ways, which nearly always involved partition and direction. Christian Borch refers to Hitler's meticulous atmospheric staging for propaganda purposes, which involved physical mass meetings.[10] Crowds brim with hyperaesthetic stimulation for both the ones participating and the ones watching over. In the kettling enclosures enforced by police during protests, the enclosure is as much spectacularised as it is its external 'free zone' of passers-by, in their turn taken up by the media and other spectacularising machines.[11] Crowds are often brought to what Elias Canetti in his seminal work *Crowds and Power* has called the point of 'discharge', namely the boiling point (for kettling) where violence erupts even in initially peaceful crowds.[12] This point is achieved through various techniques which often involve corporeal pressure by the police, and are manifest examples of Canetti's embodied and especially tactile understanding of force as the touch by the one who craves power of the

---

6   Howes, 2005
7   Howes, 2005: 284
8   Benjamin, 2002
9   Le Bon, 1960; Tarde, 1968; see Borch, 2012 for an exhaustive analysis.
10 Borch, 2014; see also Böhme, 2006. [No Böhme 2006 in References.]
11 Rowan, 2010
12 Canetti, 1984

one who is submitted, 'the central and most celebrated act of power'.[13] The kettling maelstrom of sensorial, semantic and emotional hyperactivity impressionistically serves purposes of 'public safety' but in most cases is meant as a strategy of affective intensification through the separation of an artificially delineated interior (violence) from an equally artificial exterior (peace): 'by provoking the crowd, violence is inflamed by kettling itself. The exercise of kettling is therefore incitatory in that it creates the threat in order to deal with the threat'.[14] Even without kettling, crowds waver affectively. It is a feature of crowds to be taken over by often unexamined distinctions that spread affectively. In a distinctly Foucaultian biopolitical context of crowds against alleged criminals, Moira Gatens writes, 'to be consumed by the force of such affects entails that one need not consider the causes of the behaviour that result in rape, murder, violence'.[15]

Hyperaesthesia goes hand in hand with a punctilious attempt at separating the senses. In the above kettling examples, sensory deprivation takes animal dimensions with the deprivation of toilet facilities, no space to move for the kettled protesters, and visual occlusion of the outside. Museums and art galleries are a different yet strangely comparable example of such sensory deprivation. The Aristotelian hierarchy of senses returns, where touch is considered the lowest of the orders because of its commonality with animals, closely followed by taste and smell. Jim Drobnick has studied the disciplinary technology of the 'white cube' space, namely the sparse, ocularcentric form of the standard contemporary art experience.[16] He calls them 'anosmic cubes' because of the concentrated attempts at deodorisation that take place in them,[17] geared towards the training of art appreciation in a highly visual, non-touching, aurally insulated atmosphere. An anosmic space is a ready-made that has expelled the process of deodorisation from its folds, thus making sure that its sanitisation is seen as a finished and indeed prior and permanent state of being. No aggressive chlorine smells, no open windows, no visible air-conditioning, just a cube dissociated from its social context – or at least so the story goes, since the sparseness and anosmia of the space points to a masculine, linear, white, middle-class adherence to a supposedly Spartan way of living that squeezes all its frills, pathologies and perversions in the (concealed) white closets around the

---

13   Canetti, 1984: 206. Canetti was not the only one focusing on the corporeal in crowds. Despite the main psychological focus, Le Bon's 1960 work on crowds, which is considered one of the first systematic treatments of the topic, also deals with embodied crowd movement. See Borch, 2012: 44ff.
14   Ta kale, 2012
15   Gatens, 1996b: 121. Gatens's discourse eventually points to a Spinozan knowledge that refers to the causes of behaviour. This knowledge here takes the form of distinguishing between various kinds of desire, more of which below.
16   Drobnick, 2005
17   Drobnick, 2005: 266

room. Paul Virilio refers to this when he talks about sanitisation as ideology: all in the service of segregation, ghettoisation and suburbanisation.[18] Such attempts to achieve sanitising anosmia, however, only vaguely manage to hide the aspiring *a-legal* nature of the cube: a space seemingly outside the lawscape, where the violence of legal controls has withdrawn for the sake of pure art enjoyment. This is an egalitarian space, where everyone can walk in and appreciate art. Yet, just as the various residual odours, the law is as well (or as badly) hidden: despite appearances, the white cube is a striated, controlled space full of corridors, walls, barriers, micropanoptica, pillared vistas, differentiated velocities of controlled movement and strategised pause. What is more, it is a striated space *dissimulating* as smooth, completely populated by property lines, health and safety regulations, consumer protection barriers, public morality risks, insurance dictates and so on. Still, all these are withdrawn, and what is left is one ample, white continuum.

From single-sense domination to hyperaesthetic stimulation, contemporary environments are sites of sensory extremes. A body moves along corridors of sensory direction, consciously or unconsciously obeying invitations and exclusions, sensory barriers or gestures of guilt-free overload. Electronic spaces and 'everyware'[19] construct sensorial corridors based on aural, tactile and visual stimulation in a not dissimilar way to physical spaces as Bachmann and Beyes show.[20] Social media in particular construct spatialised stimulations either through physical spatialisation, say location services, or virtual spatialisation aided by a physical spatial terminology such as 'walls', or corporeal terminology such as 'poking' and so on. Back to physical spaces of the last few years, one of the most iconic London department stores, Selfridges, has been repeating its obviously successful sales campaign, with big signs scattered around the store reading 'Buy me, I will change your life', 'Touch here', 'You did not find me, I found you' and 'I shop, therefore I am'.[21] Ironic and playful, yet squarely hitting the spot of alleviating the overspending consumer from guilt, the campaign reinforces the multisensory attack with invitations to luxuriate in the freedom of touching, smelling, feeling of what can very easily turn into a very costly experience. It can be argued, however, that this kind of 'edgy' campaign only works in some consumer societies. Indeed, the cultural base of sensory stimulation becomes very quickly obvious when, to take an example, one finds oneself in the London underground, after having been used to, say, its Madrid equivalent. One's personal space is much more extensive in London, where even at rush hour, passengers avoid looking at each other however closely they might stand, generally obey the London

18 Virilio, 1993
19 Greenfield, 2006
20 Bachmann and Beyes, 2013
21 Douglas, 2006

Transport signs that urge them to avoid eating smelly foods, endure dirty looks if they listen to their earphone music too loudly, and of course never, ever touch each other.[22] And if by accident they do touch, a highly formalised and rather empty 'sorry' comes as a linguistic *deus ex machina* and dispenses with the impropriety of the other senses.

Why is linguistic communication preferable to being touched or exposed to someone else's odour? Darwin's famous dealings with emotions and expressions points to the element of *disgust*, with its obvious links to taste. He marvels at 'how readily this feeling is excited by anything unusual in the appearance, odour, or nature of our food'.[23] The association with taste is, I think, not just an etymological coincidence (*gustus*, 'taste' in Latin). Taste is a private necessity, the one sense associated more closely with food, and as such it retains a privilege of voluntariness. One cannot always control what one hears or smells but one is expected to be able to control what one tastes. To find someone's touch or smell or even appearance disgusting is an expression of loss of control. The violation of one of the intimate corporeal cavities, the mouth, points to an undesired reduction of distance between the body and the world. This is through and through a class issue[24] that reinforces the perception that the educated classes enjoy a distance between necessity and pleasure, whereas the working masses are given to an unmediated, soil-infested satisfaction of necessity. To be fed smells, noises, bad dress sense or unnecessarily tactile attitudes makes the recipient feel less of an educated member of society and more of an animal. As Sarah Ahmed points out, 'to be disgusted is after all to be affected by what one has rejected.'[25] Unsurprisingly, the law jumps in and limits such opportunities of affect, as Martha Nussbaum shows when she examines how the law considers disgust as a reason and excuse for conduct norm-making.[26]

A couple of things become evident from the above. First, through the senses, the human approaches its elemental materiality and animality. This is often uncomfortable (especially from a class perspective) and the lawscape attempts to minimise it through linguistic prioritisation, control of sensory stimulation, or reservation of such stimulation for cases, such as subliminal marketing,[27] where one succumbs to one's senses without becoming aware of it. Senses flatten the surface between animate and inanimate. Kathleen Stewart shows how the sensorial, just as the emotional, is spatial:

---

22   See Borch, 2011b on Canetti's idea that the touch reactualises the relation between hunter and prey; also Sennett, 1970, on the fear of touching.
23   Darwin, 1965: 256
24   Corbin, 1986
25   Ahmed, 2006: 86
26   Nussbaum, 2000
27   Lakhani, 2008; Karremans *et al.*, 2006

the senses sharpen on the surfaces of things taking form. They pick up texture and density as they move in and through bodies and spaces, rhythms and tempi, possibilities likely or not. They establish trajectories that shroud and punctuate the significance of sounds, textures, and movements.[28]

This elemental quality of senses is close to a human (impression of) animality. The lawscape has a colonising relation to animality, at the same time exploiting and concealing it. It is in the interest of the lawscape to retain the illusion of centrality of both 'the white man', the empty human effigy,[29] and the colonised other as part of 'nature', therefore 'reduced' to animality. At the same time, all the hybrid, déclassé, taste-less, animal qualities in between, are being engineered towards a controlled, smooth society. The posthuman lawscape is therefore populated with what Catherine Ingraham calls 'the human animal of the post-animal world', a body 'uniquely amenable to capture' without any 'psychological resistance to the act of being housed or caged – in fact, the reverse'.[30] The posthuman lawscape is not something beyond the human but deep into human animality, in the swollen limit between the anthropomorphic animal resource and the anthropocentric human animal, creating an atmosphere of dissimulation of precisely such depth.

Second, sensory stimulation is not just about senses. Nothing is as deep as skin. I touch something, it feels good, I want it. Note the passage: I feel *it* – *it* feels good. The object becomes the subject of the sentence, and both share a sense, a direction of desire. Senses are all about emotions.[31] Desire springs between me and the thing like scented mist. The thing becomes part of my body, my body extends to the thing (and the thing becomes a Hegelian property). The feeling is shared on the skin of the lawscape, spreading like skin rash, piling up an economy of desire. In Jean-François Lyotard's *libidinal economy*, the body and the city, the one who touches and the one who is touched, share the same skin: 'in libidinal economy there is nothing but skin on the inside and the outside, there is only one monoface surface, the libidinal body is a Moebius strip'.[32] This goes beyond a material continuum of body and world. If senses are, so to speak, the libidinal formation, the locus where desire is formed, and emotions are the representation of the material desire, the *signs* of the material, then any distinction between the two is forced. As Lyotard says, 'there is no notable difference between a libidinal formation and a discursive

---

28 Stewart, 2011: 448
29 Grear, 2007 and 2011; see Braidotti, 2013 on the decline of human exceptionalism.
30 Ingraham, 2006: 85
31 See also Ahmed, 2006.
32 Lyotard, 1993: 156

formation',³³ echoing Deleuze's understanding of the common surface between the material and the discursive. Lyotard again: 'we well know that this surface is *at the same time*, indiscernibly, the libidinal skin ... and the wise flat sheet of the account book.'³⁴ Sense and emotion, desire and desire, one skin.

Albeit in a different register, the above become palpable in the work of Michel Serres, one of the great theorists of the body. In his seminal book *The Five Senses*,³⁵ Serres has carefully yet idiosyncratically assembled a contextualised analysis of the senses that goes well beyond the phenomenological and augurs, at least for Serres's subsequent work, a kind of new ecological embodiment.³⁶ One of his basic assertions is that bodies and senses are mingled (in themselves) and intermingled (hence the subtitle of the English edition, *A Philosophy of Mingled Bodies*). Serres traces the meeting of the body with the world on the skin. Not unlike Lyotard, Serres thinks of skin as the membrane that has both an inside and an outside yet whose stimulation cannot always be attributed causally to one of the two. Serres writes: 'the skin is a variety of contingency: in it, through it, with it, the world and my body touch each other, the feeling and the felt ... I mix with the world which mixes with me. Skin intervenes between several things in the world and makes them mingle.'³⁷ But skin is not only a human prerogative. Contingency is common to all bodies and elements:

> 'Contingency means common tangency: in it the world and the body intersect and caress each other ... We cannot claim to be so exceptional. We are not the only ones, surrounded by boundaries, to throw ourselves in contingency ... Everything meets in contingency, as if everything had a skin. Contingency is the tangency of two or several varieties and reveals their proximity to each other. Water and air border on a thick or thin layer of evaporation, air and water touch in a bed of mist.³⁸

This is the grand continuum of the skin, both inside and outside, intracorporeal and intraspatial.

Through skin, contingency is enacted in both senses of randomness and proximity. In *Genesis*, Serres writes: 'I am anyone, animal, element, stone or wind, number, you and him, us.'³⁹ This is a distinctly posthuman, animal

33   Lyotard, 1993: 25
34   Lyotard, 1993: 18
35   Serres, 2008
36   See Connor, 2005.
37   Serres, 2008: 80
38   Serres, 2008: 80–1
39   Serres, 1995: 35

body, in touch with its senses through the skin. At the same time, the body is the skin of the world, exposed to the noise inside and outside, 'strung tight, covered head to toe with a tympanum'.[40] The vibrations of the body are the vibrations of the world, fractals that ripple across the shared skin without discernible origin: is it the world that makes me feel like this, or is it my mood that makes the world feel like this? We can never know this. Between hearing ourselves and hearing the world, there is a black box resounding with silence. For Serres there are three kinds of the audible: the propioceptive hearing of the workings of one's body with its organs and cells and molecules; the hearing of the world's noises that 'reach the monad softly, through doors and windows', for indeed, 'being part of [it] means not hearing it';[41] and the last kind, which is the audible rupture of that which cannot be heard, 'interrupting the closed cycles of consciousness and the social contract',[42] and constituting itself as the *black box* of knowledge, which is nothing else than the ontological withdrawal of all bodies. For Serres there are innumerable black boxes that consist the body and the world, all of which allow sense to come in and meaning to come out. In French, the term used is *le sens* which stands for both significations of the word ('sense' and 'meaning'), as well as, importantly, 'feeling'. In other words, we can think of *sense* as the flow of events that enter and exit the black boxes, having submitted themselves to a process of transformation that is to remain unknown.

Serres's black box is reminiscent of what Goonewardena has called the 'urban sensorium',[43] namely the parapet that conceals political and legal engineering of the urban space from human perception. The difference to Serres is that the urban sensorium remains a thing *between* the body and the city, whereas a black box is to be found everywhere, both inside and outside of either. A black box is like the skin that neither merely stops, nor just transmits but is shared and extends either side, and whose input and output can be known and followed, but whose precise process remains inaudible, consistently withdrawn. The concept of the black box along with the preceding discussion reveals the absence of distinction between external and internal origin of occurrences, namely between feelings and senses. What is thought to come from 'inside' such as emotional states, flows and folds itself alongside what is thought to come from the 'outside', such as sensory reactions to textures, smells, sounds. Sense and emotion circulate in the black box that, like skin, extends itself in the interstices between interior and exterior and shows how the continuum is only epidermally ruptured. In that sense, one can productively read Lyotard's libidinal skin together with

---

40   Serres, 2008: 141
41   Serres, 2008: 107
42   Serres, 2008: 111
43   Goonewardena, 2005

Serres's black box, not only for their function of contingent surface, but also for the added reason that Lyotard's political position connects more readily with the discussion on engineered atmospheres that follows below. The skin in Lyotard operates as the container of libidinal intensities that mediate between emotions and systems. Lyotard's *Libidinal Economy* starts with the body like an open wound, initially splayed flat and then turned uroborously into itself in order to produce the infernal circle of the libidinal skin.[44] The Moebius strip of the skin eventually gives up itself to a theatre of binary rationality and non-contradiction, made out of its very own materiality. In that theatre, skin becomes the facilitator of systemic engineering. Desire is exploited by the system and channelled towards smooth, non-contradictory, non-conflictual, seemingly anomic spaces, in which systems reign free. This kind of channelled desire, however, as Rosi Braidotti argues, results in a 'suspension of active desire, in favour of the addictive pursuit of commodified non-necessities'.[45]

I would like to focus on the fact that the emergence of atmospheres is based on the connection between senses, emotions and symbolic meaning. To return to the first issue then, I will partly follow Lyotard in this and call these combinations *affects*. For Lyotard, and at least in this context, affects are the libidinal intensities that allow the system to direct desire. In the same vein, Lauren Berlant talks about the capturing of affect in always new combinations of desire that are already co-opted by capitalist technologies.[46] Affects are the connections between the body and the world that are exploited and channelled in a predetermined direction. Yet, there is a power in the illusion (what Berlant calls *cruel optimism*), not unlike what Deleuze calls *the power of the false*, as we have seen earlier.[47] Affective attachment to desire, however constructed and manipulated, might also carry a certain empowerment. Affect can genuinely be a mode of change and hope, as Patricia Ticineto Clough argues.[48] Even on an epistemological level, the affective turn has recast the connection between body/subject/regulation and has opened new, non-representational ways of approaching it, as Nigel Thrift has argued.[49] This ambiguity of affect, both liberating and captivating, shares some characteristics with Brian Massumi's approach, which equates intensities with affects. For Massumi, affects are 'virtual synesthetic perspectives anchored in (functionally limited by) the actually existing, particular things that embody them'.[50] This means that the body that

44 Lyotard, 1993
45 Braidotti, 2006: 152
46 Berlant, 2011
47 Deleuze, 1989
48 Clough, 2010
49 Thrift, 2007
50 Massumi, 2002: 35

embodies the affect also limits the way the affect is materialised. An affect does not have one origin – the body or the world – but is materialised as a perspective of a body. This perspective connects the body to the world (usually through other bodies). It involves the senses in the synaesthetic meaning of intermingled senses, but is not limited to them.

These definitions of affect have moved away from the more intuitive understanding of the term that identifies affect as emotion. On the surface, and from a posthuman point of view, this is appropriate since emotions merely concern the individual and miss out on the collective, the contagious and the posthuman, as Nigel Thrift shows in his genealogy of affect.[51] Still, to leave the emotional out of this altogether is also problematic, as Lisa Blackman also argues in her overview of affect as material and immaterial.[52] I am eager to include the emotional since to leave it out would confirm a distinction between the sensorial and the emotional that, as I have argued, epistemologically serves no purpose and ontologically can no longer stand before the collapse of the boundary between the self and the environment. The emotional is not only human and individual but also spatial and intracorporeal – this is the point of continuum between interior and exterior, according to Serres. Emotions are collective emergences that affect the ontology of the lawscape in the same way as symbols and senses. To include the emotional, therefore, is not a phenomenological choice but a necessary move of the new material ontology of atmospheres. Brian Massumi indeed elevates emotion to 'the most intense (most contracted) expression of capture [of affect]'.[53] Massumi's definition is usually understood in the context of the Deleuzian affect, which, in its turn, is based on Spinoza's definition of affect. Let me briefly look into them.

Spinoza's by now famous definition of affect is 'the affections of the body by which the body's power of acting is increased or diminished, helped or hindered, and at the same time the idea of these affections'.[54] Thus, an affect is idea and matter, thought and body, involving senses as well as ideas on such senses. For Spinoza there is no difference between 'passions' such as emotional states of love, hate, anger etc., and material properties, such as heat, cold, storm, thunder and so on. What counts is the affirmation of an ability of a body to affect and be affected. Every body is capable of affect if it can be affected or affect other bodies through its skin, its senses, its motions, and accordingly becomes stronger or weaker, more or less joyful. Genevieve Lloyd explains affect as 'the passage from one state to another in the affected body – the increase or decrease in its powers of acting'.[55] This definition

---

51   Thrift, 2007
52   Blackman, 2012; see also Gregg and Seigworth, 2010.
53   Massumi, 2002: 35
54   Spinoza, 2000: 164, III def. 3
55   Lloyd, 1996: 72

captures the movement within an affect, the passage from state to state that consequently leads to a greater or lesser ability to 'act', that is to move among other bodies and continue affecting and being affected by them.

This intensely relational understanding of affect has also been the basis of Deleuze's reading: 'relations are inseparable from the capacity to be affected'.[56] In *What is Philosophy?* relationality of affect is the basis on which Deleuze and Guattari feel able to distance affects from emotions: 'Affects are no longer feelings or affections; they go beyond the strength of those who undergo them ... Affects are *beings* whose validity lies in themselves and exceeds any lived.'[57] And later on, 'the affect is not the passage from one lived state to another but man's nonhuman becoming'.[58] Affects, therefore, are relational but also exceeding these relations. They are becomings (and nonhuman if referred to humans) that exceed the body that embodies them yet remain materially embedded in it by triggering strength modifications. The strength (*potentia* in Spinoza) refers to the ability for *affection*, a distinction that Deleuze retains in his work on Spinoza,[59] but somehow, and at least as a term if not as a concept, allows it to fade away later on. Thus, while affect is the ability to affect and be affected, affection is the commingling of bodies and, more specifically, the change generated in a body as a result of being affected by another. As an ability, an affect determines the strength with which a body enters an assemblage with another body, namely a commingling, an affection. It is very important that the Deleuzian affect crosses human/nonhuman lines (if there ever were any) and an assemblage constitutes the field of precisely such becomings. In that sense, bodies do not necessarily have to be categorised according to species but also according to their affective abilities – namely whether and to what extent they can form new, transhuman assemblages that respond to environmental conditions in a way that result in affirmative self-preservation.

The above understandings of affect are not necessarily compatible, and to attempt a synthesis would be neither elegant nor productive. However, I am keen on retaining several of the above elements in order to assemble an affective definition that will facilitate the discussion on atmosphere below. Thus, the material dimension of affects is beyond doubt, yet their materiality is not confined to the body that contains them. Affects pulse with their very becoming, thus relationally linking bodies with other bodies (but immanently, that is through sense *and* emotion). Affects are communicated among bodies, even though not necessarily linguistically. They are abilities to affect, and as such they find expression in the complexity of language, as well as in sensory stimulation or gesticulation towards and

---

56   Deleuze, 1990: 218
57   Deleuze and Guattari, 1994: 164
58   Deleuze and Guattari, 1994: 173
59   Deleuze, 1988: 1990

movement among other bodies. In other words, they come close to what Serres has named *sense*, namely the flow of sensory, emotional and symbolic events. In this context, I would suggest that *affect is the proliferation of sense, the mirroring of the process as capacity of affecting and being affected, the sensory excess,* in short *the multiplication of becomings in such a way that they become incalculable,* indeed atmospheric.

Through its proliferation, the affect becomes collective and ripples through other bodies with equal vehemence whether it is a smell that brings back memories and determines one's territory, a feeling of light sadness or risk from a hunter, a sudden cold draught at the back of one's neck, the sun on one's leaves. More than experience, however, an affect becomes ontologised in ways that can be directed collectively (indeed atmospherically as I show below) but not controlled by the bodies experiencing it. This is what Arun Saldanha observes when he talks about phenotypical racist patterns in the context of the Goan beach. These patterns rely on imperceptible affects originating not in the body of their emergence but the assemblage space/body.[60] Thus, capacity to be affected is the capacity of a body receiving a sensory, emotional and symbolic flow. Likewise, capacity to affect is the capacity to direct such a flow to another body. In other words, I suggest that *affect is the sensory, emotional and symbolic multidirectional flow, at the same time in and exceeding one's body, that affects other bodies.* The noun 'affect' must be differentiated from the verb. Thus, while the noun is this unhinged multidirectional flow that moves as pure ontology, the verb is the perspectival (as Massumi has put it) ability of each body to affect and be affected. The origin of affect lies neither solely in the body nor in the world, but on the skin, the material continuum between the two, the black box that conceals the process yet reveals the outcomes. Furthermore, an affect 'returns' to the body of its appearance by making it stronger or weaker. In that sense, affect is the ability to enter assemblages with other bodies. A bird that enters an assemblage with a glass panel put there to protect humans is affected by the glass, which might make it weaker and kill it or stronger in the sense that the bird might learn to avoid it. Likewise, the glass experiences affects, cracks, stains, tremors: for, as Deleuze puts it,

> why is expression not available to things? There are affects of things. The 'edge', the 'blade', or rather the 'point' of Jack the Ripper's knife, is no less an affect than the fear which overcomes his features and the resignation which finally seizes hold of the whole of his face.[61]

---

60   Saldanha, 2007
61   Deleuze, 1986: 97

As I have already mentioned, in the register that comes from object oriented ontology (OOO), Spinozan/Deleuzian bodies (including the things of the above quote) are thought of as objects, namely neither subjects/objects nor human/nonhuman, but objects that assemble into objects and relate to other objects.[62] Objects are not *objectified*. Rather they resist objectification altogether because of their ontological withdrawal.[63] What is interesting here is that the space between objects is *interobjective*. According to Timothy Morton, this space 'floats among objects, "between" them; though this between is not "in" spacetime – it is spacetime.'[64] Objects emit spacetime and create zones into which other objects are affected, becoming stronger or weaker. Remarkably, 'interobjectivity eliminates the difference between *cause* and *sign*.'[65] The sign of the broken glass is not an indication of the bird-glass collision cause, as traditional causality would have it. Rather, broken glass, collision, bird body and glass surface, as well as a myriad other objects, such as wind force, luminosity, yesterday's climactic conditions and tomorrow's glass repair, are all objects that emerge on the surface of the interobjective space as one object whose extension and intension cannot be fully grasped. Constructed causality gives way to a causality of assemblage which is never fully revealed. Causality is withdrawn, just as the objects themselves that never present themselves fully.

Another example: I walk in a supermarket and I am bombarded by an affective overload, which includes sensory and emotive invitations that produce a certain meaning for me personally – a chocolate looks inviting, awakes my desire for sugar and cocoa which is both psychological and biological, makes my body move closer to the chocolate counter, and also allows my mind to become aware of such modifications by constructing a narrative which, in one way or another, fits in my everyday experience (e.g., it is a small reward for something I did that morning). I have been affected by the flow that was efficiently directed to my body. I am aware of it, or at least I am conscious of the changes affected in my body. This is the Spinozan *idea* of the affection. But, as Braidotti points out, Spinozan knowledge is always about the body itself.[66] Conscience is only relevant to the extent that the body embodies it. The bird becomes aware of the glass and avoids it. The glass is hit by birds and becomes weaker with every crash – matter remembers, the idea of affect becomes inscribed in the affected body. The body bears on its surface the historicity of affects.[67]

At the same time, not only does an affect exceed the body of its appearance, as Deleuze and Guattari remark, but takes the body with it in excess

---

62 Harman, 2009; Bryant, 2011
63 Harman, 2005 and 2011
64 Morton, 2013: 81
65 Morton, 2013: 88
66 Braidotti, 2013
67 Razack, 2011

and flight. The body becomes a multiplicity of torn parts, a *body without organs*[68] whose 'topology is tactile',[69] aural, olfactory, visual, gustatory: 'the body exceeds the body or fails ... I feel therefore I pass, a cameleon in a gaudy multiplicity, I become a half-caste, quadroon, mulatto, octoroon, hybrid', writes Michel Serres.[70] The affect becomes a *line of flight*, a creative but tortuous path from within the body that pushes its limits further. No doubt, this line of flight might be a line of resistance against the disciplinarian striation of space, the affirmation of vitality of the body exceeded. But also, and rather unspectacularly, it can be an anticipated, planned, instrumentalised excess. It can *appear* theoretically radical and adventurous, but is in fact flattened out and indeed co-opted, which is exactly what seems to be happening when even radical theory succumbs to a rather facile and unquestioned fetishisation of the body and its rhizomatic potential. It can be engineered, politically and legally anticipated, economically desired. It can be an addictive desire that sits on top of vital desires and presents itself as libidinal. It can be presented as the 'way out' from 'life in the machine' and become simply a designed, induced emotive state. As James Nolan puts it, 'the citizen recedes; the therapeutic self prevails.'[71] In short, it can be part of an engineered atmosphere of affects.

3.2 Atmosphere

Bodies slide in corridors of affects. Affects are generated on that shared skin, caressing, scenting, beating, hunting, spreading, leaking, leaning, biting spaces and bodies and holding them together in what I call an *atmosphere*. I want to think of atmosphere as a force of attraction. Atmosphere is embodied by each body yet exceeds the body because it cannot be isolated, it is always collective and always withdrawn. An atmosphere spreads through and in between a multiplicity of bodies like a sticky substance. *Atmosphere is the excess of affect that keeps bodies together.* And, further, *what emerges when bodies, human and nonhuman, are held together by, through and against each other*. Although this definition will develop in the course of the text, it is important to begin by clarifying that an atmosphere is affect transmitted, as well as affect directed. It is both emerging and engineered.

Traditionally, and apart from studies of Earth's atmosphere from the perspectives of geography, physics or chemistry, atmospheric research has been largely confined to architecture.[72] Peter Sloterdijk's work on air, air-conditioning and atmoterror has revived the interest in atmosphere by

---

68 Deleuze and Guattari, 1988
69 Serres, 2008: 81
70 Serres, 2008: 307
71 Nolan, 1998: 6
72 See Anderson, 2009 for an overview.

extending it into critical geography, sociology, philosophy, ecology, and so on.[73] Likewise, the work of people like Gernot Böhme and Hermann Schmitz on atmospheres becomes increasingly familiar to the Anglophone world.[74] In the last few years, a flurry of activity has been generated in geography, architecture and to some extent sociology, but a great deal still needs to be done in order to get a sense, not only of what atmospheres are but more importantly, how atmospheres can be resisted, escaped, attacked. Existing research largely deals with atmospheric design matters, and even Sloterdijk's magnum opus, *Sphären I–III*,[75] is of the descriptive rather than normative persuasion, unearthing how air has been buried into the solid. Yet this unobservability has allowed it to become, at least in the last century, the main medium of population control from armed conflicts to domestic room arrangements. Even on a definitional level, most attempts at pinning down atmosphere are largely phenomenological. Take for example Böhme's definition of atmosphere: 'the common reality of the perceiver and the perceived'.[76] His attempt at leaving behind the ocularcentrism of architecture urges him to take into consideration moods and emotions but unfortunately without contextualising them in a way that breaks from the traditionally phenomenological. Thus, he writes: 'seeing is all about distance. Not immersion. The sense of being in something is "mood". By feeling our own presence, we feel the space. We feel its atmosphere.'[77] This position is problematic in many respects, and not least in that it presupposes a (human) subject and an object, thus resuscitating the Cartesian distinction phenomenology has been trying to distance itself from, as well as significantly the various racial, gender, and sexual perspectives that have the potential of radically affecting the atmosphere.[78] Böhme's 'betweeness' of atmosphere largely follows Schmitz's argument against the private nature of emotions.[79] Schmitz is responsible for the revival of the Homeric concept of emotions originating in the space surrounding the body.[80] As he writes in a previous work, 'space flows along with us.'[81] An atmosphere does not need utterances or even other human presences in order to emerge, Schmitz correctly argues. This is reminiscent of Juhani Pallasmaa's recent formulation: 'as we enter a space, the space enters us.'[82] But then, both

73 Sloterdijk, 2009
74 Böhme, 1995; Schmitz, 1969
75 Sloterdijk, 1998, 1999 and 2004
76 Böhme, 1995: 34
77 Böhme, 2005: 402
78 See also Löw, 2008
79 Schmitz, 1995: 292
80 See also Koziak, 1999; Ahmed, 2006, and also Blackman, 2012 on emotions in relation to the double of the brain and the distinction between self and other.
81 Schmitz, 1967: 3
82 Pallasmaa, 2014: 20

Pallasmaa and Schmitz continue in the phenomenological, perception-based way. Ben Anderson's suggestion of atmospheres as 'a class of experience that occur before and alongside the formation of subjectivity, across human and nonhuman materialities, and in between subject and object distinctions' redresses several important problems,[83] but remains an in-between of categories that is imperative to overcome, which, furthermore, is thought of as an experience. But neither subject/object distinctions nor dialectic in-betweens have helped much so far.

Instead, I venture an understanding of atmosphere as *an affective ontology of excess*. Although connected to bodies and their affective connections, atmosphere cannot usefully be thought of as a phenomenological entity because it does not depend on conscious apprehension. It exceeds consciousness and dwells instead in elemental materiality. Furthermore, atmosphere does not engage with the distinction between subject/object but emerges from the irrelevance of both the distinction and the terms. Atmosphere *replaces* the distinction subject/object with its mechanics of full immersion and yet withdrawal. Finally, atmosphere does not begin from one side or the other (subject/object, space/time, consciousness/body), it does not have a prescribed direction of movement, but floats on the common surface of parallel unfolding, which is neither an in-between, nor a synthesis. What is more, this parallel unfolding includes the main atmospheric paradox: atmosphere is always both engineered and emergent. Bodies constitute atmosphere and atmosphere constitutes bodies.

Atmosphere is a seductive thing, an earthly force of attraction that does not let bodies escape. Atmosphere appears as air enclosure, a *sphere* of *air* and *mist* as the etymology shows, that makes it difficult to break away and leave behind. Atmosphere attracts. Partly because of what Sloterdijk finds in them, namely the immunity and defensibility of the atmospheric space (such as the domestic unit); but also partly I think because of its pulsating velocity, its continuous yet imperceptible movement, a static yet vertiginous movement as the second etymology of the word shows, where *sphere* (in Greek σφαίρα, 'sfaira') stands for *missile* or *bullet*. The velocity of atmosphere is faster than that of individual bodies, despite the fact that bodies make up atmosphere. This is what Serres means with 'the body exceeds the body.'[84] The body is already there, before the body as it were, exposed to and consisting the atmosphere on the skin of the world. This is why atmosphere is the excess of affect. Let us recall how Massumi defines affects: '*virtual* synesthetic perspectives anchored in (functionally limited by) the actually existing, particular things that embody them.'[85] An affect

---

83 Anderson, 2009: 78
84 Serres, 2008: 307
85 Massumi, 2002: 35, added emphasis.

can never be fully captured and assimilated – it is both plural ('synesthetic') and future-tending ('virtual'). Massumi uses the 'virtual' here in the Deleuzian sense of horizon of potentiality, of not yet but potentially actualised reality. The affect's inability to be fully captured, the 'escape of affect'[86] is at the same time its ability to capture the virtual. Although firmly rooted in the *here* of the body, the affect protends to its virtual becoming. This excess, collectively yet autonomously, is the atmosphere. The way the virtual thought emerges in the atmosphere is not so straightforward. As I show below, for an atmosphere to be successful, the virtual must be actualised.

In a different yet parallel vocabulary, the lawscape-turned-atmosphere is Timothy Morton's *hyperobject* par excellence: an object (here, body) characterised by viscosity, nonlocality, temporally undulating and constantly phasing, thus always withdrawing from full ontological presence. Morton's prime example of a hyperobject is global warming. For my purposes, this is also a prime example of atmospheric engineering on all levels: gaseous, solar, geographical, scientific, economic, political, environmental. Global warming as a hyperobject is characterised by an absence of distance between the objects/bodies which are implicated in it: raindrops on our face is global warming, just as the act of flushing away our toilet waste thinking that it ends somewhere *away*. Its viscous nature means that there is no distance between here and away. Everything is within, assembling into the interobjective space yet at the same time withdrawing from it: global warming is not *here* either. 'The octopus of the hyperobject emits a cloud of ink as it withdraws from access. Yet this cloud of ink is a cloud of effects and affects.'[87] The effects and affects left behind are not causal, at least not in the sense of being merely the sign of the cause. Rather, they all constitute the hyperobject as a continuum; but they are often cut into palatable pieces of locality and present, placed at a distance in order to be apprehended. 'It's never the case that those raindrops only fall on my head! They are always a manifestation of global warming!'[88] These ruptures are also part of the object, 'are simply the invisible presence of the hyperobject itself, which looms around us constantly.'[89] In their phasing, objects appear only partially, as indexical signs of their spatiotemporal distribution, and never in their entirety. Indeed, what is there to appear in global warming, if not the entire Earth, fractal and multiscalar, apprehended by different systems, such as politics, media, science, economics, ecology, and so on, all putting between themselves and the hyperobject a desperately constructed distance of observation? How can it present itself, and to whom, if everyone is

---

86 Massumi, 2002: 35
87 Morton, 2013: 39
88 Morton, 2013: 48
89 Morton, 2013: 7

implicated in this hyperobjective implicate order of global warming? There is no one to whom the entirety of the hyperobject can be presented. Indeed, there is nothing outside the hyperobject. This total inclusiveness of the hyperobject, in combination with its spatiotemporal distribution and its self-withdrawal, is in the core of the concept of atmosphere.

Atmosphere is a process of engineering of the elements. The solid has been traditionally engineered through property lines.[90] Liquidity has been the focus of both political and legal engineering and increasingly more theoretical attention.[91] But air has received much less attention – it has been mainly forgotten.[92] I would briefly like to focus on air as an example of atmospheric dissimulation: air is engineered atmospherically both as visibility and invisibility, or what I call here, a constructed distinction between interior and exterior. Of course, atmosphere is not just about air. It is about solid and liquid too. But my mention of air here denotes this: that atmosphere is all about partitioning. As opposed to the lawscape where the focus is the interplay of simultaneous but partial in/visibilisation that allows a certain manoeuvring space for the lawscaping bodies, an atmosphere reserves no such manoeuvring space. Yet it dissimulates this absence perfectly. Air is atmosphere, but atmospheric air is asphyxiating.

Atmosphere is a rupture of air. Atmosphere severs and limits air in order to emerge. Air is placed in containers characterised by an interplay between interior and exterior, closure and openness. Perhaps the best depiction of such containers are Tomás Saraceno's art installations (Figure 3.1). In his work, baubles made of glass or plastic float about space in a web that includes other baubles, cords, as well as other organic and inorganic material. Air is not transmitted but rests safely folded in the transparent wombs resembling miniature glasshouses. These glasshouses appear perched precariously in the verticality of the vista, in a way that brings to mind Italo Calvino's imaginary city *Ottavia*: there, everything is suspended

---

90 Property law's distinct atmospheric trait is that it both partitions and invisibilises the solid, the body of the land, by relegating 'land' into a set of relations rather than the thing, as any basic property law handbook will tell you. I return to this below in the context of property.

91 Steinberg, 2013; Neimanis, 2014; Negarestani, 2008: 88,has been pointing out how the liquid (in the form of oil) is 'a narrative organizer'. The elements come together in political and legally engineered atmospheres: '… fluid or wetness, essential for blending the dust particles into one narrative with multiple undertows.'

92 Irigaray, 1983; see also Lyell, 1997: 161: 'it is time that the geologist should, in some degree, overcome those first and natural impressions which induced the poets of old to select the rock as the emblem of firmness – the sea as the image of inconstancy.'; and Neocleous, 2013: 581, in the context of air force which for him has characterised colonisation for the end of WWI, 'imperial exercise of air power and aerial exercise of imperial power has tended to be treated as a marginal note to the history of strategic bombing.'

From lawscape to atmosphere 127

Figure 3.1 Atmosphere – or Tomás Saraceno's *Biospheres* at Statens Museum for Kunst, Copenhagen 2009.
© Andreas Philippopoulos-Mihalopoulos.

over the abyss in a structure which 'will last only so long.'[93] Tomás Saraceno's work are atmospheric cities suspended mid-space, given to a temporality determined by the museum organisers, and a provisionality directly connected to the elements (the website of his project *On the Roof: Cloud City* at the Metropolitan Museum of Art in New York writes 'weather permitting'). These works do not stand for metaphors, city representations or game cities. These are cities that exist (to the extent that any city, whether imaginary or real, exists) rooted in the world.

The kind of air I am tackling here, therefore, is not a representation or a metaphor but a material reality. Through and with this air, atmospheres are engineered: a play between interior and exterior, inclusion and exclusion, reality and dissimulation. Interestingly, *atmosphere employs ontological withdrawal in order to engineer itself and its impact on other bodies*. Atmosphere withdraws from the lawscape (just as lawscape withdraws from atmosphere) by gathering around itself a singularity that relies on a dissimulation of the fact that the atmospheric singularity, just as any other singularity, remains part of assemblages and of the larger continuum. Atmosphere instrumentalises rupture in order to engineer its emergence and perpetuation. We have already seen the ruptured lawscape and the ruptured hyperobject, and we shall also see rupture as spatial justice. Instrumentalisation is neither novel, nor necessarily problematic. Nor does it become morally charged in the context of atmosphere. Atmosphere is indeed a manipulation of air (and other bodies) through rupturing but this might not be always a product of disciplinary presences, as I show throughout this and the next chapter. On the one hand, air is full of opportunities, the large openness that Luce Irigaray has imagined,[94] ready to be breathed in with future, available to be folded in the present, amenable to mnemonic bottling of the past. On the other hand, however, air is given to control, manipulation, compulsive desiring, communal identities and spatial partitioning. Air comes with property and fits snugly in rupturing partitions. For these reasons, air is a manifestation of the geological, political, legal, architectural, cultural and so on paradox of the continuum and rupture. On account of its immensity, air is regularly delegated to the position of a hanging apple ready to be harvested. Air captured is knowledge opened up. Apple in hand, and the garden blossoms to the etiolated air of the exterior. But then, no longer is there garden, and the earth becomes one shadeless surface. Air remains one.

Dividing air in airtight containers is not easy. It depends on available technology, climatic conditions and time. This is because, sooner or later, air expands, takes over, and becomes co-extensive with the air next door. This subtle interplay of partitioning yet continuing of air is the basis for the

93   Calvino, 1993: 72
94   Irigaray, 1983

engineering of atmospherics. Various technologies (legal, political, architectural) enable the construction of atmosphere. Think of the practice of nature reserves and National Parks and the way in which they artificially partition the air, thus generating a ruptured glasshouse of differentiated legal 'protection'. Mark Halsey writes:

> The islands of wildness (or, more accurately, of (re)presented beauty) which law names and places to one side, are conceivable only on the basis of an ongoing and generalised ecological violence. National Parks are carved literally out of their antitheses (the non-park terrain). In short, why privilege and preserve particular places (forests, marine environments, deserts) if not for the fact that something fundamental and enduring threatens from without?[95]

The natural reserve technology of partitioning is based on the inevitability of the exterior (in its violent, disastrous form) and the elevation of the inside to a state of museological, temporally fixed beauty. The challenge for atmospheric technologies is to maintain the (ideological or actual) airtightness of atmosphere by keeping the air in prefabricated 'glasshouses' – a significant challenge, considering that fact that an atmosphere is an affective event and as such, changeable, volatile and unpredictable. As I show below, atmospheric engineering must take place on a bed of dissimulation, for otherwise the illusion will not be complete and resistance to it will be cropping up at an uncontrollable rate. The perfectly engineered atmosphere is one that appears spontaneous (consumerism, spaces of culture), necessary (natural reserves, airports) or even unavoidable (gated communities, prisons), and above all sensorially and emotionally responsive in that it makes bodies move in pre-specified ways. This means that atmosphere dissimulates itself as *emerging and not engineered*, as we shall see in the following chapter. In most atmospheres, this can happen because an atmosphere is based on the desires of the bodies of its emergence. The coup of an atmosphere is that it generates the very affects that desire its continuation. Not only is there no way out, but significantly there is no desire for a way out.[96] This is an engineered atmosphere, one that eliminates whatever does not fit with its perpetuation, whether this is otherness, the past, *logos*, *nomos*, distinctions, the outside. The future, along with any desire for a different future, is captured – for as long as it lasts, to paraphrase Calvino. This is the time of atmosphere: a non-negotiable, constant and mostly unchangeable present that demands the total presence of bodies. As opposed to the lawscape, here repetition only yields sameness,

---

95   Halsey, 2011: 219; see also Halsey, 2006, and Kotsakis, 2011 for a similar discourse on natural reserves and the atmospherics of nature/humanity.
96   This is what Minton, 2009 means when she talks about control of fear in the city.

identity, belonging – not questioning, not externalising, not wondering about the future. This means only one thing: while relying on affective excess for its emergence, *atmosphere absorbs the excess of affect, mops around its contingent leakage, tidies up anything that can go differently, and engineers the already given direction of affects in a collective way*. Affect is desire directed, atmosphere is affect castrated, de-excessified, normalised.

The most obvious atmospheric technology is the bent glass, that thin layer of translucent skin that severs while bringing together, enabling both separateness and immersion, rupture and continuum. 'Only then can one state explicitly that existence and immersion are equally potent concepts' writes Sloterdijk.[97] For Sloterdijk, existence requires a distance of observation. Indeed, distance is a necessary rupture of the continuum. But it is also illusionary. Car mirrors know it well: 'OBJECTS IN MIRROR ARE CLOSER THAN THEY APPEAR ... "distance" is only a psychic and ideological construct designed to protect me from the nearness of things.'[98] Distance is a rupture that constructs the illusion of separateness. Think of the atmosphere constructed in traditional glasshouses or indeed any 'glass house' architecture that serves to separate interior from exterior. This is the illusion of distance at work, coupled with yet another illusion: that the continuum has not quite been broken, only spaced out. The glass allows the affective continuum to be maintained: air is see-through. Referring to a snow globe, Esther Lesley writes: 'it is and it is not, at one and the same time, distanced and close.'[99] Immersion, but also distance: a continuous interplay between rupture and continuum. The glass separates while not allowing one to forget the outside. In Saraceno's works, the viewer has the privilege of observing simultaneously the inside and the outside of a glass separation, indeed observe this very invisibilised of visibilities: the boundary. Saraceno's glasshouses emulate architectures of exclusion, whether this is a result of forceful intervention or fear of violence. In the context of the spatiality of political violence, Boano writes, 'space is produced either as a space of plenty and safety, or naught and dissolution.'[100] This either/or is a condition of spatial invisibility. But one could argue that every space has necessarily both traits. In that sense, boundaries are best thought of as *elastic* in Eyal Weizman's terminology, which does not denote flexibility or softness but rather a direct connection between spatial organisation on the one hand, and political and legal conflicts on the other.[101] The elasticity of space and spatial boundaries

---

97 Sloterdijk, 2013: 31
98 Morton, 2013: 27
99 Lesley, 2009: 517
100 Boano, 2011: 44
101 'the continuous spatial reorganization of the political borders ... responds to and reflects political and military conflicts.' Weizman, 2007: 7

might indeed play along the illusion of flexibility, mutual yielding or softness. This is how the illusion of continuum with the outside is maintained. No separation can be perfect, for otherwise there will be no need to stay in. However, and this is the great atmospheric coup, both constancy and dissolution must be located inside. Indeed, the sliver of visibilisation in the lawscape becomes state of the art atmospherics in an engineered atmosphere. In the lawscape, invisibilisation works in relation to visibilisation. Atmosphere goes a step further and constructs a totality that *excludes yet visibilises* a sliver of the exterior. Think of the case of a gated community where the residents must be reminded of the violence of the demonised outside, always threatening to break inside.

In other words, on the level of atmospheric strategy, the 'unity of difference' as Niklas Luhmann puts it, is paramount.[102] This unity of difference emerges through the glass, the surface that visibilises a difference (often illusionary) and offers to hold it together. Atmosphere constructs an artificial continuum on top of the ontological continuum. It ruptures the continuum in order to strategically present it as an interior without exterior (yet with this knowledge offered within, as I show below). The appreciation of this paradox requires a distance from the phenomenological directionality between consciousness and object. Phenomenology allows one to apprehend atmospheres from the inside and for this reason one misses out on the complexity of the continuing rupture of the air by the glass. A phenomenology of atmosphere cannot stop to think of the unthinkability of air, does not have the means to contextualise it yet leaves it outside. An approach based on an emplaced, material ontology on the other hand, allows atmosphere to emerge in both its continuity and rupture as one of the basic elements of the architecture of the posthuman.

In what follows, I would like to go through several instances of atmospherics, which, however are not necessarily characterised as atmospherics by their commentators. The purpose of this is to show that atmospherics is both more pervasive and more complex than thought. It should have become obvious by now that atmospheric engineering does not always serve benign purposes of functionality. Even when it offers such affects as calm, comfort or *Gemütlichkeit*, an atmosphere can be put in the service of happiness surveys, productivity incentives, heteronormative proliferation of social models and so on. This is the point of Laurent Berlant's analysis of the publics as affective machines of persuasion that have thematised American society through popular films.[103] Likewise, Sloterdijk's World Interior of Capital is described as Sloterdijk's *World Interior of Capital* as 'a climatized luxury shell in which there would be an eternal spring of consensus.'[104] The affectivity of luxury

---

102 Luhmann, 1998: 56
103 Berlant, 2008
104 Sloterdijk, 2013: 170

finds its most prominent form in the Grand Installation of the glasshouse of capitalism: that 'interior-creating violence of contemporary traffic and communication media.'[105] In the same way, as we know from Foucault the atmosphere of a prison serves the purposes of control. There, the withdrawal of the law is only an epiphenomenon of the density, frequency and omnipresence of the law: the air is so thick with *logos*, the space so fully striated, that the law itself becomes walls, bars, corridors, yards, time zones; on top of which another atmosphere is built: that of the power imbalances between the various bodies of guards, prisoners, prisoners among themselves, past and future, networks and substances. In discussing Hitler's campaign, Christian Borch has shown how atmospheric politics achieve specific political objectives through manipulation of air conditions and the senses,[106] reminding thus how atmospherics is an integral feature of any totalitarian regime. On a fractally different level, the Western world is its own glasshouse of atmospheric partitioning, with immigration policies that control the use of elements such as water and land in terms of spatial approaches to jurisdictional utopias, or the boundary that separates the occident from the orient, constructing both exteriors and interiors through the bent glass of religion, economy, culture and so on. Frantz Fanon writes 'the colonial world is a world divided into compartments.'[107] Racial violence has often been in the core of atmospheric engineering in the form of racial threat (when in white atmospheres) or racial discrimination and oppression (when in non-white atmospheres). Tayyab Mahmud's work on postcolonial spaces of oppression shows this amply. Slums are atmospheric constructions where 'surplus humanity' is piled up and kept inside through atmospheric techniques of accumulation through dispossession and primitive accumulation (namely Marx's concept of deprivation of the means of subsistence).[108] These techniques define the exterior of the slum as a non-possibility, thus strengthening what can be described as negative belonging, that is, belonging because of the impossibility of belonging anywhere else.

Old binarisms remain but have changed their nomenclature: thus from 'Civilised/Barbarian, Believer/Infidel, White/Black or Advanced/Primitive' to the more 'acceptable' 'Developed/Developing, Centre/Periphery, Advanced/Emerging, or Rich/Poor.'[109] Atmospherics of racial discrimination generate distinctions that in their turn contribute to the atmospherics of creation (in its colonial aspect) and perpetuation (in its postcolonial aspect) of the very international law from which they often emerge, as Eslava and Pahuja put it.[110] Mahmud takes this further: modern discourses of racial

---

105   Sloterdijk, 2013: 198
106   Borch, 2014
107   Fanon, 1963: 37
108   Mahmud, 2010b
109   Eslava and Pahuja, 2012: 2
110   Eslava and Pahuja, 2012; see also Beard, 2007 and Orford, 2009

difference and hierarchy have emerged in order to reconcile colonial domination with ideals of liberty and equality.[111] Connected to this is the recent discussion on preparedness, always linked to intense catastrophology, and originating in ways of legitimising the war on terror. Massumi has referred to these forms of intervening as *ontopower* that becomes registered as necessity on an ontological, atmospheric level.[112] Ash Amin has surveyed the current discourses of risk, resilience and preparedness under the concept of *ontological war*, in which techniques of environmental and genetic manipulation, media terrorisation, and even 'technologies of humility and the expertise and capabilities of diverse knowledge communities'[113] become atmospherically orchestrated in order to scream the grand hysterical finale of the end of the society that has to be defended. Whether it is engineering on a global scale, or localised engineering of a particular shopping mall, the affects employed and exploited are the same.

Drone atmospherics is another case of an all-inclusive atmosphere in which law becomes fully materialised and thus invisibilised. Drones act from an already partitioned air according to the specific needs of the atmosphere in whose service they are. This atmosphere reiterates notions of security, to the point that in the US military security language it takes the form of a 'counterinsurgency operation' (COIN), namely a spatial strategy of atmospheric partitioning which constructs a constellation of relatively secure territories with the ultimate aim of uniting them in an atmospheric continuum. COIN is both staffed and unstaffed, and in the latter case, it takes place through drones. COIN has four phases. Shape Phase is aimed at 'mapping the local area, weakening existing threats, building a ring of security around the area';[114] Clear Phase where 'forces enter the area as well'; Hold Phase is aimed at 'controlling the movement of people'; and finally, Build Phase, in which 'the local environment is being transformed and a new law and order system is being put in place.'[115] The same phases are predicted for drone operations, except that everything happens from a digital distance and with a clearer sense of continuum: 'the infrastructural grid of CIA compounds – termed 'the shadow' – today covers the entire world, except one blank spot in the area of Australia. Drones – referred to as 'frogs' – can navigate the entire world to the shape phase of the COIN doctrine: observing, mapping, and coding human behavior, use of space, and social interactions.'[116] This is also Mark Neocleous's point in his work on 'air power', whose main aim is 'to construct a new order' through intense, governmental policing from the air. In the context of colonial rule,

---

111 Mahmud, 2007
112 Massumi, 2009
113 Amin, 2012: 159
114 Shoshan, 2014: 89
115 Shoshan, 2014: 91
116 Shoshan, 2014: 91

he notes that this new order is constructed by crushing all rebellions against colonial rule, the destruction of indigenous modes of subsistence, surveillance and land census. Drones, according to Neocleous, continue in this mode, occupying both ends of the spectrum, from sophisticated killing machines to small technologies of battlefield surveillance. The other side is also interesting: rather than the usual videogame atmosphere associated with military drone operators, Derek Gregory shows how they are also subjected to while co-constructing an immersive atmospherics of affective corporeal encounters.[117] 'Virtual war, it seems, is less virtual than would appear at first glance' writes Caroline Holmqvist,[118] thus pointing to the increasingly co-emergent atmospheres of technological mediation and classic corporeality. This becomes even more complex when drones are seen, as Anna Leander does, as technological agents that co-produce their legality, or even the extrajudicial atmosphere of targeted killings (depending on whether they are to be considered targeted killings or extrajudicial executions), which is another aspect of the US 'drone program'.[119] As Leander writes, 'the agency of drones feeds into the controversy over the legal status of the drone program and intensifies it.'[120] This is an atmosphere that includes commercial contractors, military and intelligence services, technology, law, and so on, that manages to disorient the human input, already very weakened ('in practical terms the human in the loop may be so far removed that the legal evaluations end up resting with the drones'[121]), thus legitimising conceding the authority of legal judgment to the algorithm on the basis of which drones make decisions. The lawscape has withdrawn and the atmosphere is one of calculability, security, safety,[122] and what is more, dissimulation: '[drones] seem to work like magicians do: invisibility, silent sleek working, a clever trick and – hey presto! – the enemy has gone, suddenly evaporated in a hell of debris', as Antonio Riello writes in the accompanying literature for his art project *Dronology: A Symbolic Form Of Our Time.*[123]

117 Gregory, 2011
118 Holmqvist, 2013: 542
119 Leander, 2013
120 Leander, 2013: 814
121 Leander, 2013: 822
122 'It is hence suggested that not relying on the automated judgements would worsen the protection of civilians and increase collateral damage. To this is added that, unlike people, technology systems are not emotionally engaged and hence do not react out of fear, anger, exhaustion, or despair. On the contrary, they can be programmed to be self-sacrificing. The argument is often pushed further to suggest not only that it is appropriate to leave some important judgements to technology at present, but that even more should be left to them.' Leander, 2013: 823, references omitted. In this atmosphere, affects are mistrusted yet relying on this atmosphere is also based on affects.
123 Riello, 2014

Whether through drones, human or other bodies, an atmosphere is a powerful, shared and contagious event. In her seminal psychoanalytical work on affective transmission, Teresa Brennan has shown through clinical observation how affects are transmitted from bodies and spaces to other bodies and spaces.[124] One of Sloterdijk's main points is that the most significant military engineering development is to attack the environment, namely the atmosphere of the body that includes other bodies and spaces, rather than the body directly.[125] In defining atmosphere, Sloterdijk refers to a capitalist scenography, where 'atmospheres are made available as total settings of attractions, signs and contact opportunities.'[126] This is what Braidotti means when she refers to the 'contemporary bio-genetic capitalism' which unites human and nonhuman in a posthuman atmospherics of 'reactive interdependence.'[127] In the same vein, Marie-Eve Morin writes: 'as our management and regulation of the atmosphere is now part of our social and political concerns, the atmosphere can no longer be seen as an exterior but has effectively been integrated within the system of human relations.'[128] Timothy Choy talks about how atmospheric governmentality amounts to a suspension of ordinary forms of experience, with obvious ecological repercussions in terms of concealed differences of atmospheric pollution measurements across legal cultures.[129] Choy calls attention to air's relative undertheorisation in relation to solidity (earth, water), and draws specific conclusions on spatial inequality of income and human atmospheric permeability that point directly to a causal link between occlusion of air from analysis and atmospheric manipulation. And so, atmospheres are constantly engineered, not just around humans but *by* humans, and even more *through* our very own bodies and assemblages with other bodies. This is the complicity with the power relations that allows subjectivity, not only to emerge but importantly, to sustain power relations through desire, as Judith Butler has shown.[130] In her work, Butler has reinterpreted Althusser's *interpellation*,[131] which is the process of 'subjectification', that is to say turning a body into a mere subject to ideology. In Althusser's famous example, this turning takes place through the act of being called by the authority (police calling 'hey, you there!'), that always manages to generate a reaction (a turning) by most of us ('me?'). Butler talks about the possibility, not only of formative hailing but also *performative*

---

124 Brennan, 2004; also Mitchell, 2011
125 Sloterdijk, 2009: 48
126 Sloterdijk, 2004: 180
127 Braidotti, 2013: 49
128 Morin, 2009: 68
129 Choy, 2011
130 Butler, 1993: 121–40
131 Althusser, 2001

hailing, one that allows for alternative gestures of non-obedience. This, as I show later, might be one way of withdrawing from the way atmospheric pressure is built around and through us. Things, however, are no longer that simple. In a time of intense atmospheric engineering, Althusser's interpellation is atmospherically diffused. No one needs to call us anymore. We do it ourselves (as Althusser writes, 'freely'), constantly self- or other-checking, being interpellated not through ideology (this has been suffused in atmospherics) but of a constructed, furious desire to perpetuate the atmosphere. As opposed to ideology, atmosphere has no horizon, no external and aim. It freezes future and its 'outside' into a perpetual and inescapable now and here. It is engineered yet emergent, it is concentrated yet diffused. There is an air-written, ethereal contract between atmosphere and law: 'the contract becomes the sovereign mechanism for normalising each individual because it rearticulates, in a highly fluid and transitory manner, the individual as the agent of a choice that has been made voluntarily, as the intention of free will.'[132] One needs to withdraw from the contractual free will of the atmosphere in order to disobey.

Atmospherics is not just material. As Elena Loizidou shows in her work on Butler and Althusser, to privilege only the material body misses out on what she calls 'the figural body', namely the linguistic, discursive, symbolic body.[133] Talking only about materiality is one way of perpetuating the new atmospherics of materialism that deprioritises the immaterial to the point of allowing it to carry on colonising the material. I have touched upon this in Chapter 1, when I was referring to the need for abstraction. Both abstraction and immateriality run the same risk of being sidestepped for the grand furore of material thinking, with potentially disastrous effects. Loizidou brings to the discussion the case of Jean Charles de Menezes, the Brazilian man who was shot dead in London in 2007 as a result of a series of errors on behalf of Metropolitan Police; essentially, however, de Menezes was the victim of one grand atmosphere of racialisation based on symbolic rather than merely material features of skin colour, clothing, etc.[134] This is why it is important to define affect not only as material but as symbolic as well. Symbolic and material share the same surface and emerge in atmospherics of desire in which the distinction is neither possible nor useful. This is Nathan Moore's point when he explicitly connects image and affect.[135] Images/symbols are not representations but affective movements, indeed bodies that become stronger or weaker depending on their encounters with other images/symbols. Affect thus unfolds on the symbolic

---

132 Moore, 2012: 147
133 Loizidou, 2008: 40; see also Renée Marlin-Bennett's, 2013, work on the (human) body as information enabler, container and manipulator.
134 See also Vaughan-Williams, 2007
135 Moore, 2012

and material level in a parallel, non-causal way. Collectively, this affective movement is directed towards the emergence of the totality of atmosphere. Reza Negarestani describes this totality as extending to ideologies, gods, histories, objects and materials which escape categorisation and nomenclature, yet retain a hold onto us through our desire to maintain ourselves in their grip.[136] This reveals the way in which atmosphere folds onto itself and dissimulates itself as pure desire. This atmospheric movement, which I consider one of the most traits of atmospheres, is left out in the literature. When for example Sloterdijk writes that 'atmospheres are made available as total settings of attractions, signs and contact opportunities',[137] he focuses on the undoubtedly true fact that atmospheres are engulfing and producing of their own inescapable immanence. But in so doing, he leaves out the *excess* of the totality that is being mopped up by the atmosphere and takes the form of self-dissimulation. This is not just a theoretical point. There is no doubt that an atmosphere is a designed thing, product rather than happenstance, a legal, political, economic, architectural and so on script that fixes the presence of various bodies. But even in 'tidying up' and enclosing the affective excess, atmosphere remains, at least to some contingent extent, excessive. What is routinely omitted from the literature is the fact that atmospheric engineering regularly dissimulates itself, its excess and provenance, in order to push the engulfed bodies into further blindness with regards to the scripts that orchestrate them in the first place. In other words, one ends up with atmospheres concealing atmospheres, self-concealing atmospheres that cover up the fact that they are legally and politically manipulated, and instead successfully presenting themselves as naturalised, spontaneous, rhizomatic.

The salvaging point is odd: engineering does not necessarily work out in the intended way. Collective, inherent, emergent engineering means that atmosphere is a contingent event. This is the meaning of an atmosphere as both *engineered* and *emerging*. It emerges from within bodies, indeed the bodies that engineer it. These bodies, however, remain unpredictable, even in such an established construction as the global glasshouse. This can go either way: Sloterdijk refers to Dostoyevsky, who, even amidst his fascination with the glasshouse phenomenon, was convinced that 'eternal peace in the crystal palace would mentally compromise the inhabitants.'[138] Or it can go the other way: affects become too excessive to be controlled, and when aggregated in the form of an atmosphere, they change in volatile ways. Urban atmosphere in particular, is a risky thing, full of incalculable factors. Nigel Thrift describes cities as 'roiling maelstroms of affect',[139] while

---

136 Negarestani, 2008
137 Sloterdijk, 2004: 180
138 Sloterdijk, 2013: 171
139 Thrift, 2007: 170

Sloterdijk, in a rather more positive note finds that a city is 'a place of enhanced improbabilities.'[140] Such heightened complexity is not limited to cities. The mere fact that atmospheres are usually engineered by the bodies of their emergence makes any atmospheric engineering a contingent event, even when established techniques of complexity reduction are followed, such as common rituals that enable the assembly of 'agents of coexistence in the improbable.'[141] In other words, the affective excess of an atmosphere haunts the atmosphere and breaks it from the inside. Affective excess creates contingency and opens up a space which is used precisely in the way it was *not* supposed to be used when engineered. This can be the space of withdrawal, which as I show in Chapter 5 of this book, works against dissimulation and creates the conditions for a possible emergence of spatial justice.

Despite their precariousness, engineered atmospheres are relatively successful glasshouses of constancy and stability. The twist is that constancy and stability must be fully endorsed in the interior, even if as illusionary constructs that entail a strong declaration of how *unstable* or *inconsistent* the exterior is. In their own way, law, politics, architecture, economy and so on converge in preserving this dramatic difference between interior and exterior. This is no conspiracy: each system, to use Niklas Luhmann's categories,[142] is blinded to its outside environment, ignorant of what happens outside its limits. Law can only understand legal processes, just as architecture can only understand space and bodies through the building project. Yet, there is convergence. Systems come together and construct total atmospheres where corporeal self-discipline is distributed between the systems. Think of shopping malls that are built in such a way that one has to walk slowly, cannot easily find the way out, and is bombarded by constant shopping 'needs' offered by megashops without doors or gates but just smooth walking into; add to this the fact that one cannot stage a protest or bask or run or wear a hood or do anything other than what is prescribed; and then add what the customers expect from a shopping mall and how suspiciously any untoward gesture is seen.[143] This is the perfectly engineered atmosphere: when the very bodies police themselves, even in absence of obvious legal norms. This is what Clare Colebrook, reading Foucault and Deleuze, means when she writes, 'one is no longer governed by a tyrannical body, but is self-governing precisely through the absence of any specific norm or quality.'[144]

In what follows, I trace the main steps to engineering an atmosphere through an observation of air partitioning, with its emotional and sensorial

---

140   Sloterdijk, 2005: 948
141   Sloterdijk, 2005: 948
142   Luhmann, 2012 and 2013
143   Bottomley, 2007b
144   Colebrook, 2009: 20

affective effects. Suffice to say, however, that it is not easy to produce an engineered atmosphere. To foreshadow the below, I would like to argue that *an atmosphere is successfully engineered when it manages to rupture the affective continuum with the exterior, while at the same time reproducing it inside.* This is the geology of atmosphere: outside there is no air. In here is the only possible place to be. This means two things: first, that no change is expected, unless fully conditioned (planned, announced, smoothened out, controlled) within the atmosphere. And, second, the desire for change has been suffused within the atmospherics of either pleasure or oppression. Here, one has everything one desires for one's 'comfort': we are surrounded by the 'climatological erotics' of air-conditioning, as Mark Dorian puts it,[145] but this time as a self-loving, self-sufficient atmoporn.

3.3 Engineering and perpetuating an atmosphere

Tomás Saraceno's constructions oblige us not to forget the air.[146] *The unthinkability of air is continuously present because it is continuously interrupted*: glass partitions that allow seeing-through float about in typically anosmic gallery rooms. An exercise in how to fit the sphere in the cube, Saraceno's works revolt against the obsessive deodorisation, and bring in alien affects that would otherwise be controlled. Bruno Latour has pointed to the combination of Sloterdijk's spheres and his own networks in his essay on Saraceno's expansive architectural structures.[147] The works embody what Sloterdijk has called 'connected isolation',[148] and what here I refer to as ruptured continuum.[149] This is not an opposition but a co-existing paradox: the glass ruptures the continuum of the air inside and outside while allowing visibility. To put it plainly, all elements connect and isolate at the same time, partake both in the continuum (of the artpiece, of the air) and its rupture. In terms of matter used, the work is a prototypical employment of Sloterdijk's observation on the use of bent glass and the construction of glasshouses for modern atmospherics: 'such edifices took into account that ... the random uprooting of organisms to plant them elsewhere could only occur if the climatic conditions were transposed along with them.'[150]

Temporally speaking, glasshouses (and atmospheres) interrupt the air while promoting an encapsulated history of provenance and a situated

---

145 Dorian, 2012: 29
146 As opposed to Heideggerian dwelling, so firmly grounded on earth that excludes the unthinkability of air. See Irigaray, 1983
147 Latour, 2011
148 Sloterdijk, 2004: 255
149 Except that here there are no two atmospheres and no separation between this and that atmosphere, but a continuous rupture of becoming-other atmosphere, as I show in Chapter 4.
150 Sloterdijk, 2005: 944

capture of future.[151] The uprooting involved in the transplantation finds its equivalence in the construction of the ancient Greek city, which, Sloterdijk notes 'was a greenhouse for people who agreed to be uprooted from the modus vivendi of living in separation and instead be planted in the disarming modus vivendi of living together.'[152] Living together has always been premised on the parallel exclusion of living apart: for the Greeks it was the Barbarians, a typical identity forming exclusion (I am not Barbarian, therefore I am Greek). This can be traced forward to the urban distinction between *urb* and *suburb*, and even further in a classic cyclical form, to the *exurb*,[153] showing how distinctions are no longer simple binarisms but multiplicities, all of which need to be marked negatively and thrown outside the glasshouse. For the Old Testament city builders, however, the rationale might have been a little more complex: 'God the first garden made, and the first city Cain' writes Abraham Cowley (1618–1667) in *The Garden*. Cities are the walled spatiality of Cain against the unknown environment, be this God, other humans, or nature. Cain's divine mark, a sign of both his absolute inclusion in the divine will, and absolute exclusion from divine forgiveness,[154] meant that the first city (itself unknowingly following the principles of glasshouses) was both a way to defend oneself from divine breath and yet capture it inside the city walls. However high the walls, they contain an inside that is identical yet exceeds what is outside, not unlike a *kare-sansui* garden, whose cosmology includes both itself and the larger order of nature,[155] city and world situated either side of the bent glass are characterised by a simultaneous continuum and rupture. In their turn, continuum and rupture undercut each other and contextualise their paradox in spatialities, temporalities and corporealities deprived of the pharmakon of a synthesis.[156]

In his book *Atmospheres*, Peter Zumthor unfolds his engineering in a few positions: first, the body ('as a bodily mass, a membrane, a fabric, a kind of covering, cloth, velvet, silk, all around me. The body! Not the idea of the body – the body itself! A body that can touch me.'[157]); second, material combinations ('materials react with one another and have their radiance, so that the material composition gives rise to something unique. Material

---

151 'The air itself is one vast library, on whose pages are forever written all that man has ever said or woman whispered. There, in their mutable but unerring characters, mixed with the earliest, as well as with the latest sighs of mortality, stand forever recorded, vows unredeemed, promises unfulfilled, perpetuating in the united movements of each particle, the testimony of man's changeful will.' Babbage, 1838: 36
152 Sloterdijk, 2005: 944
153 Spectorsky, 1955
154 Saramago, 2011
155 Crowe, 1997: 15
156 Philippopoulos-Mihalopoulos, 2007a
157 Zumthor, 2012, 23

is endless.'¹⁵⁸) which have to do with what he calls 'critical proximity between materials' – neither too far, nor too close. We get a feel of what a precise science atmospherics is, especially with the following few positions that refer to the sound of space, its temperature, the way the light streams or not through, the scale of things (or what Zumthor calls 'levels of intimacy'¹⁵⁹), the beauty of form, coherence, and even what he calls 'between composure and seduction'¹⁶⁰ which refers specifically to the way bodies move in architecture. In order for bodies to be seduced, the process should be 'akin to designing a stage setting, directing a play ... direction, seduction, letting go, granting freedom ... where nothing is trying to coax you away, where you can simply be.'¹⁶¹ Zumthor's approach is a loving one, not out there to deceive. It believes in promoting beauty, form, even harmony of sorts. It is anthropocentric in the gentlest possible way. It is even intimate, talking about personal experiences in a soft way.

The whole book (which is the transcription of a lecture Zumthor gave in 2003) is an engineered atmosphere, from the feel of the book cover cloth to the smell of the paper, the subdued use of colour, the gentle font. We are in it. Any critique (anthropocentric, dissimulating of strategies, Western-centric, classist) feels out of place, cheap, facile. Yet in the middle of the book, folded almost imperceptibly, a section on what is, I think, the key to and yet the hammer against the atmospheric glasshouse: the 'tension between Interior and Exterior. A fantastic business, this. The way architecture takes a bit of the globe and constructs a tiny box of it. And suddenly there's an interior and an exterior.'¹⁶² This very distinction is how one engineers an atmosphere, yet at the same time how one might manage to come out of it, distinction as the first act that severs the universe in two. Spencer-Brown's first distinction, Luhmann's first move, this distinction is what allows one to close the book and look outside.¹⁶³

With this, I have reached the first step of engineering and the conversion of a lawscape into an atmosphere. I will not talk about *tension* between interior and exterior, first because I do not think there is tension but continuum; and second, because, despite the frequent pronouncement in this book that 'there is no outside', I want to be radical and go all the way with the dissimulation. So, *the first and most important characteristic of an atmosphere is the distinction between interior and exterior.* This distinction is the most fundamental gesture in sociologist Niklas Luhmann's systems

---

158 Zumthor, 2012: 25
159 Zumthor, 2012: 49
160 Zumthor, 2012: 40
161 Zumthor, 2012: 43–5
162 Zumthor, 2012: 45
163 Luhmann, 2002; Spencer Brown, 1969

theory,[164] who has significantly influenced Sloterdijk's work, and which can be of use here too. In Luhmann's systems theory, the system builds its identity by excluding its *environment*. The two, interior and exterior or system and environment, can be put together and observed as a unity. This means that the distinction between the two is a rupture of a continuum that serves the particular atmosphere. Once inside the atmospheric interior, one cannot see outside. 'There is no outside!' – but how different this cry sounds now. Think of how gated communities organise their space on the basis of exclusion of the outside. Often the lines drawn are arbitrary, post-facto, fed by the need for security, which has little to do with actual risk. In any case, there are no pre-existing boundaries in atmospheric engineering. Rather, boundaries are created by the passage through space, and are 'ontologically parasitic on their hosts, the entities they bound.'[165]

Add to this the second characteristic of an engineered atmosphere: that *the exterior is included in the interior*. This is often done through assimilation of concepts and techniques from the exterior. While these techniques (legal, political, technological, aesthetic) are normally contained in and by the lawscape, when employed within an atmosphere, they become atmospherically conditioned and, to the extent possible, have their contingency reduced. Techniques that can go either way in the lawscape, are given a predetermined direction (sense, meaning) within the atmosphere. In most cases, the predominant affect of an engineered atmosphere (of a gated community, of a club with strict entry requirements, of fortress Europe, and so on) is one of contentment: we have all we need right here. An engineered atmosphere is 'an enclosure so spacious that one would never have to leave it',[166] with oodles of *nomic* freedom at our disposal. Or that one *could* never leave: the diametrically opposite affect emerges in atmospheres of oppression and legal hypervisibilisation, such as kettling police strategies, prisons, or courtrooms. The *logic* presence of the law is made to be everywhere and visible, to the extent that the body cannot use the law anymore as a way of escape, protection, power. Atmospheres of oppression rely on a constant redrawing of spatiotemporal boundaries: Vaughan-Williams talks about the way mobile phone operators can be ordered to close their network, as it happened in the 22/7 London bombings. While this was in order to help the police carry out its service, it also denotes a new spatiotemporal zoning power from which no body can escape:

> such decisions are no longer localized or fixed at particular border sites in the margins of sovereign territory but increasingly more wide-

---

164   Luhmann, 2012 and 2013
165   Fall, 2005: 16
166   Sloterdijk, 2013: 175

spread or diffused throughout society: a phenomenon that might be captured by the concept of a biopolitical generalized border.[167]

The law in these cases becomes entirely assimilated by the body of the authority yielding the legitimate (or self-legitimising) force. The two atmospheres (nomic and logic) are equally enclosed, their outside always included. The greater the inclusion of the exterior, the greater the risk in the interior. In his oeuvre on community, Roberto Esposito writes: 'as in all areas of contemporary social systems, neurotically haunted by a continuously growing need for security, this means that the risk from which the protection is meant to defend is actually created by the protection itself.'[168] Esposito's fusion of biopolitics and autopoiesis echoes what Luhmann has written on the autopoietic quality of risk, namely that risk always grows autopoietically and along the means of its prevention.[169] For this reason, an autopoietic atmosphere includes and pacifies its own risks. The autopoiesis of an atmosphere reaches its apogee in the way Reza Negarestani describes the objective of any enclosure:

> to distill all cosmic processes into one unified body which is cyclically infinite yet functionally restrictive (everything must be unified). Such an environment or sphere functions as a cyclic or a spherical shape with an inner limit and an outer boundlessness.[170]

With this, we have reached the third characteristic of an engineered atmosphere: that *an illusion of synthesis is offered inside every atmosphere*. This is the main characteristic of a system's self-description, or sense of identity, according to Luhmann.[171] A system is a hermetically closed topology, constantly withdrawn albeit part of a larger continuum. This continuum, that is to say the autopoietic environment, is reproduced within the systemic boundaries. The system does not have access to the outside, except through its own construction. For the system, the system is all there is, the universe and its limits, the end of time, the perfect immanence. This is the illusion of synthesis that an engineered atmosphere emulates. For Sloterdijk, a sphere is the shared circle of humanity.[172] The fact indeed that humans dwell in a sphere *makes* them human – or this is what atmospheric rhetorics want us to believe. The illusion of synthesis as a *human* sphere is one of the triumphs of the atmospherics of humanism. Space orders nature, and makes sure that

---

167 Vaughan-Williams, 2007: 191
168 Esposito, 2011: 141
169 Luhmann, 1993
170 Negarestani, 2008: 102
171 Luhmann, 2012 and 2013
172 Sloterdijk, 1998

everything external to such nature remains external. The membrane around the sphere protects the interior from the exterior and offers Sloterdijk's much valued immunity, the main focus of his trilogy. Negarestani again: 'Air as a manifest refinement is a vision-machine through which the world *looks safe*, that is to say, already consolidated, having been forced to take the path of unification and purity.'[173] Feeling safe is a collective affect that ontologically translates in the following: any continuum needs to be ruptured in order to be tolerable. Likewise with feeling part of a 'disrupting' or 'protesting' crowd: to be inside is to be safely identifying with 'the right side'. Comfort is no more impressionistic in a kettled enclosure than it is in a shopping mall.[174] To talk about comfort in kettling sounds odd, but comfort comes from a desire to belong, especially when the delineation is unambiguous. One needs to fold the continuum around oneself in order to *feel* (comfortable, safe, belonging, present, alive, vindicated, heroic): this fold ruptures the continuum. This means that in order for the atmosphere to be perpetuated, it needs to be contained as an isolated instance of the continuum, as a rupture which refines the continuum according to the specifics of the atmosphere.

But think of Saraceno's glasshouses again: their immunity is only impressionistic. The air inside is one with the air outside. The glass sphere within the gallery cube manages to maintain its immunity, stability and independence, only if the exterior remains air-conditionally (if inside) and/or climatologically (if outside) tolerant of such separating membranes. As indeed the museum organisers put it, 'weather permitting'. Yet the illusion of immunity and stability is important and can only be maintained if the exterior is *dissimulated*, half-hidden from view, almost forgotten. The walls need to be high, the glass bullet-proof, the plastic impermeable, the cameras solidly pointing at where the action might take place, the gaze of the guards catching your movements. The future must be frozen, kept outside of the eternal present of the bubble. Is this a shopping mall or a prison? We are conditioned by the air inside, a round present vibrating with vaporous promises.

The ultimate move of such dissimulation is when dissimulation turns towards itself and doubles up: as a guarantee of longevity, *an atmosphere must dissimulate the fact that it is an engineered emergence.* This is the last characteristic of engineered atmospheres. When the atmosphere dissimulates, it does so in order to reinforce its own continuum: 'there is nothing outside' means 'there is nothing inside to differentiate it to the outside'. Dissimulation means: no one has engineered the atmosphere, no one has organised

---

173 Negarestani, 2008: 102–3, emphasis added.
174 See Lambert, 2013a: 46, on how the problematic of comfort unfolds from a Spinozan point of view: 'comfort and joy are not synonyms. We might even wonder if they are not antonyms.'

matter and non-matter to generate the atmosphere. Atmosphere dissimulates its engineering and presents itself *only* as an emergence. Atmospheric self-dissimulation means that an atmosphere dissimulates itself (as well as its origin and its non-rhizomatic nature) as non-atmosphere, as mere lawscape where there is always a possibility of non-scripted imbalances, excesses, conflicts or revolts to take place. In a self-dissimulating atmosphere, that most accomplished of atmospheres, there is nothing to go against: the atmosphere has converted itself into Quixotic windmills.

3.4 Coda: back in the room

I would like to break the atmosphere of this chapter by allowing a little bit of textual law, and specifically intellectual property law, to emerge. Think of your initial welcome to the lawscape: the music, the smell, the taste, the textures. Think of how cosy you felt. Think of your affects – you wanted to have a Coke; you had a Coke; you wanted to stay in the room; you stayed in the room. Think of the atmosphere, comfortable, safe, energising. No law, just smooth space, reassuringly urbane, tasteful yet with a hint of homebaking. At the same time, you might have realised that there is a bit of law around to protect you: you close the door behind you, this is your private space, the law protects that. You were offered the Coke, you did not steal it; you legitimately bought your iPad. The atmosphere is assembled by a safe, small measure of law, there to protect you and make you feel immune in your enclosed sphere. But look again. Or rather, smell, listen, touch again. The red and yellow colour combination is a registered trademark of Kodak.[175] The smell of roses comes from the rubber used for the floor of the room – the Sumitomo Rubbers successful application for a trademark.[176] The first notes of *Für Elise* by Beethoven have been registered as trademark by a Dutch company.[177] The iPad touch screen is part of patented technology for which Apple has been in dispute with Samsung over the past few years.[178] Finally, Coke, well! Coke is obviously one of the best examples of a fully protected product in terms of taste, appearance, logo, bottle – the whole lot. And the bonus of sorts in the room: if you were to approach the darts, you might realise that they emanate a distinct smell of dark beer. Even this combination is successfully registered by Unicorn Products,[179] a company who obviously thought that its target audience would be able to identify with it, and wanted to secure that no one else would use it.

175 Vaver, 2005
176 Schaal, 2003
177 Vaver, 2005
178 Merelle Ward, 2011
179 Schaal, 2003

To secure one's product from being used by others, and thus ending up in loss of economic activity due to copycats, is the main ground for the above kinds of intellectual protection. However, an important consequence of such protection is the control of affective movement and the production of such an atmosphere that is conducive, not only to more consumerism, but significantly to an apolitical, anomic lawscape. To control one's sensory stimulation amounts to a control of affect. The manner in which the flow of sense, emotion and symbol circulates in an atmosphere is firmly inscribed in the lawscape. Think for example of the mood-enhancing quality of colours. Kodak's red and yellow is simply one case of mood direction, Fuji (green and red, also protected) is another. Of course, these are combinations of colours rather than single colours, and that indeed makes a difference, since at least the combinations need to be manufactured. The US Supreme Court, however, when judging whether a dark shade of green used in cleaning pads could be protected by trademark, has ruled that 'nothing ... shall prevent the registration of a mark used by the applicant which has become distinctive of the applicant's goods in commerce.'[180] In that vein, the famous red sole by fashion shoemaker Louboutin has been found to be validly registered and protected as trademark.[181] Such future-arresting decisions, combined with the fact that 'there is no shortage of color-only marks on the Principal Register of the US Patent and Trademark Office'[182] justify fears of 'colour depletion' – once again, a term framed by and for the industry, rather than from the perspective of the effect it has on social atmosphere. A colour-depleted atmosphere is not monochromatic. Rather, it is an atmosphere that offers only the colours that the body (thinks that it) desires, brimming with the illusion of choice and thus concealing the fact that choice is predetermined and dissimulated as desire.

Smell is more insidious than vision. Carsten Schaall writes, 'scientists work on the calculated use of smells in cinema seats and there is already a prototype called I-Smell that spreads fragrances and smells on the Internet.'[183] The first US patent, number 1017669, was obtained in 1810, and smell-marks (olfactory trademarks) have been granted to tennis balls smelling like freshly cut grass or fuel additive that smells like grape, cherry or strawberry.[184] Although olfactory protection by trademark is harder than patent, the Dutch Court in the by now famous case of *Lancôme v. Kecofa*[185]

---

180 *Qualitex Co. v. Jacobson Prods. Co.*, 514 U.S. 159 (1995), 15 U.S.C. § 1052(f); see also 'Colours Alone Can Form Trade Mark', *EU Focus* 148, 2004: 17
181 See Sreepada, 2009
182 Starr and Bennett, 2009
183 Schaal, 2003
184 Menezes, 2010
185 *Lancôme v. Kecofa* [2005] E.C.D.R. 5, trans. A. Field, *IDEA, The Journal of Law and Technology*, 45 (1) 31–4, 2005: 32–3

confirmed infringement of copyright of the scent by another company that copied 25 out of 26 of the ingredients of Lancôme's *Trésor* perfume and sold it under a different name and for a lower price. In a rather interesting passage, the Court mentions that 'to receive copyright protection, the work does not need to be new in the objective sense. It needs to be subjectively original as viewed by the maker.' This decision presents two basic problems: one, the fact that the basis of the protection was deemed to be the scent, namely the composition of the perfume. But as it has been commented, 'if copyright protection is sought for the composition, then we are claiming protection for naturally occurring substances, or what is essentially publici juris.'[186] The second issue refers to a warped understanding of temporality: to define originality not as a question of first production but according to the maker's subjective view, as the Court suggests, creates opportunities for continuous original 'inventions' that will increasingly be depriving the atmosphere of odours found to be original according to the 'maker'. The maker, therefore, is properly speaking a demiurge of time and affect.

To register something composed by Beethoven as a trademark sounds impertinent to say the least. Yet this is exactly what a Dutch firm did when they registered the opening bars of *Für Elise*. The case reached the European Court of Justice who, at least in principle declared that sounds marks were registrable.[187] This chimed well with the roaring lions of MGM being accepted as registered trademarks in the US.[188] As David Vaver correctly points out,

> the ultimate question … is of course whether firms should be entitled to appropriate as their trademarks a sound in nature, and equally, whether they should be entitled to reappropriate pieces of famous music when the music is out of copyright.[189]

Space itself is becoming patented. In 2013, Apple successfully managed to patent a whole building: its cylindrical glass pavilion in Shanghai.[190] Most indicative of the way intellectual property protection determines the atmosphere is the cases of design law, motion marks and dress trade. The first is a particularly nebulous area of law that controls ownership of design that has a direct impact on the form, smell, taste and so on of the product, as well as the availability of the product especially when spare parts are controlled by design monopoly.[191] The situation becomes ever more complex

---

186 Menezes, 2010
187 *Shield Mark BV v. Kist* C-283/01 [2004] Ch. 97
188 *Oliveira v. Frito-Lay*, 251 F.3d 56 (2nd Circuit 2001)
189 Vaver, 2005: 903
190 Andreini *et al.*, 2013 cited in Fake Industries, 2014
191 Suthersanen, 2010

since it is much easier to register a design than a trademark or a patent. Motion marks on the other hand are multimedia trademarks that include holograms, gestures and moving image marks. The idea of protecting a gesture is quite formidable, yet several supposedly characteristic gestures, usually performed by celebrities, have been registered.[192] The 20th Century Fox Film Corporation logo with floodlights trailing back and forth across the sky is a registered multimedia trademark.[193] Finally, dress trade is a form of trademark that refers to 'the three dimensional shape, graphic design of the product and its packaging.'[194] This obviously goes further than the form and includes: 'the 'look and feel' of the product,[195] as the recently trademarked Japanese bottles of juices that imitate the feel of the skin of the fruit they include.[196] Remarkably, dress trade is successfully used for the protection of spatial layouts. Thus, not only Pizza Hut's red roof and McDonalds 'golden arches' are protected by trade dress as architectural designs,[197] but also the restaurant's 'décor, menu, interior layout and style of service.'[198] Conceivably, there is nothing stopping a restaurant from registering the effect that its lighting has on the colour of the food served or a particular kind of idiosyncratic service that will have its clients returning.

These are just some examples. As Vaver points out, 'over time there has been constant pressure from industry – note, not consumers – to widen the subject matter of protection to include as trademark virtually any perceptible feature in the sensory world that can be used to attract custom.'[199] There are myriads of other laws that contibute to the engineering of a legal atmosphere, such as planning law,[200] property law,[201] environmental law[202] and health and safety regulations (and their self-perpetuating atmospheric mythology[203]). I chose to introduce the issue of atmospheres on the basis of intellectual property law because of the fact that sensory control is direct and unmediated on the body, yet manages to diffuse and dissimulate itself. This it does in two ways: first, by targeting the

---

192 Martin, 2007
193 WIPO, 2009
194 Karki, 2005
195 Sifleet, 2005
196 Zaino, 2009; this is reminiscent of the *Jif Bottle case, Reckitt & Colman Ltd v. Borden Inc* [1990] 1 All E.R. 873, that revolved around an unregistered trademark of a bottle in the shape of a lemon and which was found still to be protected.
197 MacMurray, 2005
198 Baraban and Durocher, 2010
199 Vaver, 2005: 897
200 Valverde, 2011
201 Blomley, 2004
202 Philippopoulos-Mihalopoulos, 2007
203 Almond, 2009

environment rather than the body; and second, by dissimulating itself as desire, that is as personal preference, often by-passing the conscious, that 'demands' Kodak, Coke, Apple, or beer-scented darts. In some cases, the proffered hyperreality is superimposed on a more basic desire for, say, natural smells or tastes. This sensory desire, as Emily Grabham has convincingly demonstrated (in her case, touch), 'embeds itself into the normative fabric of the law, creating and maintaining expectations around what is proper, decent and safe.'[204] This means that the legal sensorium, Grabham continues, becomes 'detached from specific moments and mobilised within legal processes', indeed becomes fetishised by the law only to be snuggly reattached, I would argue, to the materiality of the situation in hand, claiming echoes of universality. But this is the paradox: the more universal the law, the more diffused it is. The more diffused it is, the least obvious the lawscape is. The room is just a room. The atmosphere is perfectly engineered to appear as a space filled with bodies that are guided by preference, choice, opportunity, freedom. Scratch the surface and you will discover that all these preferences are nothing but corridors of affective compulsion.

So, now we know. The atmosphere has collapsed and has given way to a lawscape that engages in its usual in/visibilisations. This is Spinoza's third perception (or knowledge, as Deleuze reads it). Deleuze's take on Spinoza's knowledge is structured around the body of a wave and a human body. While the first knowledge is one of passive encounter with the wave, the second is a knowledge of swimming, that is moving around and through the wave, and the third is the profound understanding of relations between the material and immaterial bodies involved in the assemblage:

> it does not happen anymore between the wave and myself, meaning it does not happen anymore between the extensive parts, the wave's wet parts and my body's parts; it happens between the relationships. Relationships that compose the wave, relationships that compose my body, and my skill when I can swim.[205]

Going into the technicalities of intellectual property law and how effective the above trademark registrations are when disputed, is a lawscaping technique of engaging with the law so that the law becomes spatially contextualised. This is a legal contextualisation of Spinoza's third kind of knowledge, which liberates from illusions of free will.[206] Indeed, atmosphere is a theological enclosure of guaranteed freedom, brimming with illusions (of consciousness, of free will, of finality, even of spatial justice as I show

---

204 Grabham, 2009: 350
205 Deleuze, *Cours Vincennes: Sur Spinoza*, in Lambert, 2013b: 71
206 Spinoza, 2000

below). A way to withdraw from it might be the knowledge of causes – not necessarily legal causes but deeper, lawscaping causes of desire. This, however, is an aspirational knowledge, always tending towards adequacy but never quite achieving it, despite Spinoza's and even Deleuze's conviction. The reason is not only the vast complexity that needs to be reduced, but more importantly the ontological condition of withdrawal that can never be accounted for, and which indeed does not have a place in Spinoza's world. What does, however, have a place is illusion and the power of imagination, which allows one to question one's supposed desires.[207] This is possibly a little less demanding than the third knowledge of causes, perhaps even more important than knowing the causes, since law and its practice teaches us that knowing the causes is just a way of using the tools better but not necessarily fairer. In that sense, imagination works on the level of law and its illusions.[208] This however is not the only way to withdraw from an atmosphere. As we have seen, succumbing to atmospheric excess and allowing the complexity to break through the atmosphere is one way. Actively resisting it and disobeying it, as Butler and Loizidou suggest, is another way, which, however presupposes the knowledge of causes. A change of natural or elemental conditions, a technological letting-down, or simply a happenstance are other ways. Whatever the cause, once the atmosphere collapses and the bodies withdraw, the lawscape re-emerges. One always falls back into the lawscape. But perhaps something has changed, and a locus of spatial justice has been opened. Before that, however, we make a brief bucolic excursion to the possibility of multiple atmospherics.

---

207 Gatens and Lloyd, 1999
208 Philippopoulos-Mihalopoulos, 2009

Chapter 4

A change of air
The posthuman atmosphere

Can there be a multiplicity of atmospheres? If we follow the phenomenological route, which is prominent in current atmospheric research, the answer will be positive. If we want to move away from the phenomenological, then a different mechanism must be suggested that will accommodate atmospheric change without relying on individual perception. The question is not only of theoretical interest. This chapter paves the way for the discussion on spatial justice, and sketches the terrain on which spatial justice might emerge. Without wanting to spoil the plot, it is not coincidental that throughout the book I have been referring to atmosphere in the singular. Yet an atmosphere changes, varies, and can be to some extent predicted. So how is it that an atmosphere is both singular and plural, or at least varying? Where is its otherness? And is this otherness not a threat to the legal, political, architectural and so on engineering of an atmosphere? How does this affect the already considerable atmospheric contingency? In what follows, and building on the previous chapter's finds in terms of atmospheric engineering, I am dealing with the following interrelated issues: first, whether there can be an atmospheric multiplicity; second, and related to this, whether there can be a conflict between atmospheres; and finally, how does the atmospheric paradox of emergence yet engineering take place?

In order to look at this more closely, I move the discussion onto a posthuman lawscape that exemplifies the indistinction between the various bodies that constitute it. From this, an atmosphere arises, both emerging and engineered. The main question is whether the particular atmosphere, potentially placed in opposition and even conflict to another atmosphere, constitutes an ontological atmospheric plurality or not. This Chapter's lawscape is the Northern Italian region of the Triveneto with its nomadic *transhumance*. The term comes from the Latin *trans*, meaning 'across', and *humus* 'ground', and denotes how flock and shepherd move nomadically from one place to another in search of food. In other nomadic pastoralism cases, this usually happens twice a year, summer and winter. The Veneto pastoralism, however, is unique in the world.[1] The practice of transhumance takes place year-round, a *moto perpetuo* that embodies movement to the point of fixing it. The shepherds of the Veneto have no base, even a semi-permanent one.[2] Rather, they territorialise a large and complex mountainous area, engineering an atmosphere of independence, self-sufficiency, and legal autonomy – all this on top of the area's overabundant state regulatory regime that controls who, where, when and how one moves across the local patchwork of private and public property, environmental protection boundaries, political sensitivities, economic zones between industry, agriculture and pastures, network of roads, permits, exchanged favours and local connections. Every time the flock crosses the plains, a new circle of desire is carved in the area. This area is identically delimited yet differently territorialised every year: differently superimposed (temporally and spatially), differently read by the flocks, at different speeds and with different moments of confluence and conflict with the system of property around it. Is that two atmospheres, namely transhumant and sedentary? Further, does the transhumant atmosphere split into human and animal atmosphere?

Our guide to the area is Valentina de Marchi's text *Hunger for Grass* ('Fame d'Erba'),[3] currently the definitive study on the Triveneto pastoralism. In the autumn of 2008, de Marchi spent approximately five months with the shepherds, preceded and followed by shorter trips to the area. I have subsequently visited the area and tried to trace at least some of the loci of the text. But this is not easy. I remained an outsider to a tight atmospheric construction. I should have been prepared. The very first sentence of de Marchi's book reads: 'If one day, upon reading these pages, you feel the need to meet one of the shepherds and their flock I talk about here, remember that it will not be easy to find them.'[4] Intractability is described in the book as a way of life dictated partly by the nomad's choice (who often would give the wrong directions in order not to be found) and partly by the legal regime that obliges the nomad to remain outside. The transhumant flock withdraws from the atmospherics of private property and state control, and generates its own atmosphere. The transhumant atmosphere emerges from human and nonhuman nomadic bodies, and distinguishes itself from the outside by means of an affect: hunger. Hunger becomes the flock's territorial refrain, its visceral code that distributes space in one circular songline.[5] This is an immanent legal regime of territorial presence

1   Aime *et al.*, 2001; Blok, 1981
2   See Gagnol and Afane, 2010 for another connection between spatiality, (in)justice and pastoralism.
3   De Marchi, 2009a. De Marchi's texts are in Italian and all translations and possible mistakes are mine.
4   De Marchi, 2009a: 8
5   Refrain is employed by Deleuze and Guattari, 1988. See above, Chapter 2 for its connection to territory.

that defies property lines while silently drawing its lines of withdrawal, hushed flows of pause and revolt carved in the fabric of the earth. This legal regime is intricately linked to its space of appearance. It is a proper lawscape-turned-atmosphere.

4.1 The Earth that moves

Immerse yourself. Imagine a large flock that goes on for as long as the eye can see. As many as 2,000 sheep flood the narrow valley, moving slowly like a woolly glacier, a carpet spread over the surface of the earth. There is nothing outside the sheep. The earth is defined by their bodies. Amidst them a shepherd, immersed, in some ways powerless, a master dominated by its own affective lawscaping. The strongest, most prominent affect here is hunger. Hunger is the alchemy that fuses the assemblage shepherd-flock in a continuous presence. It makes the ground move always forward and always in the singular direction of desire: 'you never turn back with the flock.'[6] Yet affects are not singular. They proliferate and vary according to the body that embodies them. This is the hunger of the sheep but also the hunger of the consumer to be fed with the meat of the sheep. Both hungers are part of the assemblage. The closeness between the shepherd and the animal, however, does not exhaust itself in the capitalist understanding of mass meat production. It goes further and pre-dates it. As I show below, the process of rearing is personal, intimate, fused with the process of identity formation for the shepherd. It qualifies the connection between the human and the animal by allocating power in a radically different way than, say, in an industrial production unit. Here, the apexes of decision-making are spread across an immanent surface. The flock's hunger is an affect shared by the shepherd, thus removing the centre of decision-making from the consciousness of the shepherd's mind and over to the space taken up by flock and shepherd. As Hayden Lorimer writes, 'ultimate authority does not rest with [the] herder: Mikel's insistence is that "Sarek [the chief animal of the herd] will decide" and all others [shepherd included] follow.'[7] Just as any affect, hunger is transmitted among the various bodies, fleshing out both their sociality and their corporeality: 'the term "transmission of affect" [is meant] to capture a process that is social in origin but biological and physical in effect.'[8] While I would be reluctant to retain the above causal differentiation, Brennan's quote affirms that affect, as I mention in the previous chapter, is a sensorial, emotional and symbolic flow, straddling the material and the immaterial. Thus, the materiality of

---

6 A shepherd in De Marchi, 2009a: 200
7 Lorimer, 2006: 498
8 Brennan, 2004: 3

hunger has a sensorial and emotional purchase from which it cannot be meaningfully distinguished. It follows that an affect is idea and matter, thought and body, involving connections between bodies as well as ideas on such connections. Hunger as affect makes the flock weaker or stronger, pushing it into encounters with other bodies, testing its power and increasing it or decreasing it accordingly.[9] At the same time, an affect is both immanent in the reciprocity between bodies and space, and excessive in relation to them. Hunger is immanent yet exceeds the body of its emergence: the sheep's hunger exceeds the body of the sheep, catching into its affective net the body of the shepherd and the consumerist body of the meat-eater. All the bodies of the assemblage feel the hunger.

This is a clear manifestation of the posthuman trait of the lawscape. To recall, the posthuman lawscape referred to in Chapter 2 is about two things: first, the inclusion and understanding of relevance of nonhuman bodies within the assemblages of various bodies that trammel and co-produce the lawscape; and, second, the understanding of the human no longer as pure (whatever that might be) but as a set of intensive and extensive qualities that can be natural/technological, non/human, in/organic, im/material and so on, with which the human gathers into spatiotemporal assemblages. The shepherds have a name for it: 'the malady of the sheep' (*la malattia delle pecore*),[10] a malady where species meet (to spatialise Haraway's title[11]). The preposition 'of' connects ambidirectionally shepherd and sheep (Who has the malady?), upsetting thus the predetermined hierarchy of direction or consciousness in the phenomenological sense. Affective ability, rather than species, is the prime criterion for categorising bodies – namely how they move and stand still, and whether and to what extent they can form assemblages that respond to environmental conditions in a way that results in conative perpetuation. Hunger is the principal affect of the transhumant body that pushes the latter into constantly new encounters with other transhumant or sedentary bodies, namely other flock-shepherd assemblages, local farmers and their properties, tourists and their routes, consumers, researchers, abattoirs, the police with its own pronounced set of normativities, and so on.

I have described the lawscape as a spatial manifold.[12] Indeed, the Triveneto transhumant space is neither an abstraction nor a linear, measurable jurisdiction. On the one hand, it is a concrete mountainous terrain

---

9   To recall Spinoza's definition of affect: 'the affections of the body by which the body's power of acting is increased or diminished, helped or hindered, and at the same time the idea of these affections.' Spinoza, 2000: 164, III def. 3
10 De Marchi, 2009a: 33
11 Haraway, 2008
12 Above, Chapter 2.

with a Mediterranean climate.[13] On the other hand, it is an embodied, manifold space. The shepherd's knowledge of the space is an extension of his body,[14] his corporeal mark in space itself: 'the way in which the shepherd knows and sees the world is inseparable from the physicality of being in the world, and thus, from his identity.'[15] The physical and experiential dimension of this knowledge cannot be stressed enough. When faced with maps of the territory, at worst the shepherds would not even look at them and at best they would be unable to read them.[16] This is not ignorance or lack of skill. Rather, the shepherd cannot read a map because he is not moving on it with his flock. He is disembodied in front of the map. The flock is the negation of representation, embodying in its body an ontology of nomadic presence that cannot be traced, either on the map of the territory or the map of the earth. Indeed, as I show below, shepherds habitually cover up their traces on the ground in an attempt to remain untraceable. Yet, a mnemonic trace is carved, palimpsestically superimposed in the way they deal with space: 'the shepherds read the territory as a continuous and contiguous space, consisting of pasturable, prohibited and transitory areas. The nomads need a system of fluid spaces, consisting of decentred areas, not excessively exploited or controlled.'[17] This diagrammatic understanding of the connection between one's body and the space in which the body is situated has nothing representational or metaphorical about it. Rather, it is one of these occasions where a body's movement becomes tautologous with the space of movement, constructing thus a vibrant idea of its materiality. Space and body become the lawscape by determinining immanently the movements inside.

How can this body-space continuum fit in with the classic property regimes of land division and functional exclusion? How can the flows of affect work when confronted with the various stoppages of the law? To put it in Deleuzian parlance, how can the smoothness of the affect mix with the striation of the law? There are two answers to this: first, affect is not just smoothness, nor law is just striation. Just as law is both striation and smoothness, oppression and protection, division and confluence; in exactly the same way, affect constructs its own legality of movement, its own law that chops up what appears as affective smoothness into an infinity of

13   Verona, 2006: 14
14   It has always been a man, as I show below. The gender issue of transhumance is brought forth by De Marchi, 2009a, and creates an imbalance between the nomadic (husbands) and the sedentary (wives and rest of family). There is another atmospherics to be studied in the sedentary women's world, who meet up with men at specific points in the manifold. It is issues of this sort that destabilise any romantic impression of the transhumant.
15   De Marchi, 2009a: 201
16   De Marchi, 2009a: 201
17   De Marchi, 2009b: 5

corridors of imitation, waves of contagion and behaviours of conflict that are variably directed by an abundance of social, political, legal (in the narrow sense), architectural and so on pressures. *Affects are also subject to law. Indeed, to push it further, law is an affect, carried in bodies and the distances between them.*[18] In the Triveneto case, hunger dictates a variety of routes and a variety of velocities. The region is fully striated by the affective desire to be fed. Some routes slide in more readily while others require different moves. The flock reaches some pastures quicker while others dally and end up becoming redundant. The carpet of the earth is broken up in several parts, each part a separate flock, each one pulsating with the same desire to be fed, surrounded by a shepherd who summons the ground in the hope that his flock will arrive first, before it goes to the next shepherd and ends up brown and downtrodden. This is the transhumant lawscape, fully legalised through its own bodies and spaces, its own movements, its own priorities, conflicts and antagonisms.

The second answer to the question on the conflict between affect and law is a little more surprising. At first instance, the flock tramples over private property, moves across boundaries and zoning and protected areas. It moves against the law, it tresspasses, it allegedly causes harm to private property.[19] Private owners shoot at them; police officers discourage them from moving along their routes; motorways sever their routes, and industry kills their pastures. Amidst all this, there is a twist: counter-intuitively, hunger has nowhere to hook on when the ground is smooth, earthy and eaten up. In order to be even momentarily sated, hunger requires boundaries erected by the world: private property to tread on and sedentary institutions and people that refuse passage. As De Marchi writes, 'private property today is what basically guarantees freedom of passage for the shepherd',[20] and allows shepherds to graze more easily than state-owned property, ecological parks or other public enclosures that are heavily policed. *The transhumant stasis needs walls to turn against.* Private property has been successfully engineered to be part of the atmosphere that includes transhumants and sedentaries. Private property has been *included* in the atmosphere as the atmospheric exterior, as we have seen in the previous chapter. However, and for purposes of spatial justice, to stay at this would not be enough. Spatial justice is not a condition of arrival but of departure. Transhumants have organised themselves atmospherically, but they need to keep on withdrawing in order for spatial justice to keep on emerging as a question, as I discuss below. Yet, the atmospheric dissimulation is nearly perfect: the lawscape has become atmosphere! The

---

18 See Moore's, 2007, point about affect as law through the image of law.
19 See Karplus and Meir, 2013, for a similar perspective on the conflict between nomads and sedentaries.
20 De Marchi, 2009a: 239

lawscape dissimulated itself atmospherically, a wolf in sheep's clothing: law and space become absolutely absorbed by the body of the transhumant and its atmospheric rather than 'legally imposed' position. There is no other law but the law of hunger.

'It is realistically impossible to respect all social rules and imposed institutional arrangements':[21] de Marchi refers to the way the exterior is being incorporated inside an atmosphere. Inside all is *nomos* – law as *logos* has been included inside the atmosphere as the *exterior*. Transhumant bodies operate 'under the illusion that the government is removed and cannot harm them in any way.'[22] This illusion accentuates another illusion, namely the feeling of freedom that a shepherd passionately cultivates: 'we shepherds are free, we pay no tax on the territories we cross' says Paolo; 'I chose to become a shepherd in order to be free from bosses. Up in the mountain we are really free, no one breaks our balls up there', says Matteo, another shepherd;[23] finally Gianni: 'the animal is our sole boss. We shepherds do not want another boss above our heads.'[24] This is the power of atmosphere: only from these dizzying heights can property be appropriated and illegality turned into a legitimate tool. At the same time, this is atmospheric dissimulation at its most ironic: in the atmospheric immanence, there is no law. Everything is air, rarefied and free. Yet, *everything is free means everything is dissimulated*. The lawscape remains everywhere but its dissimulation as atmospherics is complete. Even morality does not manage to break through the atmospheric glass. Take for example the inevitable question of sexual intercourse with animals: very quickly, moral criteria imported from outside are applied by the shepherds themselves, and while on the one hand they are keen on shedding this image and react in a defensive way at its mention, on the other hand they strongly try to maintain the allure of tradition best exemplified in the archetype of the old shepherd seeped into nostalgic pastoralia, which, in a round-about way, also implies a certain sexual intimacy with the animal. This is the self-dissimulation of atmospheres at work on all levels. As de Marchi pointedly observes, 'they *act out* their own identity',[25] completing the atmosphere as a reflexive situation, yet revealing the fact of its being directed, indeed scripted by the law of the very actors whose bodies make up this atmosphere.

In the nexus between affect and law, which is at the same time a conflict and an atmospheric inclusion, the concept of *territory* emerges, further linked to *territoriality* through animality. Territoriality is the ethological concept for a spatiality appropriated by an animal's behaviour. Territory becomes

---

21 De Marchi, 2009a: 33
22 Marx, 1996: 124
23 De Marchi, 2009a: 64
24 De Marchi, 2009a: 123
25 De Marchi, 2009a: 181, emphasis added.

territoriality through hunger. Without negating property as such, the flock maps out a territory on the basis of the flock's territoriality, which, as Andrea Brighenti points out, is virtual: 'an *imagined* entity, a space that is carved out, excerpted and circumscribed in view of a set of tasks to carry out.'[26] This virtual territory is superimposed, moves alongside and internalises the system of private and public property relations, its very presence a performative assertion of the inadequacy of property as a total description of space. Even though transhumance relies on private property, it does not claim property ownership but property possession, or to put it as Agamben has, *use*.[27] While ownership claims a temporal chunk closely resembling or at least aspiring to permanence, in possession, space is carved in temporal slices solely determined by the affect of hunger. Use, on the other hand, becomes co-extensive with the conative existence of the body. It is not exclusive but is enclosed. The same applies to the categorisation of the sheep as wild or domesticated. The two categories slide alongside each other: although clearly a domesticated animal, a sheep in a flock is at the same time a roaming animal, thriving, as Mussawir puts it, 'upon relations of escape and capture' and having 'no use for property relations.'[28] Yet property is being reappropriated by the flock as territorial mapping, and escape by the shepherd as a standard dissimulating strategy. The domesticated is getting wilder.

To recall, I have defined an atmosphere as the excess of affect that keeps bodies together, and what emerges when bodies are held together by, through and against each other. The assemblage flock-shepherd, its position on the fabric of the earth, static and revolting against the movement of the world, produces an affective excess that holds bodies together while offering the immanence of the atmospheric interior. Affect is continuously generated, in excess yet from within, and atmosphere is there to mop out the excess and arrange it back into immanence. This is the paradox of emergence and at the same time engineering of atmosphere at work. Removed from the shopping mall or airport, the transhumant atmosphere is both more intense and more diffused. Atmosphere is air, breath, exhalation, ozone, earth. It is dung, wool, grass, fear, desire, death, sickness, loneliness, freedom, defiance. It is, perhaps most prominently, the smell of the *carretto*, the folding home of the human and the beast. The carretto is what most accurately represents transhumance today,[29] and an echo of that briefest discussion on whether Palladio's villas were made to accommodate humans and animals together.[30] The atmosphere of the *carretto* is the atmosphere of *becoming* (becoming animal, space, death), the fold that folds into itself: a

26 Brighenti, 2006: 68
27 Agamben, 2013; see also Keenan, 2010, on possession.
28 Mussawir, 2011: 67
29 De Marchi, 2009a: 194
30 Ingraham, 2006: 18

mobile home on two levels that the shepherd shares with the most vulnerable of the flock, the newly-born and the sick. Animals on the ground floor, humans on the upper floor, folded into each other in a contiguous continuity of nomadic monads.[31] Noises circulate, smells percolate, movements shake the two levels of the human and the animal and the bed becomes grass, the grass becomes plate, the plate becomes door: 'there is no space in which the limits between human-animal are rigid.'[32] The ultimate limit also collapses and in the *carretto*, one becomes death: 'there is no taboo towards dead animals. Animals and their death are experienced as continuous with the human world, the food or the inhabiting space; a continuity that the ones who are not shepherds find disconcerting.'[33]

The indistinction between subject and object, already at work in the posthuman lawscape, becomes fully fleshed in the transhumant atmosphere. The indistinction reaches the point of making any terminological difference redundant, and replacing it with one body assembled by the shepherd/flock/space/time. Transhumant atmosphere allows one to remove oneself from the quest for identity, experience, phenomenology of otherness and such old semantics, all of which still residing in the subject-object distinction. Once again, this is emphatically not a suggestion for a further in-between, namely the idea that there is a third space between the various dualistic configurations of values. As I have mentioned, and I will do so more analytically in the context of Chapter 5, while a third space is useful as a mechanism of avoiding the oscillating eternity between oppositional values and their consequent prioritisation between them, it has also been fetishised to the extent that it is now seen more as a synthetic panacea rather than a way of revolting against this dualistic, synthetic logic. For this reason I have been describing atmospheres as an inclusive immanence that can never be seen as a synthetic totality on account of its excess. Immanence is not a smooth, free-flowing plane. On the contrary, it is full of ruptures, counter-flows, affective excess.

Yet atmosphere relies on an *illusion* of synthesis, itself in turn relying on the intimacy of affect. Every shepherd knows each and every sheep of his flock in its singularity, either through how it looks or how it grazes: 'the competent shepherd always walks in the middle of the flock, vigilant and attentive to an animal troubled by an embedded thorn, an inflammation in its hoof, indigestion or pregnancy.'[34] This is quite a feat, considering that most flocks consist of between 1,000 to 2,000 sheep, and points to the intensity of affective knowledge between the bodies, the excess of which generates an atmosphere so cohesive and so intense that it easily gives the

---

31 Deleuze, 2006
32 De Marchi, 2009a: 35
33 De Marchi, 2009a: 35
34 De Marchi, 2009a: 198

impression of total synthesis. Of course, it never is so. The affective nature of atmosphere means that it is never stable but in a constant state of becoming on various levels of intimacy and institutional relations. So, we have the paradox that, on the one hand, the Triveneto transhumance manages to escape the grip of a legal atmospheric script (that typically includes division of territories, policing of movement, accuracy of registration for taxation purposes, and so on), while, on the other, it harbours intense conflicts and disagreements inside.[35] The latter manifests itself as mistrust among them, the need for isolation in order to escape surveillance, the spatiotemporal spreading over the territory of the various flocks and so on. But even conflict is understood as a naturalised part of the illusion of atmospheric synthesis. When de Marchi asks about the relations with other shepherds, she gets the following telling response: 'we cannot agree on anything … He goes his way, I go mine.'[36] Tension and internal conflict abound to the extent that any political transhumant presence oscillates from the imperceptible to the fundamentally fragmented.

The need for atmospheric synthesis becomes a question of identity. As Simone the shepherd says, 'whoever asks for permission to enter the fields is not a real transhumant. The real transhumant never asks, just enters.'[37] Asking for permission to enter would be a concession to the outside of the atmosphere, something that the elusive understanding of the transhumant identity does not accept. Property belongs to the invisibilised outside. It comes in only as a slither of visibilisation of distance: the transhumant shepherds employ the distinction between *baio-a*, to refer to themselves, and *paor* to refer to the non-transhumant, the sedentary. As De Marchi writes, 'although also local, a *paor* is mostly a land owner, and as such he retains the control, the right and thus the power. The transhumant, on the other hand, is aware of the fact that he is an *intrusive alterity*, a passage through territories that do not belong to him and in areas that are removed from his own original community.'[38]

The identity of the transhumant is not understood dialectically in opposition to the sedentary. In fact, one can barely talk about identity. It is more appropriate to use the Deleuzian/Guattarian term *haecceity*, which, as we saw in the context of the lawsape walk in Chapter 2, refers to the understanding that one 'consists entirely of relations of movement and rest between molecules or particles, capacities to affect and be affected.'[39] Haecceity is identity seen as hybrid collectivity that does not focus on the individual but on the connection of the individual with other bodies in the

---

35 De Marchi, 2009a
36 De Marchi, 2009a: 220
37 Simone, a shepherd quoted in De Marchi, 2009a: 224
38 De Marchi, 2009a: 182 (emphasis added).
39 Deleuze and Guattari, 1988: 262

broader sense of the term. Haecceity is the assemblage between one's body, other bodies, the space in which one is moving, the body of law that determines one's movement: 'The street enters into composition with the horse, just as the dying rat enters into composition with the air, and the beast and the full moon enter into composition with each other.'[40] Space is not the container, the background or even a factor of identity. In haecceity, space is part of the continuous becoming, the ever changing composition that sucks in, spits out and becomes other. The transhumant haecceity is a composition of intensive qualities such as 'knowledge and technique, gestures and glances ... perceptions and tastes',[41] and extensive qualities such as the way affects circulate between shepherd and flock, or shepherd-flock and territory, which in its turn includes other bodies. Yet, haecceity is also a fixing of ontology as limitless. A body's ability to couple with other bodies reveals at the same time the lack of an outline, and a totality that can never be fully revealed. While not finite or unchanging, haecceity fixes the totality of the body as a *singularity that is permanently withdrawn*. Seen in the above way, haecceity conceptually facilitates both the spread of the assemblage and the isolation of the ontological withdrawal. This is the reason for which the 'real' transhumants generate their own legality, neither lawful nor unlawful, but parallel to the regime of state law. In the same way, the transhumant identity/haecceity is not a dialectical result between transhumants and sedentaries. It embodies no final solution between the arrogant Cain and the humble Abel. The synthesis becomes parallelism; the book becomes a pile of bodies. Just as in José Saramago's novel *Cain*, the farmer Cain abandons the sedentary identity for a nomadic haecceity that fuses with the rest of the Biblical world;[42] in exactly the same way the transhumant haecceity is nomadic: 'by dint of not moving, not migrating, of holding a smooth space that [the haecceity] refuse[s] to leave ... Voyage in place.'[43]

Everything moves within immanence. But immanence is a continuum, static in its rush, immobile in its frenzy. Have a closer look at the flocks: on first impression, as soon as hunger is sated, the flock moves on. On closer inspection, the flock does not move. The flock is always there, a static immanence that hugs the earth. Shepherds and their flocks are static, unchanging, insulated, immanent. They are bodies within an immanent atmosphere, participating in it while co-constructing it. What moves instead is the whole plane, sliding noiselessly underneath thousands upon thousands of legs: a precipitous conveyor belt that often exceeds 400 km, offering to the sacrificial altar mountains and valleys, rivers and banks, open

---

40   Deleuze and Guattari, 1988: 262
41   De Marchi, 2009a: 185
42   Saramago, 2011
43   Deleuze and Guattari, 1988: 482

state property and private agricultural enclosures, all in the pursuit of fresh terrains of grass. The transhumants are static nomads, pivots of pulsation on the earth's sliding corridor. But should not nomads move? Is this not the definition of a nomadic smoothness as opposed to logic striation? Not quite, Deleuze and Guattari say explicitly, 'it is false to define the nomad by movement.'[44] Nomadic movement is 'immobility and speed, catatonia and rush, a 'stationary process', station as process.'[45] The nomadic movement is the 'absolute movement', or the speed in which the body 'in the manner of a vortex' swirls in palpitating stasis. The body carries the smoothness within, withdrawing into a *stasis,* namely both pause and sedition, both rest and revolt.[46] To stand still is the feature of the transhumant. The withdrawal from mobility to stasis is a short one: the shepherd knows how to leave 'but at the same time his journey is stationary, in place.'[47] Stasis is a pausing revolt, a stop against the current of the world, anti-heroically, even mischievously as I show below. The transhumant is an *unter* rather than an *übermensch,* hiding in silence while the world moves around.

Shepherd, flock, space, time make up the immanence of the atmosphere. In its interior, the atmosphere pulsates with tension, conflict, catenation of moving and stopping, of siding with and moving away from. Shepherd, flock, territory, seasons: an assemblage of bodies, enclosed in its immanent expanse, an immovable presence on the surface of the world, distilling and mapping the world according to its atmospherics. This is a directed, atmospheric engineering which, paradoxically, emerges from its own bodies: both engineered (by hunger) and emergent (through the bodies), the transhumant atmosphere adheres to the characteristic moves of atmospherics: an arbitrary distinction between inside and outside on the basis of identity (we are transhumants, not sedentaries), an inclusion of the outside (in the sense of state law of property and zoning that threatens our lifestyle), an impression of synthesis (in the sense of conflict between us that translates into fierce independence from each other; but also, as I show below, in the sense of isolation as the only way to avoid surveillance), and a self-dissimulation of the atmosphere itself (silence – see Section 4.4). Flock and shepherd stand counter to whatever movement might be imposed on them by invisibilising it and concentrating on the movement

---

44   Deleuze and Guattari, 1988: 420
45   Deleuze and Guattari, 1988: 421, referring to Heinrich von Kleist.
46   Douzinas, 2012: 36: 'Stasis is a strange word. It means, first, the upright posture, standing tall and serene, holding your stance. This first meaning is associated with the meaning of the English word stasis as stillness or immobility. But the Greek stasis, in one of those tricks of the cunning of language, means also sedition, revolt or insurrection, the opposite of stillness … It brings together place … and demand (to stand up together opposite and against).' It is worth noting that revolt might not be the opposite of stillness after all.
47   Deleuze and Guattari, 1988: 421

of the earth, itself following the elemental desire for food. In the atmospheric interior, there is no state law of property and boundaries, just affective desire in the form of hunger. The desire to be fed forces flock and shepherd against the world and within a 'lawless', intimate atmosphere of affect.

4.2 Ruptures in the service of atmosphere

Let me return to the questions that opened this chapter: can there ever be more than one atmosphere? And can there be a conflict of atmospheres? From a phenomenological point of view, there can be several atmospheres: yours, mine, hers, theirs; different moods, different times; different angles, different floors, even different bathroom layout.[48] But if one wants to avoid the narrowly centralised human focus, one needs to construct atmosphere as a geological affective totality that covers all available surfaces and that can only be apprehended in its withdrawal, as a never-entirely-emerging hyperobject. Anything outside the atmosphere is already part of it, just as atmospheric gas expands and takes over all available space. *Logos* is part of the transhumant atmospheric enclosure, but significantly in a position of exclusion in the imaginary exterior. Private property becomes included, even co-opted, as part of the transhumant atmospheric ruptures. Conflict becomes a matter of internal becoming-other. Atmosphere is an excessive enclosure that borders with its infinite surroundings, and whose lines of flight trammel it from within, constantly tensioning it, often to the point of breaking it. As already mentioned, atmosphere is an autopoietic system at its most affective. An autopoietic system constructs itself and the elements of its construction.[49] It is hermetically closed, yet it reconstructs its environment inside its boundaries. An autopoietic system, just as an atmosphere, does not know any boundaries. The whole autopoietic environment, whatever is 'outside' the system, is potentially the system's past, present and future topology, always included within.[50]

Bodies in the service of atmosphere: atmosphere constructs its internal oppositions in order to carry on spreading. Ruptures in the service of atmosphere: atmospheric conflict is the multiplicity of internal distinctions (interior/exterior) within an atmosphere. To go *against* an atmosphere is an internal move, from within the very atmosphere, causing the atmosphere to speed up its becoming and move to a different configuration of elements. This is still within the control sphere of atmosphere, forming part of atmospheric economy. It is a voluntary act of openness. Negarestani says: '"I am open to you" can be recapitulated as "I have the capacity to bear your

---

48 Borch, 2011a
49 See Maturana and Varela, 1972, and Philippopoulos-Mihalopoulos, 2007a
50 Philippopoulos-Mihalopoulos, 2014

investment" or "I afford you".⁵¹ As we have seen, conflict, disagreement and differences can all be part of the atmospheric economy, already built in the materials of the engineered project. This means that any difference is already atmospherically included. To put it simply, ontologically there can be no two atmospheres at the same time. What happens instead is that an atmosphere becomes-other atmosphere, thus accommodating change, plurality, conflict. The atmosphere moves its topology while changing. An otherwise peaceful meeting where all goes according to the agenda and where the atmosphere is one of relative predictability and possibly boredom, can have its atmosphere phenomenologically ruptured by another atmosphere which might involve someone who, say, driven insane with boredom, would start shouting and hitting the ones present. Or, if a change of local access laws affords a more unproblematic passage for the transhumants, then the atmosphere changes in ways that radically resemiologise questions of topology, identity, affective spreading. In both cases, one could say that there is an atmospheric rupture. But the rupture would only be impressionistic, for the previous atmosphere (peaceful meeting or difficulty of access) no longer exists. Indeed, *there has been no rupture, merely a becoming-other atmosphere*, a rhizomatic moving along of the atmospheric topology, an amoeba-like expansion and contraction of atmospheric boundaries. This is what Esposito describes: 'The most complete normative model is indeed what already prefigures the movement of its own deconstruction in favour of another that follows it.'⁵²

How does atmospheric becoming operate? Atmosphere is affective excess collected, transmitted, directed, and even conflicted. Teresa Brennan has shown through clinical observation how affects are transmitted from bodies and spaces to other bodies and spaces.⁵³ This is Sloterdijk's point when he refers to Gabriel Tarde's concept of imitation.⁵⁴ Affective imitation spreads like grass among the bodies and spaces of these bodies. It has also been put as contagion of susceptibility.⁵⁵ Although much more aggressive sounding, the point of contagion is that no one can be left outside an atmosphere: one is already part of the atmosphere, co-constructing it with any reaction one might have: bodies in the service of atmosphere. The constructed distinction between interior and exterior soon becomes flooded by the atmospheric synthesis of total inclusion. Think of how transhumant identity impressionistically distinguishes itself in direct opposition to the sedentary,

---

51   Negarestani, 2008: 198
52   Esposito, 2008: 188
53   Brennan, 2004: 170. See also Mitchell, 2011.
54   Sloterdijk, 2004, and Tarde, 1903
55   Le Bon, 1960; see Borch, 2012 and Blackman, 2012.

when reality is much more complex.[56] The initial distinction between 'us' and 'them' (interior and exterior) is put in the service of a synthetic atmosphere of inclusion. Even when these atmospheric parameters change, and people who do not traditionally come from a transhumant family become accepted as transhumants,[57] the synthesis of the interior/exterior proves resilient. The transhumant atmosphere becomes-other while still remaining the same. Think of Saraceno's bubbles mentioned in the previous chapter, and how their glass partitions allow the air to be visibilised, its presence not forgotten. In exactly the same way, the affective excess is the momentary visibility of that modulation, the passage from one atmosphere to another. Just as the unthinkability of the air is constantly present because it is continuously interrupted, in the same way atmospheric becoming renders the atmosphere ontologically present through the epistemological emergence of atmospheric difference and conflict.

At the point of becoming-other atmosphere, an atmosphere becomes visible, ontologically vibrant. *This vibrance becomes epistemologically accessible as rupture, or different atmosphere.* This vibrance may be understood and indeed strategically engineered in order for bodies to be encouraged to withdraw from the atmosphere. I am not referring here to the possibility of revolting against an atmosphere. This might be a way of withdrawing from an atmosphere but also falling into the becoming-other of the same atmosphere. An atmosphere of conflict and ideological charge is as much an engineered atmosphere as one of oppression or consumerist safety. What is needed, as I show in the following chapter, in order to break away from the atmospheric spell is a corporeal withdrawal from atmospherics *in toto*. This is much harder than it sounds. Atmosphere is a continuum. Atmosphere does not stop. It only momentarily ruptures in its passage to becoming-other atmosphere. The epistemological tremor produced, just like any other rupture, can become strategically employed by atmospheric engineering. However, there is always the hope that the crack that makes the atmospheric excess whistle furiously like an overboiled kettle, could turn into spaces of withdrawal, possible re-emergence of the lawscape, and even questions of spatial justice. This is the topic of the next chapter.

---

56 The parallelism between transhumant lawscaping practices and private owners or the state is not seen as different atmospheres but as parallel lawscapes that contribute to the emergence of one atmosphere. The atmosphere is one of external difficulty and internal cohesion on all sides. The atmosphere of opposition to the other side is what allows the inside of the atmosphere to engineer itself in such a watertight manner. If things change on either side, if a natural disaster makes transhumance impossible, or in any other case of change, the atmosphere will become-other.

57 Which happens occasionally, as de Marchi, 2009a and 2010 writes, but without fundamentally challenging the atmospheric distinction, not because of transhumant tolerance of difference but because of the deep-rootedness of the atmospheric synthesis of inside and outside, inclusion and exclusion.

To answer, therefore, the larger question: atmosphere is always in the singular. The rupture (with the constructed exterior) has been brought inside, folded in, and from in there, it continues to separate thinkable from unthinkable, this air from that air, the illusion of the inside from the illusion of the outside. Negarestani refers to this paradox as the Epidemic Openness, or the 'Pandemic Horror, the horror of the outside emerging from within as an autonomous ... Insider and from without as the unmasterable Outsider.'[58] The hope is that this openness is both part of the economy of atmosphere *and* 'a stealth strategy.'[59] In the following section, I examine how this inside emergence can be used as a stealth strategy through yet another flying visit to Triveneto.

4.3 'I don't know'

The flock is wrapped in silence. Two thousand sheep evaporate, leaving behind a static atmosphere. Every sound has been absorbed in this carpet of dissimulation: secret language, mental mapping, little tricks, fraud, topographical evasion, all in the service of the vanishing of the flock. The magician looks into the hat and hides his face behind a sonorous yet muffled 'I don't know'. The latter is the 'canonical response to all questions considered dangerous.'[61] Questions that span from the apparently innocuous enquiry of how many sheep they have, to where they are going next, where they were earlier, or what route they normally follow. If 'I don't know' is not enough, they will lie yet in a seemingly absent-minded way: 'Where are you going with the sheep?' 'This way' you say, and instead you are going the other way.'[62] Their dissimulation technique is to create an atmosphere of intractability, silence and withdrawal that enables them to carry on looking for grass. On the symbolic level, the atmosphere is also generated through the excess of miscommunication. They often pretend they only speak the local dialect and do not understand Italian. They keep on answering in a hazy, generic way. They adopt a specific body posture, which discourages questions or comments and physically protects them from exposure:

> If you keep your head low, you cannot even hear what the others say. Anyway, they rarely say anything but calling us idiots or stupid. The

---

58   Negarestani, 2008: 198–9
59   Negarestani, 2008: 198
60   A shepherd quoted in De Marchi, 2009a: 231
61   De Marchi, 2009a: 231
62   De Marchi, 2009a: 220

shepherd does not even answer, he walks away mute. *Once I have eaten, it is goodbye.*[63]

In the above quote, the affect of hunger is so obviously shared among the assemblage that when the shepherd talks about himself, he actually means his flock. Once I have eaten, I move away silently, with my head down. You will rarely see a bell on my neck – we do not do it in this area, it is too noisy. If there are non-transhumants around, I will speak to my colleague in *gergo*, our own secret, flexible language that keeps on changing according to the situation. If I have caused some sort of damage to the crops while grazing, I will walk away lightly and silently. If you come after me, I will manage to hide my traces, all eight thousand hooves, all two thousand jaws, I know how to do it, I have been doing it forever, you will not be able to trace me. If you come perilously close to me, I will swap my *carretto* with a nearby shepherd so that you are fooled into thinking that you were looking for him. By the time you discover our little trick, I will be elsewhere, hidden in a different fold of the earth. I am not like you, I am half animal half human, a hybrid of continuous becoming, a pause on the fabric of the earth that makes the earth slide underneath.

Stealth can be used strategically. The transhumant body withdraws by dissimulating its presence on all affective levels. As I show in the next chapter, withdrawal is the condition for the possible emergence of spatial justice. Withdrawal, however, is a corporeal move away from an atmosphere. Withdrawal does not partake of the atmospheric becoming, does not facilitate the vibratory passage between atmospheres. It either rides that passage or simply constitutes its own. The immediate consequence of withdrawal is the return to the lawscape where affective negotiations can begin anew in order to reorient legal and political realities, with the ultimate aim of spatial justice.

In the transhumant case, however, we have something at least impressionistically different: here, withdrawal is employed in order to *engineer an atmosphere*, rather than to move away from it. In other words, withdrawal is the main strategic move through which the transhumant atmosphere is engineered and perpetuated. It is, properly speaking, an atmosphere of withdrawal. This gives rise to three observations: first, withdrawal is not an isolated move. Every body withdraws. The transhumant withdrawal moves away from a state atmospherics of private property, boundaries and zoning that includes the transhumant as a fixed, controllable presence. In that sense, and on a first level, the transhumant withdrawal might be seen as a

---

63 De Marchi, 2009a: 230, my emphasis. 'The nomad opts for stupidity, causing the ground to rise up to the surface. This is not conditional upon the existence of the state – the nomad is not the antithesis of the sovereign, but is, rather, a force in his/her own right.' Moore, 2012: 140; see also Moore, 2010.

step towards spatial justice. Second, although the transhumant withdrawal generates an atmosphere which, at least from a certain perspective looks like an atmosphere that could give rise to spatial justice, *no atmosphere can generate spatial justice*. Spatial justice emerges at the point of withdrawal from an atmosphere. So even if we could say with certainty that spatial justice did emerge as a result of the transhumant withdrawal from state atmospherics, we cannot say with equal certainty that spatial justice is still *here*. Spatial justice is a constantly negotiable rupture, constantly claimed and constantly moving. Spatial justice might emerge from within an atmosphere but only because a movement of withdrawal would have taken place that has moved the body *deeper* in the atmosphere, back in the continuum of the lawscape. For if lawscape is a continuum, atmosphere is the partitioning rupture of the continuum, separated from the lawscaping continuum by means of brittle glass, furious desires, elastic walls of dissimulation. These partitions need in their turn to be ruptured for spatial justice to emerge. Deeper in the atmosphere might mean: ride the atmospheric excess. Or, take distance. Wake up. Dream. Imagine. Question. Think again. Hesitate. Transcend. Return. Any desire that allows the body to have an *encounter* with the whole comfortable atmospheric illusion, whether this is enabling or oppressive, is a way *deeper* in the atmosphere and, therefore, *out of it*. So, spatial justice might emerge if the transhumant flock withdraws from the 'comfortable' atmosphere of stealth and lands itself on the lawscape of state law, thus attempting to reorient the lawscape. Or it might emerge if a private landowner withdraws from the received idea that transhumants are detrimental to local agriculture and attempts to reorient the given lawscape with regards to the transhumant passage. The point is this: *even spatial justice becomes fixed within an atmosphere*, even a 'good' atmosphere. Nothing can escape the atmospheric fixing, the absolute present of eternally actualised possibility. Spatial justice must be negotiated constantly and atmosphere does not allow this. An atmosphere brims with spatial justice but the thing from its illusion is often impossible to tell. This is why it needs to be constantly checked against the assemblage of its emergence.

This leads to the third observation: withdrawal does not have an a priori moral characterisation and therefore privileged access to spatial justice. Even as an ontological movement, withdrawal is not above being strategised and manipulated. Withdrawal can be put in the service of the emergence of an atmosphere that encourages consumerism, or one where a more 'interesting', 'exoticised' way of life is generated. In a more sinister way, withdrawal has been used in order to deliberately exclude the South Israel Bedouin tribes from legal protection by dissimulating them according to a romanticised nomadic stereotype that does not correspond to their reality, as Ronen Shamir shows.[64] For these reasons, I am resisting

---

64   Shamir, 1996

romanticising the transhumant any more than, say, the shopping mall. I am too conscious of what Moore writes: 'the danger is that the sovereign and the nomad can begin to operate as another double bind.'[65] I want to resist falling in this trap of sovereign presence and sovereign opposition. Sovereign is the body that withdraws, as well as the body that generates the atmosphere. Deleuze and Guattari talk about withdrawal as a movement that takes with it the whole parapet of legality, propriety, property, and exposes the violence of heroic exposure.[66] Is a transhumant shepherd a heroic figure? Hardly, especially when seen from up close. Trickstery and dissimulation are not the staple of heroism. In Hollywood parlance he is an 'anti-hero'. Or simply a body that pulsates with animal law, that is a law that does not subscribe to the given prescriptions of legality but withdraws from it, creating its own atmosphere. Still nothing romantic about it. The anti-hero is unreliable, untrustworthy, arrogant, contradictory, antagonistic, petty. He keeps women aside, as mere support staff for his anti-heroics.[67] He breeds animals, not out of ecological dedication but for their meat. He is fundamentally indifferent to the needs of other human bodies. The shepherd's social position is particularly complex, with sedentary farmers and other residents trying to block his passage, anathematising him as nuisance, or proper catastrophe and accusing him of various countryside plagues, such as flora damage, desertification, theft of grass, trespassing and so on – for all of which he might indeed be responsible, although he vehemently denies it. To the extent that he can, therefore, he remains silent, invisible, away from open vistas and regular human society. He becomes his beast more and more, never totally one or the other, and this becoming never following one or the other direction only. The sheep becomes human as much as the human becomes sheep under the expansive roof of the *carretto* and on the elemental surface of the earth. Pretty image, pretty sentence even, but this kind of romanticism worked better in the nineteenth century.

In other words, transhumants have fully engineered the transhumant atmosphere and keep on perpetuating it through the mechanisms we saw earlier. Before, however, we rush into condemning transhumant or other atmospherics *in toto*, we need to remind ourselves of this: that *all engineered atmospheres rely on emergence*. This is arguably the most paradoxical aspect of atmosphere: how can it be both an emergence *and* a centrally manipulated human feat of engineering? In his *Laws of Imitation*, Gabriel Tarde analyses the charismatic leader and how such a presence can kick-start a wave of imitative somnambulism: 'society is imitation and imitation is a kind of

---

65   Moore, 2012: 140
66   Deleuze and Guattari, 1983
67   De Marchi, 2009a writes extensively on family structures and gender relations that generally follow a patriarchical pattern.

somnambulism'[68] – somnambulism on a geological level, where the whole earth moves according to a desire dissimulated as gravity.[69] The answer to the question therefore can be found, once again, in the process of *dissimulation*. We have already seen dissimulation at work in the previous chapter, where the steps of engineering an atmosphere were described. Let me look at it more analytically. An engineered atmosphere dissimulates two things: its continuum with its own variable, changeable potential (indeed, its own variability *in toto*), and its engineered origin. The former is dissimulated as absolute stability, eternal *present*, fully actualised virtuality, predictability and therefore security of expectations: nothing to wait for, this is the best/worst possible world. Nothing can or will change. The latter is dissimulated as originating rhizomatically in the manifold of bodies and spaces, namely the im/material and directionless expanse of potential causes and effects,[70] as opposed to directed political and legal action that aims at achieving specific goals in indirect, environmental ways. Both these dissimulations aim at only one ultimate thing: the self-dissimulation of an atmosphere as a non-atmosphere, as air continuum, even as lawscape. If successful, the generated atmosphere precludes any possibility of way out, resistance or even reaction to it.

What happens when atmospheres 'place the individual in a circuit of feeling and response, rather than opposition' as Clare Hemmings argues, namely in a circuit with no apparent exodus, and no allowance for a space of negotiation?[71] Quite simply, the desire to find a way out is minimised and so is the possibility of spatial justice. This is the same for all atmospheres, whether of the comfort of belonging, or that of conflict or legal oppression. In a conflictual atmosphere, conflict is internalised and naturalised, pushing bodies to desire more conflict. Once again, this is not necessarily a 'bad' atmosphere. Conflict is often necessary in order to move forward the atmospheric becoming, and in the process to open up spaces of renegotiation. The problem is when atmosphere already includes and operationalises conflict as a way of dissimulation of its political direction. In such cases conflict *must* carry on, not because of an ultimate goal of justice but because it is in the service of the atmosphere. Conflict for the sake of conflict, but even further: conflict in the service of the atmosphere. In a legally oppressive atmosphere, namely atmosphere that emerges as *logos* in its full materiality of camps, prisons, forced deportation and so on, there is nowhere *else* to go. Every body must move along the corridors prescribed

---

68   Tarde, 1903: 81
69   Negarestani, 2010, talks about the war against the sun, or in our terminology, how to distinguish between desire to maintain 'comfort' and desire to withdraw. War against the sun is the only way that will liberate the earth from its 'heliocentric slavery'.
70   Deleuze and Guattari, 1988
71   Hemmings, 2002: 552 in the context of affect.

by the law, and any opposition and search for withdrawal is drained out of the body. *Lasciate ogne speranza, voi ch' intrate*.[72] There is no outside in the atmosphere of legal oppression, the present is eternally fixed as fully as the ultimate cry of *not yet* or *no longer*.[73] But we forget: there is nowhere *logos* without *nomos*. The continuum always includes both. Yet *this* is dissimulated.

Paradoxically, what entraps also offers a line of flight. If it is law's dissimulation that offers no exodus, it is law's re-emergence that might offer it. This might sound conservative, even reactionary but is not meant in this way. It is, however, partly a rehabilitation of the law. A psychoanalytical approach to affect reveals another common ground between law and affect. For Brennan, affect is a judgment.[74] This judgment manifests itself physiologically in the body, and, I would argue, also in the environment, since as Brennan admits, 'there is no secure distinction between the individual and the environment.'[75] Yet, judgment is prioritisation, reduction of complexity, distinction.[76] Judgment is a line drawn through space and across air, separating here from there, element from element, body from body. It is a move that moves bodies (including one's own) from one side to the other. It is indeed the glass that partitions this from that air, the illusion of inside from the illusion of the outside. An affective judgment, therefore, operates as separating, a gesture that distinguishes, first, an origin from a destination, and second, a direction. In other words, an affective judgment separates, first, the subject from the environment, and second, the direction of affect. This does not mean that the subject is necessarily 'I'. I can be the destination of the affect ('me'), the end point of its direction. In that case, the subject becomes the environment, to whose atmospheric influence I succumb.

This mendacious construction of a dual position in the emergence of atmosphere ('me' and 'it', both potential objects of the sentence) is a residue of what Brennan calls 'the foundational fantasy', namely the desire for separation from the environment that 'founds the subject in its sense of superiority.' The fantasy itself is 'a passionate judgment directed toward that other in order to maintain a certain relational position.'[77] It is clear that we are still operating on the surface of dissimulation:[78] body and world are one, yet their distinguishing is, once again, a strategy of self-preservation and affirmation of difference. The breath of the world is shared among the

---

72 'Abandon all hope, ye who enter', Dante's famous door sign.
73 Braidotti, 2007
74 Brennan, 2004
75 Brennan, 2004: 6
76 Luhmann, 1998
77 Brennan, 2004: 111
78 Albeit on a fold: dissimulation of dissimulation.

breathers, yet each breather stands aside, circulating air between her and the world in a private communion: 'air disrespects borders, yet at the same time is constituted through difference.'[79] Air's aggregate nature forces one into a distinction that generates identity. The same applies to space, earth and water, cities and concert halls, prisons and valleys, supermarkets and carton boxes. In the whirls of the one atmosphere, a body emplaces itself with the normative force of desire. This is, properly speaking, the 'dissection [découpage] into comparable and countable parts' of the skin of the world as Lyotard puts it,[80] the effect that capital has on affect, law has on territory, and anticipation of the future has on the earth. This is the *logos* of spatial striation,[81] the rationale and the word that chops up territories and properties, inclusions and exclusions, the inside and outside of the walls.

The same issue arises in the totalising effect of atmosphere and the way affective judgments of exclusion spread imitatively. In engineered atmospheres, the judgment is pre-given and its direction predetermined atmospherically and therefore preconsciously. At the same time, it is continually reinforced through the bodies of desire themselves, be this inappropriately casual clothes, bad table manners, cockroaches in the kitchen, beggars in the Olympic city, protesters at G8 summits, buskers in the mall. Yet, and here comes the line of flight from within: amidst all that legally-directed affective judgment, a different kind of movement might emerge, or at least one that takes a different direction, one that withdraws from the wave of engineered imitation.

This is the space of spatial justice, as I discuss in the following two chapters. A withdrawal, a movement *away* from the totalisation of an atmosphere is possible, and indeed happens regularly. 'We are always on the way to withdrawal.'[82] Of course, the more controlled an atmosphere is, the harder it will be for such a withdrawal to become aware of its necessity. Senses, emotions and symbols are hard to wake somnambulists. Yet, it can happen, especially between the 'cracks' of atmospheric engineering, the space of the inevitable risks when faced with the heightened improbability of the endeavour, or the not always smooth passage between the various atmospheric becomings. Withdrawal is not a passive movement but one of 'surfing' opposition: find the cracks and ride them, blow up the affective excess, glide on the atmospheric movement of lawscapes, dwell on the contingency of atmospheric dissimulation. Or even more radically, as Negarestani puts it, accept that 'openness emerges as radical butchery from within and without.'[83] As I show in the following chapter, however, even this

79 Choy, 2011: 165
80 Lyotard, 1993: 44
81 Deleuze and Guattari, 1988
82 Negarestani, 2008: 50
83 Negarestani, 2008: 199

withdrawal ends up in the lawscape; perhaps a differently oriented lawscape, a momentarily spontaneous, orgiastic, unplanned lawscape, but nevertheless a lawscape. This might not be such a bad thing after all: law is both *logos* and *nomos*, striation and smoothness, bitumen and breath. Law tears down walls as well as erects them, digs corridors of forced movement as well as passages of side movement. To be critical of the law *in toto* means nothing. What law orders in its dissimulation can be demolished in its re-emergence. Affect as judgment takes the form of both engineered atmospherics that guide desire; and the spontaneous, uncontrollable emergent atmospherics of changeability, the rhizomatic moving against engineered atmospherics, and the construction of an atmospheric becoming-other.

Chapter 5

The rupture of spatial justice

Another space: imagine entering a concert hall and seeing that someone is sitting in the seat that you have already booked. You have the ticket to prove it, and what is more, the particular seat is your favourite in the whole hall. This little piece of paper you are holding is your guarantee of your claim, and you only need someone to mediate in order for you to get your rightful seat. You find an usher who confirms the seat and, escorting you, asks to see the ticket of the person currently seating in your seat. But, lo and behold!, this person has a ticket with the exact same seat number as you do. What do you do?

The answer is the focus of this chapter: a definition and elaboration of a new, fully spatialised and corporeal understanding of spatial justice. Spatial justice could be the most radical offspring of law's spatial turn. Instead, in the literature it remains a geographically informed version of social justice, a slightly trendier conceptualisation that casts sideways glances to its surroundings.[1] But should spatial justice really be seen as social justice with space as 'add-and-stir', namely without any fundamental difference to the generic concept of social? And, perhaps even more outrageously, does this mean that (social) justice can afford *not* to be emplaced and geographically informed? Or could it be the case that the majority of the existing literature on the subject has made some politically facile assumptions about space, justice and law, thereby subsuming the potentially radical to the banal? In this chapter, I suggest that the concept of spatial justice is the most promising platform on which to redefine, not only the connection between law and geography but more importantly, the conceptual foundations of law.

This chapter attempts two things: the focal point is to suggest a conception of spatial justice that derives from the lawscape. Such a conception cannot rely on given concepts of distributive or social justice, since it can no

---

1 Namely, a branch of justice that refers to the distribution of benefits and burdens in society, organised by social institutions such as property, state, organisations, etc. For a classic discussion see Miller, 1979, who, however, distinguishes between legal and social justice, whereby the former is defined in the positivist way of punishments and restitution.

longer be constructed on processes of consensus, rational dialogue, renegotiation of territory, demos, agency or even identity formation. Instead, the concept of spatial justice put forth here is informed by the concepts of lawscape and atmosphere, in relation to poststructural, feminist, postecological and other radical understandings of emplacement and justice. The lawscape and its atmospherics demand a rearticulation of justice as something that might – just might – emerge from within the spatiolegal operations. Spatial justice opens up the space of conflict between the various bodies that dwell in the lawscape. Its manifestation comes in the form of an embodied gesture, that of *withdrawal*: a body withdraws from the atmosphere and returns to the manoeuvring space of the lawscape. This is the only way in which a body can question the emplacement of itself as well as other bodies: by withdrawing from an atmosphere of fixed positions. At the same time, this is the only way in which the law can generate justice: by withdrawing before the demands of justice while retaining its position as the main means in which justice can be achieved. This geography of withdrawal is not metaphorical: it is only by understanding a unique corporeal emplacement that spatial justice can be adequately sketched.

But please do not withdraw from your claim to the concert seat yet. Stay a while longer.

5.1 An aspatial spatial justice

Spatial justice emerges from within the lawscape. This is not so hard to conceptualise, since as I have argued, there is no outside to the lawscape. Everything is a matter of spatial positioning (even in abstraction), and every spatial positioning is potentially or actually controlled by law. There is nowhere else for spatial justice to emerge. By linking it to the lawscape, spatial justice sees both its elements of law and spatiality fleshed out, and is epistemologically claimed by a spatially-turned legal discipline. So far, the concept has been used by a political and geographical discourse that could not engage with the technicalities of spatial allocation, parcelling and categorising as prescribed by the law. Yet, the concept and praxis of justice is intimately linked to the law. The mechanisms and processes of the legal system attempt to provide for solutions to conflict in a manner that is as consistent as possible. The 'crowning' of this legal mechanism is justice, rather famously accoutred with the (legitimate) violence of the sword and the objectivity of the blindfold.[2]

If spatial justice is to be claimed by the law, it can no longer remain the lukewarm hybrid of socially 'just' spectres, distributive justice wish lists, neoliberal articulations of participation, parochial territorialism and geopolitical analyses. These too are relevant and have been very useful in

2   Goodrich, 2014

pushing the spatial justice popularity. But as a product of law's spatial turn, spatial justice is now forced to incorporate the characteristics of the former and to push them to the extremes of an embodied, spatialised and symbolic gesture, all the more challenging because of its claims to possibility and applicability. The kind of spatial justice put forth here relates, at least to some extent, to the irreducibility of one's corporeal emplacement in the lawscape. This means that spatial justice is the ultimate expression of one's spatial and legal claim to a unique corporeal position which by necessity excludes all others: *spatial justice emerges from the fact that only one body can occupy a specific space at any specific time*. In other words, *spatial justice is the struggle between bodies to be in a specific space at a specific time* (Figure 5.1). Spatial justice is neither a simple question of local democracy, nor however a utopia, something-to-come, a messianic solution. It is at the same time less and more mundane than the above. Because of the way defined here as an open corporeal gesture, spatial justice brings to the fore the relevance, in addition to spatiality, of corporeality and more generally materiality.[3] However, these material bodies, as I have already mentioned, are not individual human bodies in the way generally understood. Rather, they can be a collectivity, a flock of sheep, a fleeting community clustered in a lift going from the third to the fifth floor, or a boat of illegal immigrants. A body is understood as an assemblage of various conditions and materialities – we are not isolated things; skin does not separate us from the world; otherness is not over there but very much with us, on us, in us. A body as an assemblage is a reality of contemporary physics: 'the Einstenian object isn't a unity ... the pencil you are holding in your fingers is only a rigid extended body on account of a false immediacy. Nothing in the universe apprehends the pencil like that, really. Not even the pencil apprehends itself like that.'[4] Rigid separation is an epistemological construct, often a necessity, according to the foundational fantasy of distinction between self and environment, as Teresa Brennan writes.[5] The pencil melts into your hand along with the cup of coffee which is both standing next to you and flowing through your body, the air in the room in which you read this, the impact of the news you heard earlier on about China and your mother's imminent telephone call: an assemblage. It is this assemblage that claims space in the particular time. Even better, we are talking about the space *of* the particular assemblage, namely assemblage *as* space. Only this spatiotemporal assemblage is here. Next to you, there is a different assemblage, with its own spatiality and temporality, but again, linked to your assemblage, a different assemblage emerging *ad infinitum*.

3 See above Chapter 2.
4 Morton, 2013: 62
5 Brennan, 2004 – see Chapter 4 above.

Figure 5.1 Spatial justice – or Cai Guo-Qiang's *Head On* at MOGA, Brisbane 2013. © Andreas Philippopoulos-Mihalopoulos.

    This is not an attempt at trivialising the phenomenon. Spatial justice can only emerge on a flat surface. The latter is not meant in the sense of egalitarian distribution or constant flows, but as an ontology that resists pre-formed hierarchies and boundaries across species or kinds of materialities.⁶ This is the only way in which existing and historically ossified paradigms can be reconsidered, and a new form of material considerations can take its place. Yet, and this is the problem as we have seen with most flat theorisations, they do not account for existing and continuing power disparities. Flat ontologies may indeed be flat, if one constructs them so, but they are also regularly made to tilt this and that way. Horizontality is often compromised, either in the form of atmospherics, where the sphere of inclusion occludes the way out; or even in the mesh of the lawscape, where different visibilities bring forth different interests. In the lawscape, the law emplaces the spatial desire to be *here* and translates it into legal claims, argumentation, decisions (themselves structured according to a certain understanding of power). It opens it up to legal solutions. Spatial justice, however, does not offer a solution, at least not in the form of an a priori, normative answer, or indeed a resolution to the conflict in the sense

6 Marston *et al.*, 2005

of justice done. Spatial justice discounts the possibility of ever arriving at 'the' solution. There can never even be *a* solution. There is only a 'final solution', with the all-too familiar results. The discourse of a 'solution' is perhaps a necessary rupture but its persistent mythologisation ends up occluding the project of actually working on the problems themselves. Solutions are just ways of wrapping up the problem. Every 'solution' is co-opted, every end-project is always already part of atmospherics, we need concepts that address the problems, not solutions that promise to fix.[7] For this reason, the concept of spatial justice takes the form of a question mark. It operates from within the material and symbolic space of the lawscape and demands continuous assessment and negotiation of where one positions oneself (where one finds oneself positioned) and the responsibility of situatedness. The suggested definition points to an unresolved situation, just like the initial incident of the theatre seat. In leaving the question open, the definition I put forward here suggests that there are no easy solutions to what I consider arguably the most important current political and legal challenge. The open definition attempts to highlight a problem of spatiality and temporality that goes largely unnoticed or subsumed in other, grander and perhaps more easily understood narratives of historical claims, national identity, sovereign control, or even property boundaries and questions of self-determination. Focusing instead on a concept of spatial justice as stripped-down and as basic as the above, allows us to see the issue for what it is, namely a corporeal gesture across space and time that crosses, while also relying, on legal boundaries.

Both in name and inference, spatial justice has been widely employed in the recent literature that attempts to deal with concepts and practices of justice from a geographical perspective. This engagement is in fact a predominant feature of the much-hailed 'spatial turn', which, as I have commented in Chapter 1 of this book, has been noted in disciplines as widely varied as politics, theology, history, art,[8] and of course law. The spatial turn is often seen as a reaction to the textual turn that deconstruction established, and an emphatic focus on material rather than linguistic processes. This, however, may not be so straightforward, since it is indeed within or at least following an all-embracing, all-destroying textual turn that a spatial/material/corporeal turn could be fully felt.[9] At the same time, spatial and textual, as I have mentioned, share the same surface. It is to a large extent,

7   Such a concept can be a solution but 'the problem to which it corresponds lies in its intentional conditions of consistency and not, as in science, in the conditions of reference of extensional propositions. If the concept is a solution, the conditions of the philosophical problem are found on the plane of immanence presupposed by the concept, and the unknowns of the problem are found in the conceptual personae that it calls up' (Deleuze and Guattari, 1994: 80).
8   Warf and Arias, 2009
9   Doel, 1999

therefore, within the ambits of the prior textual turn that the following dealings with spatial justice are both to be read and possibly critiqued. From a wide array of cameos, I have chosen to focus on only a handful of perhaps the most representative dealings with spatial justice, which will serve as a brief contextualisation of my attempt to define the concept.

Perhaps most promisingly, David Harvey's pivotal work on justice and space augured a new way of conceptualising geography's dealings with justice.[10] Along with Lefebvre, Harvey brought justice home by radically re-reading Marx's spatial credentials and by linking them to urbanisation. He begins by focusing on the futility of defining space in any *a priori* fashion since 'space becomes whatever we make of it during the process of analysis rather than prior to it.'[11] When he engages with space's definitional parameters, Harvey distinguishes between the 'complex, non-homogeneous, perhaps discontinuous' *social* space, and the *physical* space in which, as he says, the engineer and the planner typically work. Harvey posits what he calls the *pattern* as a bridge between these two spaces, or more specifically the *production* of the pattern by the social space and its superposition on nature (the physical space).[12] This pattern, however, is resolutely described as human (whether in its capitalist or utopian collective manifestation), thereby ignoring ecological, technological and other production processes that eschew the 'social' in its narrow, anthropocentric description. Likewise, the direction in which this pattern is imposed is unilateral, namely there is no exchange of processes, no mutual constitution. There is, in short, a double and parallel focus on a human process of analysis on the one hand, and the social/physical distinction on the other. This narrow and somewhat idealising human focus, reproducing images of absolute and centralised human control, is characteristic of the whole work, and, I think, the breeding ground of problems with regard to the role of space in the discussion. The main consequence of such a division is a deprioritisation of physical spatiality in preference to a processual, social, human (but not phenomenological) understanding of space. This becomes particularly obvious in the discussion on urbanism, where cities are emphatically not seen just as 'a set of objects arranged according to some pattern in space.'[13] While no one would disagree with this, the case remains that there is only a marginal role reserved for those objects or indeed for that space in Harvey's analysis. Something for example that has eschewed Harvey's attention is that not only social space but also physical space can be described as 'complex, non-homogeneous, perhaps discontinuous', and as such the locus of processes that are produced outside the narrowly defined

---

10   Harvey, 1973
11   Harvey, 1973: 13
12   Harvey, 1996
13   Harvey, 1973: 302

social confines. Indeed, this way of manifold thinking, already discussed in Chapter 2, banalises the distinction between social and physical space.

The discussion is further removed from spatiality when the concept of spatial justice is dealt with – indeed, not as *spatial* but as territorial distributive justice. This not-so-subtle difference has been traditionally ignored by the ensuing literature that carries on seeing Harvey's work as central to the notion of spatial justice.[14] The difference between spatial and territorial is not straightforward,[15] but it is safe to say that territorial is not just spatial. Which means that territorial justice is spatial only to some extent. When coupled with the adjective 'distributive', even if decoupled from Rawlsian references (as Harvey makes clear in his critique of Rawls's justice), territorial justice becomes less physical and more processual, less spatial and more social. Thus, faithful to its title, Harvey's book does not profess to offer a conceptualisation of spatial justice, but a spatial perspective on social justice.

Edward Soja, on the other hand, has consistently taken to the term in earnest.[16] In *Postmetropolis*, Soja explicitly and repeatedly refers to 'spatial justice and regional democracy', very rarely dealing with either of them separately and hardly ever distinguishing or defining them. The book came out in 2000, and one would be forgiven to assume that by then the term was clear beyond doubt. Despite such an impression, there has never been a systematic discussion on the concept and Soja's book has not done much to address this gap. What is clear instead is that he too is dealing with social justice from a spatial perspective. Thus Soja states: 'I do not mean to substitute spatial justice for the more familiar notion of social justice, but rather to bring out more clearly the potentially powerful yet often obscured spatiality of all aspects of social life.'[17] This is an exercise in degrees of spatial involvement then, a bringing into light of the spaces in which social (in)justice can be located while conceptualising itself and the world. However, there is nothing that enters a new dialogue between what is space, what is justice and what is the thing that can be brought out of their obscure communion. In his recent book *Seeking Spatial Justice*, Soja has attempted a more grounded theorisation of the concept, explicitly linking it to spatial considerations.[18] He writes: 'justice, however it might be defined, has a consequential geography, a spatial expression that is more than just a background reflection or set of physical attributes to be descriptively mapped.'[19] And later, 'spatial justice is not a substitute or alternative to other forms of justice but rather represents a particular emphasis and

14 See Hay, 1995; Dikeç, 2001.
15 Brighenti, 2006
16 See Soja, 2009.
17 Soja, 2000: 352
18 Soja, 2010
19 Soja, 2010: 3

interpretive perspective.'[20] From that point onward, Soja performs an excellent strategisation of the concept with regards to projects and movements, but fails to show what is new or potentially radical about spatial justice. He carries on in the vein of the above quotes, which perceive geography as a 'consequence' of justice, implicitly reiterating in this way that there can be no justice which is not spatial. So what is particular about spatial justice? If it is just a 'particular emphasis and interpretive perspective', it has inferior purchase than something as mobilising as Henri Lefebvre's famous but admittedly rather sketchy *right to the city*,[21] and in that sense it might even be considered obsolete (or just another semi-revolutionary idea); while at the same time, by thinking of it as a 'perspective', we return to a solidly anthropocentric and phenomenological, consciousness-centred conceptualisation of spatial justice. While this might serve some social mobilisation purposes, it fails to engage with the structural basis of the problem of exclusion from justice. This can be simply put as follows: the above definitions of spatial justice construct an illusion of human agency and its consequent ability to control and influence the spatiotemporal conditions in a direct way. We only need a hero. Or even better, we are all heroes inside. But as Martina Löw writes,[22] the best that social agents can do is *reproduce* space and its qualities, whereas social structures and systems actually *produce* it. This is not a refusal of individual or collective action in the form of questioning, resistance, revolt or revolution. It is, rather, a sobering call for contextualisation of heroics within the spatiotemporal atmospheric parameters in which they emerge.

A valiant attempt at defining the term, or at least at engaging with the problems of definition, comes from Pirie in his article 'On Spatial Justice'.[23] He begins with what he calls 'formal juridical notions of justice' which he summarily leaves aside in order to focus on 'less legalistic specifications of justice, that is, in social or distributive justice.'[24] In fairness, however, he engages in a brief yet thorough, consistently amusing and sporadically scathing overview of conceptions of justice, from Rawls's 'three lexically ordered principles of justice' to theories of subjective justice 'so as to sidestep high theory',[25] without however at the same time sidestepping some well documented reservations about the value of conferring to justice the role of 'the supreme arbiter of distributions.'[26] He infers out of this 'intractable literature on justice' that 'the burning issue to be confronted in presenting a

20 Soja, 2010: 13
21 Lefebvre, 1996, and also Dikeç, 2002; Mitchell, 2003; Harvey, 2004; Layard, 2012; Butler, 2013a
22 Löw, 2008
23 Pirie, 1983
24 Pirie, 1983: 465
25 Pirie, 1983: 467
26 Pirie, 1983: 468; e.g., Kamenka and Tay, 1980

case for a concept of *spatial* justice is whether the kinds of question posed can be answered without recourse to objective or subjective criteria of *social* justice.'[27] Indeed, Pirie suggests, not without a certain disappointment, that 'the term "spatial justice" once again appears as shorthand for the phrase "social justice in space".'[28] This is another manifestation of the vacuous transdisciplinary exercise in 'add space and stir'.[29]

Pirie put forth a challenge for a formulation of spatial justice that would rely on 'an alternative conception of space itself.'[30] Mustafa Dikeç's work on spatial justice explicitly addresses this challenge.[31] His analysis, however, does not employ an alternative conception of space but Balibar's concept of égaliberté, namely a universal, transindividual politics of emancipation. This relies on a practice of collective struggle whose presence remains unquestioned both by Balibar and Dikeç. The problems with such grand assumptions aside for the moment, Dikeç's use of space remains lacking in terms of theorisation in an otherwise engaging analysis of spatial politics, with the risk of misaddressing the challenge of spatial justice.

The problem with the above analyses – and in full awareness of an arguably unjust generalisation – is that the spatial remains an adjectival context, a background against which considerations of the surrounding space are thrown into relief along with the 'obscured spatiality of all aspects of social life.'[32] Society is reinstated in its primacy, the human subject never abandons *his* enlightened central perspective, and the usual rationalising political processes are applied even if presented in their revolutionary variation. However valiant and necessary in view of their spatial openings such endeavours may be, they lack some fundamental attributes: first, they lack a radical vision that would responsibly mirror the current societal conditions and state of thinking. Thus, harping on relics of modernity such as the fixed spectres of identity, community, demos, popular will and consensus, purposefully ignores the discrediting of such fictions (one look at electoral procedures or the way the world is organised transnationally should be convincing enough) and perpetuates without questioning their supposed relevance. Second, by insisting upon an anthropocentric specificity of resource distribution, the existing spatial justice discourses constitute a blatant marginalisation of the current radical thinking on the fluidity of the boundary between human and natural/artificial/technological. In a typical sidestepping of what has originated and further developed in feminist thinking, spatial justice reasserts itself as a human (namely, masculine) need.

27   Pirie, 1983: 471
28   Pirie, 1983: 471
29   Ellem and Shields, 1999
30   Pirie, 1983: 471
31   Dikeç, 2001; see also Dikeç, 2002, which sketches a spatial praxis with regards to Parisian suburbs.
32   Soja, 2000: 352

Third, current theorisations of spatial justice fail to deal with the spatiality of space, those characteristics of space that render space the awkward, angular, unmappable, unpredictable factor that it is; on the contrary, there is a constant and unsurprising double retention. On the one hand, there is the retention of the relevance of time. At best, time is thinly dissimulated as the supposedly spatial metaphor of a line on which justice eventually arrives: but the linearity of the metaphor betrays the lack of its spatial credentials. As I have extensively discussed in this book, space is not a line defined by two points but a manifold plane of disorientation. In that sense, no theorisation of spatial justice that I am aware of has managed to disengage itself from the given priority of time, the concept of struggle towards a better and fairer future horizon conditioned by an interminable waiting. On the other hand, there is a retention of a conceptualisation of space as a measured and measurable factor, given to Euclidean properties and legal appropriation. But this is the idiom of a spatial turn that is turning away from its spatiality and deeper into what I have called in Chapter 1, a geographical dimension. The term 'geographical' in this context denotes the description, the *graphein* of the earth, the dealings with representation rather than the violence of an affective event.[33] Fourth, current discussions on spatial justice are moralising. They claim universality through their particular emplacement. They talk in the name of the deprived and the dispossessed. They essentialise individual needs. In the usual schizophrenia of the utopists who claim to be pragmatists (and take offence if they are thought of as utopists), propagators of spatial justice claim to have a solution to the problem.

Finally, from an epistemic point of view, spatial justice seems to be a battle for geography to assert the centrality of space – and while this is laudable, it sadly stops there. Once asserted, geography and geographers carry on the way they begun, without an observable transdisciplinary leap and without the kind of theoretical enrichment that one coming from the discipline of law would expect. This by no means implies that law is theoretically richer. It is simply an observation on the missed chance of a new locus of transdisciplinary encounter between law and geography which has the potential of being theoretically fecund but which seems to have left geography with only another catch-phrase, vacuous and repeating of already well-articulated thinking. But at least geography did take up the challenge of flirting *en masse* with justice, a concept which, however summarily, is acknowledged as a partly legal concept. Law has hardly deigned to experiment with its spatiality in any way as extensive as this.[34]

The concept of spatial justice suggested here attempts to address the above shortcomings. It specifically posits itself as non-anthropocentric,

---

33   Delaney, 2002
34   All writings on spatial justice come from geography. There is a minimal engagement with the concept from law – see indicatively Davoudi, 2005; Béland and Lecours, 2005.

spatial rather than crypto-temporal, and fundamentally amoral (but ethical instead, namely contextualised rather than overarching and overseeing the moral landscape[35]). It also attempts to be postdisciplinary, yet emerging from that fleeting space between law and geography. At the same time, however, the concept is aware of its limitations. Epistemologically, it can never do away with the incommunicability between epistemes. Ontologically, and in the same vein, it can never be the solution. Spatial justice cannot bring about better identities, more organised popular will, broader consensus, healthier or richer developing countries. Nor can it do away with time and its fundamental role in conceptualisations of justice. In fact, it specifically does not attempt to do the latter. Instead, it simply posits a shift of emphasis, and a temporary one at that. The best it can hope to do is delineate the problem, initiate a discussion on the conditions, acknowledge the hitherto marginalised spatial factor: in short, while acknowledging and working through the impossibility of a solution, spatial justice brings forth the conditions of such an impossibility, thereby allowing a flicker of possibility to stream through.

5.2 The desire to move, the desire to stand still

As I have shown in Chapters 1 and 2, contemporary thinking has developed radical ways to conceptualise space. A concept of spatial justice that chooses to ignore these developments can only be inscribed within a narrow epistemic opening – from law to geography and vice versa. This opening, however daring and indispensable, is not enough for a radical semiologisation of spatial justice. It *might be* enough for a renewed interest in understanding local conditions in the sense of contextualising difference, forming legal subjectivities, and even achieving a clearer practical understanding of fair distribution of land and resources. However important, these understandings can already be achieved through the existing vocabulary of either law or geography. In other words, the space in which such a conception of justice is to be located is already found, explored, colonised. The quest now is for a different kind of space conceptualisation that will allow itself to be folded in an understanding of justice and thus radically alter the latter from within.

Let me start with the most obvious of assertions: justice cannot afford not to be spatial. At its most basic, the adjective refers to the emplaced, geographically specific conditions in which justice is to be considered, planned for, constructed, imagined. In its turn, this merely points to the need for considering the particular when legal justice is administered. This is what Goodrich, Douzinas and Hachamovitch mean when in one of the founding texts of critical legal theory they write 'the justice of judgment

---

35   See Philippopoulos-Mihalopoulos, 2009.

will depend on law's answer to the unique and singular demands of the person who *comes* to the law.'[36] There is a movement *to* the law, an embodied displacement of 'the person who *comes* to the law' in order to find justice that must be taken into consideration by law. The movement is specific, spatially inscribed, bodily demanding, and expresses a 'here' that deserves to be heard at least as loudly as 'now'. *Justice cannot be just if it is not spatial.* This does not deny the universal appeal of the concept, or its potential universal application. Being spatially emplaced does not exclude the universal. On the contrary, it reinforces it by grounding it on the particularity of the monad: for how else can the universal be if not by resonating in every particularity?

The above paradox is one of many that characterise the concept. In every respect, spatial justice emerges in the middle of other concepts, disciplines, desires, bodies. 'The middle is by no means an average; on the contrary, it is where things pick up speed. *Between things* does not designate a localizable relation going from one thing to the other and back again, but a perpendicular direction, a transversal moment that sweeps one and the other away, a stream without beginning or end that undermines its banks and picks up speed in the middle.'[37] The middle offers no originary starting point to set the process off, no destination to give direction. History is corporeally and spatially inscribed in the folds of the matter from which spatial justice emerges.[38] Spatial justice is located in the middle of the simultaneity of the spatial manifold, those 'delicate milieus of overlapping perspectives, of communicating distances, divergences and disparities, of heterogeneous potentials and intensities.'[39] Right in the thick of this, the various directions are nested: the body desires to move, the body desires to stand still. I want to be where you are, exactly *there* exactly *then*. Or, I want to stay put, exactly *here* exactly *now*, in exclusion of everyone else. The motives (greed, attraction, hunger, possessiveness, territorialism, reterritorialisation) can be put aside for the time being, for what is important is the desire to move or to stay still. But how different are these two desires? How different is the desire to striate from the desire to perpetuate smoothnesss?

We have seen in the previous chapter how the transhumant *stasis* is both sedition and sedimentation, revolt and stillness. Stasis takes place in smooth and striated spaces, requiring both while lodging itself in the middle. Striation and smoothness are never encountered in isolation but always interfolded ('smooth or nomad space lies between two striated spaces'[40]).

---

36   Goodrich *et al.*, 1994: 22
37   Deleuze and Guattari, 1988: 28
38   See e.g., Vergunst, 2010, or Gibas, 2013, on how spatially embedded history becomes part of corporeal movement in space.
39   Deleuze, 2004c: 50
40   Deleuze and Guattari, 1988: 424

The body itself, sometimes striated, sometimes smooth, generates its own space and moves with, towards, through, parallel to or against other bodies. Or space itself, folded with the body and part of the broader assemblage, enables the body to slide or to get boxed in, to move in *glissando* or in convulsions. This can be the static movement of the nomad, or the striated, pillared movement of the state. The difference is not unlike that between a game of chess and a game of Go:

> in chess it is a question of arranging a closed space for oneself, thus of going from one point to another... In Go, it is a question of arraying oneself in an open space, of holding space, of maintaining the possibility of springing up at any point ... The 'smooth' space of Go as opposed to the 'striated' space of chess. The *nomos* of Go against the State of chess.[41]

In both games, the board folds in around the piece and determines individual will and freedom (which, in turn, is behind the choice of the board, which is behind the board itself, *ad infinitum*). Static or spreading, deterritorialising or colonising, desire for space generates spaces of desires. Each body wants to carry on being and becoming. *Spatial justice is the quest for a conative space, that is a space where an autopoietic, self-maintaining body can unfold.* Whether on paths of striation or points of nomadic smoothness, a passage is performed. There can be no moral judgment on the continuum of spatial horizontality: this judgment belongs to time, historical responsibility, confirmed causalities. Spatial movement cannot contain moral judgment because there is no depth in its haptic horizontality, no visible horizon on which the origin or the destination can be carved. We find ourselves thrown here, emerging human bodies and ethnic groups and animals and industry and oceans and cyborgs, and in that simultaneous horizontality, we desire. This is the beginning of spatial justice then, right in the middle between striated and smooth, the desire of bodies to claim space. At the point of tremor on the passage between crossing over and staying put, spatial justice emerges. But this is not a tranquil passage. When the lines conflict and the bodies clash, when a geopolitical presence is not tolerated, when two peoples are forced to 'share' the same space at the same time, when the industry moves into the forest, when the ship moves into the fish stock, when the tsunami moves into the village: there is conflict. Spatial justice is the movement out of this conflict while delving deeper into it, in the thick of the lawscape. It is the excess whose line of flight returns in the middle, right where it begun, in the thick of *logos/nomos*.

---

41 Deleuze and Guattari, 1988: 389

In their own way, both *logos* and *nomos* attempt to trace desire. Striated movement is controlled, but always contains resistance, revolt, indifference, other routes. Smooth, nomic movement is not controlled by the state but by its own stasis, its own *non*-movement, its swirl. Thus, when Deleuze and Guattari explicitly equate the nomad with the in-between (the 'intermezzo'),[42] they seem to encourage the location of justice on precisely this nomadic in-between, the nomad herself. However, I would suggest that the in-between of relevance here is precisely the in-between of the in-between as it were, the *inter-* in the intermezzo: rather than just of the nomad, the space of justice is located in the fold between smooth and striated, nomad and state, *nomos* and *logos*. Justice needs the calculation of the law and the there-inscribed possibility of passing over to the smoothness of *nomos*. Justice relies on law's calculability,[43] that is to say on the lawscape, as it would rely on a trampoline which enables one to jump higher, but on the understanding that one will return to the trampoline (only to jump again, and so on). In the same way, spatial justice is in the middle of desiring bodies, of conative spaces, of lawscaping practices, of atmospheric enclosures.

A necessary clarificatory digression: it is important not to conflate spatial justice with the in-between in the sense of third-space, third term, the other of the dialectic, or Hegelian synthesis of thesis and antithesis. To start with, neither *logos* nor *nomos* occupy convincingly either of the two positions. As explained, they are constantly interdigitated, not dependent on each other but folded within each other, so that there can never be a smooth or a striated space as such. At the same time, the emergence of spatial justice is not a way of conciliating the two, or finding some neat synthesis of how the inner strife between *logos* and *nomos* is to be played out. This easy association points to a larger, deeper problem. By now, the various attempts at carving a space in-between have become central in their marginality (if the pun is allowed) and consequently have lost their radical potential. In-betweens have become co-opted hubs of capitalist pacification. We can all fit in the limbo of the imaginary outside, which assuages the damned when the latter are faced with the impossibility of a Hegelian synthesis. By fitting in the third space, however, we are embodying the synthesis at which dualisms precisely aim. Marcus Doel writes: 'the addition of a third space does not disable dualism: it merely opens the metaphysics of binary opposition to a dialectical resolution of contradiction.'[44] The production of an in-between is the dreamland of the dualist contradiction, in which both thesis and antithesis retain a stake while ostensibly becoming surpassed, resisted, refuted. The fluidity of the in-between quickly solidifies despite intentions

---

42 Deleuze and Guattari, 1988: 419
43 Derrida, 1992
44 Doel, 1999: 120

to the contrary. How often has the world witnessed much-hailed in-betweens becoming simplified spaces of oscillation, political negotiation, circumstantial manoeuvring and self-congratulatory compromise? Are we not governed by self-propagating in-betweens, be this socialism, neoliberalism, sustainability, human rights, international law, happiness indicators, big society resuscitations, ethical consumerism and so on? Governed by distance rather than affinity, void rather than continuum, antithesis rather than a non-relativistic multiplicity of positions, dualism's deeper problematic structure is not exclusion but hyper-inclusion. The in-between becomes normalised as the inclusion of resistance, the important third party that brokers the agreement, the centre of power where power is diffused and control appears dissimulated as non-power. The in-between is the apex of atmospherics.

All this, however, is of lesser importance when compared to what I consider to be the main pathology of the in-between: by offering a solution to the problem of dualism, an in-between reasserts and firmly establishes the dualist position. Even when the in-between genuinely manages not to become a synthesis, but stubbornly maintains its focus on the emergence of a differentiated space, the initial rupture is taken for granted. Interestingly, the proliferation of in-between spaces has done little to alleviate the initial dualism. Rather, the various in-betweens form part of the phenomenon of the politics of centre/periphery, whose aim is to reduce the antithesis into an in-between, which ultimately, however, turns against itself and self-cannibalises in a cloud of political apathy. In short, by accepting the possibility of an in-between, one accepts the ontological priority of the dualism and its subsequent epistemological necessity of a synthesis, however materialised. In fighting Hegelianism, the in-between manages to be Hegelian through and through.

For the above reasons, spatial justice must not be reduced to a position between *logos* and *nomos*, striated and smooth, or indeed one body and another. Even if the apparent duality of the two bodies can be salvaged quite easily through the clarification that we are always talking about multiplicities, and in most cases the body in question is a collectivity claiming space from a multiplicity of collectivities; even in this case, the expectation that justice is the solution to the legal problem of conflict subscribes to the classic dualist thinking of synthetic judgment. Avoiding dualism and its implicit expectation of third-space solution, has proven to be hard even for Deleuze and Guattari, despite their frequent and explicit denunciations of dualist thought. But one has to resist falling into the trap. This should be fought even on the level of instrumentalism: to use dualist thinking in order to counterattack it (indeed what Deleuze and Guattari seem to be doing in *A Thousand Plateaus*) can also be found guilty of espousing dualism in the sense of the a priori separation. Indeed, more than dualism itself, the problem is the acceptance of a rift that needs/must/can be mended.⁴⁵ Thus, both aforementioned foundational problems of dualism risk re-appearing from the back door. So, rather than synthesis, spatial justice is an emergence (which means, it lies beyond prescription, controlled mechanics and systematic articulation of the result). Rather than originating in a dualism, spatial justice emerges from within a multiplicity (which means, it is not an oscillation between two opposing poles, but an often arbitrary picking of various positions that form a surface on which one moves). Rather than an outcome in the sense of causal link between legal and corporeal movements, spatial justice resists causality. Further, it also resists attribution, namely post-facto causal linking that takes place on a virtual plane, itself potentially co-opted by its own striation. Finally, spatial justice emerges properly speaking *in the middle*.

Unsurprisingly, one must be careful not to confuse the middle with the in-between. The middle is the space of revolt against the usual tools of origin, centre and boundary: 'One never commences; one never has a tabula rasa; one slips in, enters in the middle; one takes up or lays down rhythms.'⁴⁶ Just as the grass has no one root, central part or limits to its expansion, in the same way to begin in the middle is to find oneself folded between the multiplicity of the world without a discernible origin, a specific centre and determined territorial limits. This is the challenge of spatial justice: to negotiate a contested space without taking recourse to origins, central commands, outlines. To be thrown into the mobile multiplicity of the grass is to follow the blades waving in the wind: spatial justice is required to understand bodies as posthuman assemblages, that is with their political, personal, legal, religious, technological elements, moving on a continuum of differentiated power relations.

Grass is opposed to the tree with its defined root, trunk and volume: 'arborescent systems are hierarchical systems with centres of significance and subjectification.'⁴⁷ While this is one side of the continuum, the other side is its constant rupturing, partitioning, lawscaping, tilting. Deleuze and Guattari urge us to 'make rhizomes, not roots, never plant! Don't sow, grow offshoots!'⁴⁸ Rhizomes specifically encapsulate the ideas of horizontal, trans-species, heterogeneous growth. Not a linear, vertical construction but a surface where any modulation is absorbed, closed in and eventually spread in lake-like smoothness. However, rhizomes have been routinely fetishised in the literature as the way to guarantee openness, flexibility, flatness and contingency. But rhizomes are also co-opted, overcoded and used

---

45 On this see Dolphijn and van der Tuin, 2012, and especially Chapter 6 'Pushing Dualism to an Extreme'.
46 Deleuze, 1988: 123
47 Deleuze and Guattari, 1988: 16
48 Deleuze and Guattari, 1988: 24

in ways that go against the very idea of rhizome.[49] Received legal histories, prefabricated political positions, historical origins and facts that have been maintained as affects of spatial and temporal nostalgia or claims for reterritorialisation make use of rhizomes and their affective way of spreading. We are all encased in atmospherics of legal and political engineering which spread imitatively, rhizomatically. The space in the middle is not always open and possible. It often succumbs to the rhizomatics of atmospheres. This is the struggle: the middle is neither necessarily 'good' nor 'bad', positive or negative. In its point of folding, the tilting surface gathers speed, becomes vertiginous, with bodies sucked in or centripetally deracinated. It is not an easy space to be in, and no readily available moral hook is there to orient us. Rather, the space in the middle is precisely *in the middle*: neither this nor that side; but then again, not a boundary and therefore not flanked by sides. Likewise, it offers no direction: just as the leaves of grass move with the wind, the space in the middle consists of the encounter between the grass and the wind. Finally, the space in the middle offers no chronology and no external causality: all is interfolded in simultaneity and immanence. The wind becomes the grass. The grass becomes tomorrow's grass. Its beginning is in the middle, in the space of here, manically flapping around its movement.[50]

Yet, the middle can be thought of as the space of struggle. It is a space of *encounters* with other bodies, a space in which one's body affects and is affected by other bodies. An encounter for Deleuze pushes the encountered parties off their comfort zone of categories and identities, and throws them in a 'mad becoming.'[51] The grass becomes wind and moves along the wind's breath, and the wind becomes grass and spreads itself on the ground: becoming itself is pushed deeper in the middle, as it were. To quote Elizabeth Grosz, 'one "thing" transmutes into another, becomes something else through its connections ... Fingers becoming flowers, becoming silver, becoming torture-instruments.'[52] This is the crux of the struggle of spatial justice: to allow becoming to take over without, however, succumbing to atmospheric demands that make one desire spaces not needed for one's conative maintainance. False needs make for false becomings. In encountering other bodies, 'the body exceeds the body or fails' as Michel Serres writes.[53] Bodies need to move, to exceed their own

---

49 Michulak, 2008; Joseph, 2013
50 Deleuze and Guattari, 1988: 10: 'The orchid deterritorializes by forming an image, a tracing of a wasp; but the wasp reterritorializes on that image. The wasp is nevertheless deterritorialized, becoming a piece in the orchid's reproductive apparatus. But it reterritorializes the orchid by transporting its pollen. Wasp and orchid, as heterogeneous elements, form a rhizome.'
51 Deleuze, 2004a: 141
52 Grosz, 1995: 184
53 Serres, 2008: 307

body, not in order to colonise other spaces but *to not fail.* This movement is static, fully corporeal, fully spatialised, aiming to pass into another state of the body, in a differentiated assemblage, in a different lawscape.

I conclude this section perhaps counter-intuitively by quoting from Maurice Merleau-Ponty's phenomenology of the body.[54] For Merleau-Ponty, the world and the body are contiguous. This contiguity finds its expression in the concept of *flesh*, namely the chiasmatic, reversible and elusive folding between self and other.[55] Flesh appears in Merleau-Ponty's unfinished final book *The Visible and the Invisible*,[56] which contains his eventual reply to Cartesian dualism and Hegelian synthesis, and more particularly in his final notes which were never to take the form of a full text. (But perhaps this is the destiny of all battles against dualism: sketches of flights, parenthetical half-sentences that resist the structure of textual dualism between communicating and understanding.) The following quote is from his working notes on the book:

> Position, negation, negation of negation: this side, the other, the other than the other. What do I bring to the problem of the same and the other? This: that the same be the other than the other, and identity difference of difference – this 1. does not realize a surpassing, a dialectic in the Hegelian sense; 2. is realized on the spot, by encroachment, thickness, *spatiality*.[57]

The paradox of spatial justice can never be resolved through a tidy theoretical synthesis, but through a continuation of the ambiguity that is 'realized on the spot, by ... *spatiality*', namely a superposition of desire onto itself in a spatiality that contains the impossibility of simultaneous emplacement of same and other. Spatiality is both the cradle of the claim and the locus of resolving it. Rather than being subsumed to a solution, this resolution escapes the oscillation and moves on a line of flight of difference of difference. This is the space of emergence of spatial justice. Its thickness and its continuous desire for superimposition, escapes the dualism. It does not fly towards a third space, but remains still, withdrawn from surpassing, emplaced 'on the spot'. Spatial justice can only emerge right *here*, as a continuation of the negotiation yet withdrawing from it. Space is layered, superimposed, thick: spatial justice allows one to find another layer of

---

54 My frequent urges throughout this book to move away from phenomenology notwithstanding, I find that Merleau-Ponty's corporeal-spatial thought has moved in new directions that attempted and sometimes succeeded in doing away with some of the problematic phenomenological distinctions.
55 For an application of flesh to law and space, see Trigg, 2008
56 Merleau-Ponty, 196858 See Chapter 2.
57 Merleau-Ponty, 1968: 264, original emphasis.

negotiation by remaining immanent, emerging in the continuum of the lawscape. This is the gesture of spatial justice: we move by standing still. We are changing layer.

5.3 A rupture in the continuum

The lawscape is a continuum.[58] The multiplicity of lawscapes are folded into the continuum of the lawscape. Each lawscape fractally reproduces the in/visibilisation of the lawscape, itself an interplay between *logos/nomos*. There is no essence here: lawscape is the manifold emerging from the in/visibilisation of law and space, like a sound equaliser screen whose values go up and down depending on the song without, however, a main button to control what gets in/visibilised. Lawscape is an emergence from the various micro and macro in/visibilisations performed by the bodies that constitute the lawscape. Space and law are fully embodied in the bodies of their emergence. Bodies carry law and space, indeed generate law and space, through their moving on the lawscape. Just as any surface, the lawscape is tilted. Bodies fall more readily onto other bodies, and the sliding is more easily allowed by the terrain. Some bodies weigh more than others, are more powerful than other bodies and force the latter out. The conflict of bodies carries on, mostly on unequal footing. We owe this awareness to the new feminist thinking that has come from a material engagement with bodies, elements and spaces, and which the given flat unhierarchical surfaces of new materialisms were once again reproducing some masculine imaginary. Astrida Neimanis puts it excellently when she says that the 'challenge to a hierachized ontology of bodies cannot, for posthumanist feminist positions, result in a flat ethics: non-hierarchy does not mean homogenization; difference and differentiation are still fundamental and necessary facets of embodied being and relationality.'[59] Posthumanist feminism sees the continuum as a tilted surface where bodies have unequal power. But, as Spinoza reminds us, we ignore what a body can do.[60] We do not know the body's limits, capabilities, power. We can try and predict how it might affect and be affected, but cannot tell for certain. There is always space for initiative, waking up, revolution, deceit, vengeful memory or plain inertia. There is always room for surprises, for ruptures in the continuum.

Spatial justice is a rupture of the atmosphere. It springs up, sometimes invited sometimes not, always in rupture to the atmosphere and always from within the lawscape. It is a rupturing continuation. It does not come from outside (what is that?), it does not transcend any lines, it does not make the lawscape wait. *Spatial justice is always here, always now, always from*

---

58   See Chapter 2.
59   Neimanis, 2014: 14
60   Spinoza, 2000

within the law, embedded in the lawscape. At the same time, spatial justice also ruptures the continuum of the lawscape. Spatial justice reorients the lawscape, and so doing, ruptures its given orientation, its 'order and orientation' to recall Schmitt.[61] A clarification is needed here: all ruptures are ontologically part of the continuum. We have already encountered several kinds of rupture in this book: ruptures as folds within the lawscape that envelop bodies, occluding them while unfolding their potentialities; ruptures as dissimulations of the lawscape as atmospherics; ruptures as proliferations of the lawscape that lead to a superimposition of lawscaping; ruptures as ontological conditions of withdrawal. Whatever the mechanism of rupturing, they all share the following characteristic: they distinguish between an interior and an exterior. This is not the same as the inside and the outside. Within the continuum (which is a continuous inside or outside, depending on what you will), flutterings of exterioriority are created because of the *necessity* for an outside. Referring to Deleuze's work on the fold, Elizabeth Grosz writes, 'what is truly radical in Deleuze's understanding is his claim that this outside must be thought itself, or perhaps even *life* itself.'[62] The outside conditions the inside, displaces it, renders a pure effect of the outside's contortion. The outside is the virtual condition of the inside, and both are immanent to the real, according to Deleuze. Yet, the only way that the outside can become ontologically relevant is through the *exterior* that a rupture produces, notwithstanding the fact that a rupture is only a mechanism of *illusionary* breaking up of the continuum. The outside only becomes relevant through the illusion of distinguishing between interior and exterior. Spinoza has shown in his treatise of fictions that these are necessary actions that promote understanding through a continuous check between imagination and rationality.[63] Illusionary ruptures are extensions of imagination (a form of knowledge for Spinoza) that allows a body to deal with its ignorance of conditions. In the context of hyperobjects (the continuum in our case), Timothy Morton writes: 'the gaps and ruptures are simply the *invisible presence* of the hyperobject itself, which looms around us constantly.'[64] I would go even further. The hyperobject looms *through* us rather than simply around us. We bear the hyperobject in our bodies and our assemblages with the world. We are the hyperobject. We embody and generate its continuum and its ruptures and we determine when its continuum might be too much for our bodies to bear. We invisibilise the very continuum by rupturing it: we partition air into glasshouses, fluid into exclusive zones, ground into territory. We chop up the lawscape in manageable bites of ruptured continuum: rooms,

---

61 Schmitt, 2006
62 Grosz, 1995: 134 added emphasis
63 Spinoza, 2000; see also Gatens and Lloyd, 1999, on the role of imagination.
64 Morton, 2013: 76

routes, holiday spots, militarised borders, retreats, friendships, relationships, future, memory. We need ruptures.

This need only vaguely dissimulates a larger need: to partition the ontological continuum into epistemological ruptures. Atmospherics do precisely that: the creation of the illusionary exterior is sometimes more important than that of the interior. We hide behind small dents on the surface of the continuum in order to pretend that we are outside. 'There is no outside! But we forget this ... How lovely it is that we forget!' sings Nietzsche.[65] We need to forget and so we bring epistemology forth. The most fundamental question, 'How do we know what we know?', is answered by partitioning our observation plateau from the observed, bringing forth the parallelism between matter and idea, observed and observer. We the observers/partitioning line/they, the observed. We are at a distance, so we have an overview. But Niklas Luhmann cautions: every observer is always already observed.[66] Systems observing observing systems. All observation is self-observation, and all observation is a fractal part of this continuum that connects while keeping apart. The physicist David Bohm calls it *implicate order*.[67] Deleuze and Guattari call it *the plane of immanence*. Luhmann calls it *world society*. Spinoza calls it *nature* or *god*. There are many names for this. What is important is that most of these, at some point sooner or later, succumb to the need of (mostly illusionary) partitioning. This is what Timothy Morton calls *the rift*: the rupture between the ontological continuum and 'its appearance-for another thing, or things.'[68] All ruptures are part of the rift, a defence mechanism that we, conscious human beings, need in order to deal with the immensity of immanence of life.

Ruptures allow the impression of solid emplacement. For this to happen, however, ruptures need either to be collective or to mobilise a collectivity. Gatens and Lloyd call this 'the collective dimensions of imagination.'[69] This perhaps does not need clarification, in view of the fact that a body is always a collectivity of material and immaterial bodies, while at the same time is part of ever larger collectivities. To recall, a body is both collective as an assemblage and withdrawn, and its presence in other collectivities indisputable, indeed ontologically necessary, while at the same time never whole. The mobilisation of collectivity needed for rupture (and spatial justice as rupture) does not deny withdrawal but *mobilises*, strategises it even if not always in an obvious manner. The mobilised withdrawal of collectivity does not have to be a 'united front' or a worked-out juridical agreement. It can be, and increasingly is, an emergence that lasts for as long as it does.

---

65 Nietzsche, 2005: 175
66 Luhmann, 1998
67 Bohm, 2002
68 Morton, 2013: 78
69 Gatens and Lloyd, 1999: 39

Ruptures are by definition relatively short-lived. They allow a body to hook on them, fold into them, suspend its movement or simply hide behind them. More importantly, they add verticality to the continuum. They open up levels that were not immediately apprehended. Ruptures visibilise a different sedimentation of the lawscape – still part of the lawscape but perhaps slowed down, covered up. In rupturing the lawscape, spatial justice brings the lawscape forth, turns it inside out, throws its entrails into view. It reassesses it while reasserting it. Spatial justice remains an emergence, just as any rupture – it does not provide a position *against* the lawscape but a position of reorientation from within. It employs rupturing but does not aim at dismantling the lawscape. It might occasionally offer the *illusion* of an opposing position to the lawscape, and this might well be enough. This is similar to what Saskia Sassen has called *new assemblages*, namely temporal re-organisations (re-orientations?) of territorial fragments under different legal logics.[70] Whether this is new is debatable. Such assemblages take place at all times, sometimes registered sometimes not, sometimes revolutionary but often just for the system's continuation that needs to reinvent itself. Whatever it is, as I show below and have repeatedly asserted, there is nothing outside the lawscape, except another lawscape, itself part of the fractalised lawscape. Spatial justice is a rupture that delves deeper into the lawscape.

Not all illusions are the same. Atmosphere is the grandest of all illusions, engulfing one fully and stifling even the desire to get out. Not only because of the directed affects, the invisibilisation of the lawscape and the occlusion of the law, but because of its ample offering of spatial justice. *Atmosphere brims with spatial justice.* Every body is emplaced in full accordance to its desire. A body wants to consume and feel safe in the mall. A body wants to carry on fighting for freedom on the square. A body feels safe in a natural reserve. A body wants to belong in a glasshouse as vast as the word, as continuous as life, as opaque as freedom. A body is even made to believe that a camp is the only place where it messianically belongs. A body is offered all that. Space is generated as a direct embodied desire to be just *here*. This is the city of justice where the future is captured in an eternal present, the past is spatially engineered to tilt in a specific way, and the bodies move or stay still according to their desire. This is the grand atmospheric illusion, where desire is one, undifferentiated, decontextualised, floating in the formaldehyde of stability. This is the atmospheric stage where the body becomes the Lyotardian moebius strip that vertiginously turns around itself and the theatre where its desire is being compulsively performed.[71] Bodies become isolated in atmosphere, only communicating

---

70 Sassen, 2006
71 Lyotard, 1993; see Chapter 3.

with the grand glasshouse around them and made incapable of apprehending their position in relation to their assemblage. This is the atmosphere of Human Rights, of Sustainable Development, of the Rule of Law: all of them vast glasshouses echoing the isolation of the bodies trapped within, unable to communicate with each other but merely mediating their breath through the atmosphere. Atmosphere removes responsibility. You did not find me, I found you. Blame me. This is what the man on the telephone, pretending to be (and demanding to be called) 'Officer Daniels' in Craig Zobel's film *Compliance* says to the fast-food chain manager when he makes her strip-search her employee, allegedly suspected for theft. This is what participants to the Milgram experiment do when asked by a doctor to administer a fatal electric shock to the people next door.[72] Responsibility is removed from the assemblage and passed over to a centralised point of atmospheric engineering. Yet, at the same time, responsibility is seemingly fully accepted because of the driving desire that feeds in the conative nature of atmosphere. I want to carry on, says the body. I want to sexually assault her, thinks the man who is put to 'guard' the suspected fast-food employee, when 'Officer Dean' encourages him to do so. The psychoanalytical Father appears everywhere in the atmosphere, but nowhere entirely, nowhere totally visible. He appears in small, bite-size effigies, easy to ingest but with gaseous aftermath. Atmosphere is the perfect enclosure because it denies the possibility of fighting through the rupture of spatial justice. In offering the illusion of spatial justice, atmosphere excludes spatial justice, retaining its allure for as long as possible, as long as the bodies desire it. Atmosphere stops the most important emergence: that between the lawscape and spatial justice. One needs to break the atmosphere in order to move and to begin claiming one's position. *Escaping an atmosphere is a gesture not of direct conflict (to recall, conflict is inscribed within the atmosphere), not of generating another atmosphere (this is mere a passage between atmosphere, an engineered becoming) but the counter-intuitive gesture of withdrawal.*

This does not mean that atmosphere is always incarceration. There are atmospheres that are conducive to the emergence of different lawscapes. Think of the Triveneto transhumants and the way the emerging atmospherics were a defence against the surrounding lawscape of regulation, categorisation, spatial partitioning. The bodies worked from within their own lawscape, one of withdrawal, spatial multiplicity, animality and hunger, in order to engineer an atmosphere that *excluded* the non-transhumants. Is that a 'good' atmosphere? Yes, no. It serves the purpose of the transhumant, but it occludes the efforts of the state and private owners to control them. Is that 'good' overall? I am not sure and perhaps I am not interested in asking this question, since it is patently the wrong question. Every 'good' is to be seen in its own context and own position in the assemblage. So, just as

---

72 Weizman, 2012

Human Rights, Sustainable Development and the Rule of Law made an appearance in the preceding paragraph, in the same way they can emerge here too. This is neither subjective, nor individualistic. It is emphatically collective and cannot be captured by subjective apprehension. The judgment must take into consideration the way it affects the assemblage as a whole. Private owners or the local police are also part of the assemblage, but they are positioned further away from the atmospherics of the transhumants. This means that each body has a responsibility of position with regards to the assemblage of which it makes part.[73] This is Spinoza's answer to what is good and evil.[74] Nothing overarching, universal, a priori. Always a question of power, in the sense of following one's *conatus*, itself always contextualised in the broader collectivity, rather than trying to find a categorical imperative and stick to it. Contrary to usual understandings, responsibility does not emanate from the subject or even the body of the subject, but from the body's situatedness in the assemblage, its orientation and possibilities of movement or stasis: no categories, no individual understanding, no moral impedimenta. Responsibility is always situated, micro-felt and micro-acted, in the space within and among the bodies. Always mediated by the larger lawscape in which one is positioned, and judged according to the strength that the bodies acquire at every encounter with other bodies: neither cultural relativism and easy pluralism, nor uniformity and centralised control. There is judgment that unfolds slowly and locally, yet holding onto the vast ontological continuum in which it takes place.

This is why we should not rush into judging, say, the shopping mall, consumerism or need for security as 'bad' atmospherics. Each assemblage is different. Shopping malls are found to be a good place for women in Muslim countries because they make them feel safe;[75] consumerism animates cities and is often at the basis of what urban life is; zoos do preserve species and further research;[76] gated communities might allow one to breathe, however air-conditioned the air inside. The issue is of course always beyond the atmosphere and behind the need for atmospherics: why should women not feel safe outside the mall in the first place? Why should there not be more free urban activities? Why should there be a need for gated communities? All are true but all point to a different reality, a reality elsewhere, a transcendental hope in time or space. This is a widely accepted understanding of justice: nothing less than perfection, heaven, finality. But here I am arguing for something else, something that cannot wait, that is acted upon here and now. Spatial justice is not about another world. It is about reorienting the current world in order to allow a

---

73 See Bennett, 2010, who sketches a similar withdrawal from potentially problematic assemblages.
74 Spinoza, 2000: *Appendix to Part I*
75 E.g., Erkip, 2005; for non-Muslim countries see Lewis, 2007.
76 Although see the political and legal repercussions in Braverman, 2013.

different lawscape to emerge. It is about withdrawing while asserting our position in the world.

5.4 Withdrawal

Spatial justice unfolds through withdrawal. Contrary to what might appear, withdrawal is not a moral indictment, nor does it refer to withdrawing from the claim of the other body to occupy the same space at the same time. Withdrawal is not yielding to the other body, just as spatial justice is not a retreat. As I have said, spatial justice is not a solution but a question that tries to address the problem of the impossibility of simultaneous emplacement of more than one body in the same space. It begins from the violence of space: all space is violent on account of the fact that only one body (assemblage) can occupy a specific space at a specific time. This is the mirror concept of what Leopold Lambert calls the body's necessary radical choice: every body must necessarily occupy a space at every time.[77] Every body is obliged by its nature as both intensive (symbolic and ideal) and extensive (corporeal and spatial) to occupy a certain space at a certain time. Bodies partition the spatiolegal while inhabiting it. But not all parties, or more broadly all bodies involved, can find justice at the same time. Justice partitions itself over time, but can only emerge at a specific space. The violence of space requires a different understanding of justice, one that can accommodate different corporeal movements.

Spatial justice unfolds through the withdrawal from an atmosphere. Each atmosphere includes and constructs disagreement, conflict, even resistance, often put in the service of the atmospheric conatus itself. Spatial justice presupposes a withdrawal from such a construction and importantly from atmospheric becoming-other. Withdrawal leaves behind the full enclosure, with its co-optations, inclusions of exterior, and actualisation of virtuality. But where does one fall into, when one withdraws from an atmosphere? Unsurprisingly, one lands in the lawscape, simultaneously the same that gave rise to the atmosphere, and a different one; part of the lawscape continuum yet ruptured in order to create difference. This lawscape might be the locus of the hope of emergence of spatial justice, provided that in withdrawing, one *reorients* the lawscape, thus changing the register of conflict. *Spatial justice emerges when a withdrawing body passes into a differently oriented space in which the register of conflict might change.* There is no guarantee and no prescription that spatial justice will emerge, even if the lawscape is reoriented. Spatial justice is entirely contingent upon the violence of space, brimming with unequally strong bodies vying for the same spatial emplacement, same resources, same privileges, same temporalities.

---

77 Lambert, 2011

Eyal Weizman's work on the architecture of Israeli occupation in Palestine focuses on what he calls elasticity, namely the way matter and space are imprinted, formed by and continuously altered by politics and law.[78] Elasticity does not mean flexibility. The structures reproduce the standard distribution of power among parties. The Separation Wall is the most iconic structure of elasticity. Far from being elastic in the standard sense of the word, it intransigently accommodates the oppressive and exclusionary desire for a sense of safety akin to a gated community of Israeli atmospherics. To withdraw from this atmosphere and reorient the lawscape of its emergence, is one of the hardest spatial exercises that anyone could engage in. Yet, through his architectural practice, Weizman attempts just that:

> the question is whether there is a mode of action that might contain the possibility of a break rather than the constant elasticity of material organisation and political evolutions … how this interaction on the same plane can actually create a new plane, can lead to a transformation, or a phase transition – that is, something beyond the rules of the game that already exists.[79]

The question is how to withdraw from the elasticity of atmosphere that inscribes history on stone according to the desire of its participant bodies, (whether Palestinian, Israeli, geographical, geopolitical, resource-driven, historically-originating, human and nonhuman) and move to *another* plane, a new lawscape that is reoriented in such a way that transformation might be possible. This is the movement of spatial justice, a transformation rather than mere amelioration, but a transformation that can only happen through reorientation. Jean-François Lyotard has put it memorably with his concept of the *différend*: 'A case of *différend* between two parties takes place when the "regulation" of the conflict is done in the idiom of the one of the parties, while the wrong suffered by the other is not signified in that idiom.'[80] Withdrawing from the atmospheric plenitude of military oppression, biopolitical control, spatial partitioning and symbolic/linguistic propaganda aims at a new game, with new rules and new hopes.

Withdrawal has nothing passive about it. As it becomes obvious in the above case, withdrawal requires an articulated strategy that may be risky and often desperate. Leopold Lambert's work on Palestine is a first-hand look at the way bodies are made to move in the occupied space. Lambert describes the incident of the set up of a tent village by Palestinians in an area in East Jerusalem, which the Israeli government appropriated in order to build 3,000 new housing units.[81] The fragile, temporary nature of the tent village

---

78 Weizman, 2007
79 Weizman, 2010: 277
80 Lyotard, 1988: 9
81 Lambert, 2013b and 2011

contrasts with the solidity of nearby Ma'ale Adummim, the biggest Israeli settlement in the West Bank. Lambert writes: 'such a dichotomy indicates the asymmetric forces involved between state-organized militarized operations of claiming a land and an immanent encampment in which the determination is affirmed only through the presence of bodies.'[82] In the tilting lawscape surface, bodies move as best as they can, drawing on any assemblage resource they can, locally and globally. The encampment is a movement of withdrawal, attempting to cast itself on a reoriented lawscape that captures the attention of the media: 'the group of Palestinians was not really interested in creating a new village, but rather in resisting the law according to which their own land was to be withdrawn from them.'[83] Withdrawal as resistance indeed, but there is more here. There is an active withdrawal, a spatial repositioning, however temporary and fleeting. The terrain here is not justice in general but *spatial* justice: the tents reorient the lawscape, encouraging the world media to take photographs. The tents versus the solid, castle-like settlement is the orientation of the new lawscape. Unsurprisingly, the tents were presently removed by the Israeli military as a security threat.[84] Spatial justice did not emerge.

Or did it? Spatial justice is the continuation of the conflict but on a different level, a different register. The 'success' of withdrawal is measured by the degree of reorientation of the lawscape. Lefebvre refers to this when he talks about 'trial by space':[85] each assemblage (in Lefebvre's terms, 'groups, classes or fractions') mark their 'legitimacy' through their production of space. 'Ideas, representations of values that do not succeed in making their mark on space, and thus generating (or producing) an appropriate morphology, will lose all pith and become mere signs.'[86] The production though always occurs 'through confrontation with the other values and ideas it encounters there.' This means that any reorientation of space can only occur through repeated encounters with other bodies. Such encounters take place on the space of withdrawal from the atmospherics of the existing lawscape. No doubt the new lawscape needs to prove its relevance. But spatial justice as an emergence takes place regardless of the validity of the new lawscape. It is enough to reorient the lawscape towards its new validation. Vassiliki Katrivanou and Bushra Azzouz's documentary *Women of*

---

82 Lambert, 2013b: 71
83 Lambert, 2013b: 71
84 In an excellent theorisation of the Israel/Palestine crisis in terms of architectural space, Lambert, 2011, has shown both the current use and the potential use of architecture and spatiality as a weapon, that is as a means of reorienting the lawscape. Lambert puts forth a spatial idea of justice as necessity, which finds its correlate in this text, in the context of the discussion of the ontology of withdrawal.
85 Lefebvre, 1991: 416
86 Lefebvre, 1991: 416

Cyprus captures the feelings of women on both sides of the 1974 Cypriot partition, namely the North (Turkish) side and the South (Greek) side. This revolved greatly around the notion and practice of home both as a nationalist strategy and a corporeal affect. We witness for example how, in the 2004 UN referendum on the proposed reunification of the island, the Greek side government used the concept of home as the main symbol of safety for the (Greek) land posed against the potential threat of a Turkish invasion. The material aspect of home, its walls and fences, becomes a hardened geopolitical line that speaks directly to the bodies of the Greeks: indeed, the argument against the reunification was an affect, namely fear of the other side. Likewise, when the referendum results of the Greek side were revealed to be overwhelmingly against reunification, the Northern Cypriot Turkish women interviewed in the film would not hide the fact that they took it not so much as a national but as a personal wound: a body being branded with a thickset *OXI* ('no') that perpetuated their exclusion from a land which, at least for some, used to be and still is thought of as home: bodies here, homes there, thrown apart by a barbed wire that keeps on prickling skins and walls.

The gesture of withdrawal takes place, not on the border where one shows one's passport, but in the affective burden of such a crossing especially since the one that crosses even for a brief visit risks being called a traitor by one's peers. The film begins with a Greek woman crossing over in order to 'return' briefly to her 'home' namely the place she grew up and from where she was exiled after the partition. Her moment of withdrawal, and affective judgment, is when she caresses the face of a young Turkish settler (settlers came from Turkey and are in some ways the underclass of both native Turkish and Greek Cypriots). This gesture legitimises the settler and crosses a taboo line that later caused her considerable opprobrium. This was an embodied withdrawal from the atmospherics of geopolitical lines.

Every conceptualisation of home, however felt, lived, remembered, reminded, related, constructed, instrumentalised, is a direct connection between the body and the law. The law is used by both sides in order to deal with the question of spatial justice the way I have defined it here: that both you and I want to be at the same place at the same time. Our bodies want to occupy precisely the same space, whether this is material or constructed: your home is my home. Home is made of skin and tar. There was a moment in the film where spatial justice emerged – a justice not to come but right here, populating and opening up the space. A Turkish Cypriot woman crosses over to the South side and is welcomed by a Greek family now living in her previous home, physically hugging her and making her feel 'at home'. They were both making somewhat grandiloquent claims, conceding at least emotional property of the home to each other: it is your home, no it is your home. Both parties were retreating from the here of the home. Yet at the same time, they were rather amusingly suggesting that they could essentially occupy the empty house next door, and make it her new/old home – a little gesture of withdrawal from the law, a going-against the law and an opening up of a space of justice right here, next to the lawful property. In withdrawing from the given law, spatial justice claims a space that transgresses the barbed wire and brings together bodies, spaces, gardens ecologies, discourses.

The women were claiming the law as a way to come out of their nostalgia, either in order to move beyond the traditional notions of what home is, or to return to what they always thought of as their home. The women in the film embrace the law as a corporeal right just as they embrace each other tightly; they chop the law to pieces and serve it at lunch with people from the other side; they step on it to cut lemons from what used to be their own garden; they ask the law for a pardon, a gesture of reconciliation. The law is not simply the law of property that distributes or the right to vote at the referendum, but an embodied affect, a feeling formed in the space between the caressing hand and the stones around it: I belong here. For the woman in the film, the law is luminous and bituminous at once, and whatever happens, they return to the law.

In perhaps the most telling scene of the film, two women are bent over a coffee cup reading their future – a universal act of divination in Turkey and Greece and one of the most gender-characteristic moments of the film. The coffee cup contains their gazes, both focused on a space of justice rooted in the here, the coffee marks and the interlocking hands; yet at the same time looking out, towards the other side that has already crossed the wire and has moved here, right where they were sitting. The cup was the new lawscape, reoriented towards a different register. This was a female register. This was a future register. The house was renamed yours and mine, the cup was the space of a future that resonated right here. Space bleeds into time (it never managed to detach itself from it): the success of reorientation is measured temporally by the way the reoriented lawscape manages to repeat itself across the continuum. To recall, in the second chapter, I referred to the time of the lawscape as one of repetition of difference of each lawscaping event across the lawscape continuum. When reorientation is repeated across other lawscapes, in their turn reorienting themselves, then the *present* spatial justice comes to its own and ripples across the continuum.

In *Production of Space*, Lefebvre speaks about the 'moment' of justice as the discontinuity between monadic judgments, namely judgments made in the context of enclosures. Lefebvre uses Leibniz's monad, the perfect closure, the house without windows,[87] disconnected from other monads yet all part of the grand monadic nomadism, the perpetuity of life's motion.

---

87 The Leibnizian figure of the monad, in Leibniz, 1992, has been used by many but see notably Deleuze, 2006 and Tarde, 2012 for similar takes to Lefebvre's. The difference between these approaches is the position of the outside. In Leibniz, the outside is inside. In Deleuze, it becomes a radical fold; in Tarde, it maintains a certain outsidedness.

This is consonant with the definition of a body that I have been sketching in this book. For Chris Butler working on Lefebvre's relevance for the law, the moment of justice 'opens out of the possibilities inherent in the continual need for acts of judgement.'[88] This is important for two reasons: first, the emphasis on the spatial and temporal discontinuity of justice, as opposed to the narrative of solution entrenched in most accounts of justice, rendering it thus the only appropriate continuation to a struggle; and second, the importance of micro-judgments. These judgments must be understood as non-moral, affective, spatial, temporal, embodied, part of the ontological condition of a body as both connected in the form of an assemblage and isolated in the form of withdrawn autopoietic entitites. Such embodied 'moments of justice' are micro-ruptures of the grand atmospherics. As I have said, spatial justice is a reorienting rupture of the lawscape; to this, we need to add that it is also a series of disconnected, monadic affective judgments, a continuum of ruptures as it were. As we have seen, according to Teresa Brennan, affect is judgment. This judgment places distance between monads, takes place in the space between, and unfolds in withdrawal. Whether there is a possibility of spatial justice depends again on the reorientation of the lawscape.

The connection between justice and judgment is fraught. Deleuze puts its thus: 'there exists a justice that is opposed to all judgment, according to which bodies are marked by each other, and the debt is inscribed directly on the body following the finite blocks that circulate in a territory.'[89] This might sound brutal, and it is indeed inspired by Antonin Artaud's theatre of cruelty, as well as a Nietzschean understanding of violent justice as corporeal compensation of the debt inscribed in bodies. However, it is arguably less cruel than the machines of judgment that fix bodies in inescapable hierarchies that reproduce either the oppression of a better place outside, or the atmospherics of lack of debt and responsibility. Negarestani argues for the 'necessity of a philosophy of cruelty in the wake of an interminable cruelty',[90] structuring ontology *in toto* as a mesh of elaborate cruelty, which cannot anticipate less cruelty. For Negarestani, following Deleuze, 'in the wake of the philosophy of cruelty, ethics can return to the mathesis of the problem once again wherein the problem is not determined by its solution or conditions but by its capacity to generate fields of the problematic.'[91] This is the point of justice's emergence in the middle, that space of struggle that is significantly not a space of *judgment*, of secure values, of fixed constructions and of solutions. The middle is the space *between* and *of* the bodies that determine the negotiation (and not the solution) through their

---

88   Butler, 2013b: 6
89   Deleuze, 1997: 127–8
90   Negarestani, 2009: 79
91   Negarestani, 2009: 82

positioning. Deleuze's critique addresses a particular judgment: the divine, top-down, unmediated judgment, what we could call here atmospheric judgment. Recall the film *Truman Show* where, when Truman apprehends the atmosphere of enclosure and perfection and tries to *withdraw*, the voice of his 'creator', the eternal Father, god – namely the reality show director – booms from high up the set dome, simultaneously reinstating and allowing the atmosphere to become one of terror.

This is the judgment which Deleuze criticises. But there is another judgment, a moment of judgment indeed, which, in its small reorientation of the lawscape, manages to transform it: in what I always thought was one of cinema's most moving moments, Truman lifts his hand and touches the 'sky'. The 'sky' turns out to be an immense sheet of artfully painted paper, covering the inside of his perfect globe, the atmosphere of the glasshouse where even the glass was made invisible. The moment of affective judgment that took place between Truman and the paper sky is one of a series of judgments inscribed between the bodies of the film. The final justice moment is never given. We do not know what is outside the paper sky for Truman. There is no outside also means: what takes place here counts. Each affective judgment between Truman and the lawscape around him is completely devoid of the divine, and is instead mundane, everyday, insignificant. Each little withdrawal from the atmospherics of the American Dream is a moment of spatial justice. The connection I want to suggest between judgment and justice therefore is one of extreme banalisation for both. Judgment is not divine, and justice is not a solution. Both work in a relation of ruptured continuum, monadological moments of a nomadic universe, nothing leading anywhere certain but, under the right conditions of withdrawal, all leading to another level of conflict, where things can be negotiated and fought over through a different distribution of power.

Deleuze and Guattari understand the difficulty of withdrawal. They reserve it for their favourite figure, the Nietzschean schizo:

> the schizo knows how to leave: he has made departure into something as simple as being born or dying …These men of desire – or do they not yet exist? – … must reinvent each gesture. But such a man produces himself as a free man, irresponsible, solitary, and joyous.[92]

This celebration of justice is no doubt violent:

> a schizorevolutionary type or pole that follows the *lines of escape* of desire; breaches the wall and causes flows to move; … good people say

---

92 Deleuze and Guattari, 1983: 131; see also Arsic, 2005, for a fascinating Deleuzian reading of Thoreau.

that we must not flee, that to escape is not good, that it isn't effective, and that one must work for reforms. But the revolutionary knows that escape is revolutionary – *withdrawal, freaks* – provided one sweeps away the social cover on leaving, or causes a piece of the system to get lost in the shuffle. What matters is to break through the wall.'[93]

Desire captures; desire allows to withdraw. One withdraws by acting on desire, while at the same time acting against desire. These are of course different desires: desire to withdraw, to reorient the lawscape in a fairer way. And desire to stay in, in the familiar discomfort of the atmosphere, or the comfort of familiarity. Risk is absorbed by atmosphere: both risk as lack of security (namely, danger), and risk-taking as an activity that requires withdrawal (from the comfort zone). *Withdrawal is self-withdrawal*: a body withdraws from the space of its own desire, the one that keeps the body atmospherically conditioned. But since it is always the bodies themselves that constitute an atmosphere, to withdraw from an atmosphere is always an act of self-withdrawal. An atmosphere is inscribed on the body in the same way as the body inscribes the atmosphere: through desire. Withdrawal requires two distinctions: first, the distinction between the actual and the virtual, namely the potential that conditions the actual, and the actual in hand. This virtual is none other than the illusion of the outside, or the creation of an exteriority through rupture. This exteriority pulls the body towards it, while at the same time demanding of the body to reorient the exteriority along with all the bodies that contains (and constitute it). It requires a counterposing. Elizabeth Grosz writes: 'the activity of desiring, inscribing bodies that, though marked by law, make their own inscriptions on the bodies of others, themselves, and the law in turn, must be counterposed against the passivity of the inscribed body.'[94] The law is carried in bodies, inscribed or even embodied, eating through their entrails, partitioning their future, bottling their past, flattening their present. Bodies are law, and when they withdraw, they need to reorient themselves against their own passivity which refrains them from withdrawing and keeps them placed in an atmosphere. To withdraw is to counterpose one's body against the body of one's desire.

The second distinction is between the freedom associated with two distinct desires: on the one hand, freedom as denial of desire, and on the other, freedom as immersion in desire. This is the hardest distinction for the one who withdraws: to tell the difference between desire and desire: one cocooned in her own embodied atmospherics; the other coming through the vibrations of the assemblage in which she participates and which defines her. Colebrook, reading Deleuze and Guattari, writes: 'it is in

---

93 Deleuze and Guattari, 1983: 277
94 Grosz, 1995: 36

this excessive consumption that the body of the despot turns the circulation of goods into a means for establishing his precedence.'⁹⁵ But in atmospherics of desire, as Colebrook remarks, the despot is none else than the very same body in which desire emerges. The desiring body is equally the capitalist despot and the body that wants to carry on its conative striving. This is the crux of Spinoza's ethics: how do we know whether our desire is essential or false. His answer is that we should only go for the kind of desires that allow one to preserve one's *conatus*, one's desire to carry on being and becoming, and becoming stronger as a body through encounters with other bodies. We do not desire something because it is good for us. Rather we judge something to be good because we desire it. Telling the difference between desire and desire (joy and sadness, in Spinoza's terminology of affect) is a question of knowing the *causes* of something: why is it that we are desiring what we are desiring? How good is it for us in relation to the assemblage to which we belong? There is nothing selfish or subjective here, just a confidence in a type of knowledge that can access the deeper causes of desire.⁹⁶ On the level of the law, 'it is the distinction between grasping law as arbitrary command and law as knowledge that marks the difference between human freedom and human bondage.'⁹⁷ This is also the difference between identity and haecceity: identity is determined by external causes without knowledge of how they determine us; haecceity is determined by internal causes, by joyful affective connections with the assemblage that, with us, becomes what it is. It is not easy to know or to determine the causes at all times. This is Spinoza's definition of freedom, namely to become strong in the knowledge of the causes for which one desires what one desires.⁹⁸ But to withdraw from questioning such distinctions would be to succumb to an atmospherics of no-choice. To withdraw from atmosphere, on the other hand, is a different kind of collapse of distinctions: wall-breaching, piece-losing, flow-moving withdrawal that embodies the violence of spatial justice, a double violence. First violence: I withdraw from the space in which I am. This violence reveals the stronger and demands a differentiated but horizontal responsibility. And then, the second violence: I move on a plane of simultaneity, not on *terra*

---

95 Colebrook, 2009: 1896 'this, then, is that human freedom which all humans boast of possessing, and which consists solely in this, *that humans are conscious of their desire and unaware of the causes by which they are determined.*' Spinoza in his letters, 1992: 286, translation modified, emphasis added.
97 Gatens, 1996b: 116
98 Spinoza, 2000; see Kisner, 2011 for a connection between freedom and 'autonomy', which is not a Spinozan concept as such but it is interesting here because it points to a monadology of bodies; but also Deleuze, 1988, and Gatens and Lloyd, 1999, for their different and indeed much more complex take on Spinozan freedom as a fundamentally difficult and even dangerous process that deals with layers of knowledge that cannot be so easily distinguished.

nullius. My withdrawing displaces others around me, behind me, undesired, unplanned, invisible. And they in their turn continuously displace others and others than the others. The excess ripples through the plane but never dissipates, never truly escapes. An immanent withdrawal that does not move but rushes through – a *stasis* in the double sense of the word as both pause and revolt. Deleuze and Guattari again on the schizo who knows how to leave 'but at the same time his journey is stationary, in place.'[99] This static, immanent withdrawal is the gesture of a fully immanent justice. Its beginning, if there were to be such a thing in a continuous movement, is always in the middle, in the inner folding between, on the one hand the incalculability of the plane,[100] and on the other, the calculability of law, the pillars of striation. Spatial justice emerges from within the lawscape and ends up in the lawscape.

Igor Stramignoni invites: 'here, as elsewhere, one must start from *some* linear, measurable, calculable space.'[101] This somewhat Derridean formulation posits the beginning of the calculation of the reorientation from within the oppressive linearity of the known. Jacques Derrida famously said, 'incalculable justice *requires* us to calculate. And first, closest to what we associate with justice, namely, law, the juridical field that one cannot isolate within sure frontiers.'[102] I have read Derrida's position as a call to begin from within the law – indeed the lawscape – in order to negotiate its reorientation towards spatial justice.[103] But one goes back. Or indeed further. This is the crux of the mechanics of spatial justice – and possibly its most disappointing feature. Spatial justice ends up in the same 'linear, measurable, calculable space', the same lawscaping continuum, which is at best differently oriented. There is nothing outside the lawscape's incessant calculation of in/visibilisation, its strategy of atmospheric horizon, its own self-withdrawal. Lawscape is both the way out towards spatial justice, and the way in, to the micro-judgments of spatial justice. This is the inevitable distance from the permanence of Levinasian ethics: after justice, there comes the law. The law regulates the way to justice, in a constant oscillation that dictates the withdrawal of law before justice and equally the withdrawal of justice before law. *Law is the necessary precondition of spatial justice.*

---

99 Deleuze and Guattari, 1983: 131
100 See also Levinas, 1978 and his spatial treatment of justice in terms of those nearby and far off. On this, Manderson, 2005b.
101 Stramignoni, 2004
102 Derrida, 1992: 28. Note that the calculation for Derrida refers not to law (the calculable par excellence), nor to justice, but to their relation.
103 See also my analysis in Philippopoulos-Mihalopoulos, 2003 where justice is a suspension (another form of rupture) of the law, yet a return to it. To put it in Vismann's words, 'the primordial scene of the *nomos* opens with a drawing of a line in the soil. This very act initiates a specific concept of law, which derives order from the notion of space' (Vismann, 1997: 46). See also Luhmann, 2004.

Recall the role of private property in the transhumant movements in the last chapter: private property with its limitations, boundaries and exclusions, was being co-opted by transhumants in order to achieve their own atmospheric withdrawal. They would do this by quietly *stepping on* private property, since state property was either too fenced-off and protected or filled with highways, police and natural parks. Let us take the moment of stepping: a withdrawal from the atmospherics of private property certainly. But a moment of spatial justice? Not so certain. It could well be so, if private property is not atmospherically included as the exterior of the transhumant. That is, if the act of stepping on private property constitutes a withdrawal from atmosphere *in toto* and the remergence of the lawscape as reorienting, renegotiating space. Recall Cyril in Chapter 2, one of the students stepping on the grass near the House of Parliament in London: a slightly different stepping on, pure encounter with the atmospherics of a spatial partititioning, a corporeal withdrawal from it and a visibilisation of the whole lawscape. The connection between lawscape, atmosphere and justice is not straightforward. Not a facile 'law-leads-to-justice', but a complex becoming, a potential passage from law to justice, from boundaries to boundlessness, via ruptures; and at the same time, reassertion, reconsideration, palimpsestic building and sedimentation. Spatial justice leads to a new lawscape, itself always bound by the need for rules, by understandings of distance and propinquity populated by affective judgments. Withdrawal can only take place away from the atmosphere and inside the lawscape, hammering right at its foundations and up to its turrets. In reorienting the lawscape, one does not move outside. There is no better place outside. There is no better law, better society, better justice. It is all part of this surface on which withdrawal moves: a stratum that simply shifts rather than escaping to an imaginary exteriority. If spatial justice is to take place, withdrawal must be kept immanent. The space of justice is located here, deeper into the lawscape. This is not simply a manipulation of the law, a new interpretation or a legal stuttering. Nor does it mean that one has to work with the system. Rather, it is a denial of the law, a questioning of its relevance, its validity and even its lawfulness. Withdrawal rides the waving banner of the unutterable legal paradox: is the law lawful?[104] Take the example of a revolt. Illan Wall writes on the Tunisian crisis:

> on the streets of Tunis, Sfax and a variety of other cities, for months 'the people' cried '*dégage*' (clear out, get out). This was, first and foremost, a simple refusal of the distribution of people and things – a refusal of the situation as a whole.[105]

104 Luhmann, 2004
105 Wall, 2011

A refusal which, however, was expressed in the specific corporeal movement of clearing out from the streets. They went *elsewhere*. Where? Revolts work from outside the law in that they assume the difference in materiality between 'us' and 'them'. It is a necessary rupture, an inclusion of the exterior. Yet whatever change takes place with a revolt, it will have to be within the lawscape. Just as engineered atmospheres make use of the foundational illusion of the difference between individual and environment, in the same way withdrawal needs to make use of a similar rupture: 'this is a different lawscape! There is no law here, no old hierarchies, no telling-us-what-to-do!' The policeman is not us, and the fellow citizen is not them. The fact that the materiality of the lawscape is one with the materiality of, say, the judge, the police, or the *ancien regime*, makes it imperative that the continuum be ruptured *if* the overcoming of the (present) lawscape is to take place. It *must* become 'us' and 'them', however mendacious, transparently illusionary this rupture of the continuum might be. And subsequently, this reoriented lawscape has to be constructed as separate and new, in full dissimulation, for otherwise *withdrawal has nowhere to withdraw*. so spatial justice *through* and *in spite of* the lawscape. Lawscape reunited, lawscape dissimulated. Which lawscape is that? A brand new lawscape? No doubt, but also a very old lawscape. 'You have to keep enough of the organism for it to reform each dawn.'[106] Speak the law's language, enter the law's dreams, touch the law's extremities. Revolting is withdrawing, but withdrawing is immanent, reorienting, future-looking, constructing. One cannot achieve justice by revolting alone.

As I have already mentioned, withdrawal is not an isolated movement. First, it is a judgment, and thus part of a series of judgments that may lead to spatial justice. Withdrawal comes with its own historicity of previous spaces of withdrawal, as we have seen in the case of the transhumant withdrawal in the previous chapter. Second, withdrawal is often collective or at least involving an extended assemblage (*pace* Hollywood hero). It is a shift that mobilises the space, the bodies on this space and the legality that trammels their connection.[107] Research on the way Filipina domestic workers withdraw from the oppressive regime of Hong Kong employers by taking over the city every Sunday is instructive.[108] The city becomes an extended picnic mat, rolling over steps, pavements, tram stops, multinational corporation lobbies, shopping centres. The sheer flood of bodies everywhere is a repetition of

---

106 Deleuze and Guattari, 1988
107 Munro and Jordan, 2013
108 Hou, 2010 peculiarly has a photograph of the phenomenon on its cover but no coverage inside. I find interesting the idea that withdrawal of that sort does not even need to touch the linguistic level in order to be as effective as it is. See, however, Law, 2002 on the reclaiming of public space in Hong Kong by Filippina workers which, Lisa Law argues, makes for a contested transnational space.

joyful withdrawal, where bodies collectively appropriate the urban space and populate it with another lawscape. The collective nature of withdrawal also means that a withdrawn body can also affect the way other bodies are being positioned. This is more complex than it sounds: one can withdraw, not solely with the aim of one's own hope or strategy of spatial justice, but someone else's, indeed *for* someone else. This is what Irus Braverman's research on Israeli withdrawal from Israeli government policies has shown.[109] Withdrawal has far-fetching effects. It trammels the continuum riding on its tense ruptures. It resonates across in chaotic yet strategisable ways. This is not just a question of choice but of responsibility. In a posthuman lawscape-turning-atmosphere, human abilities to affect the way assemblages move are often more pronounced than nonhuman ones, and this is the reason for which the examples of this chapter necessarily focus more on (post)human bodies. Of course faced with natural elements, fuzzy causalities, technological meltdowns or simple chance, these abilities are weakened. At the same time, in issues such as climate change or biodiversity, the human effect is immediately and scientifically tangible. Placing the responsibility of indistinction in the context of the era of the Anthropocene, *reinforces* rather than alleviates the human responsibility to avoid situating oneself anthropocentrically, while at the same time potentially weakening the agentic ability of the human to intervene.[110] Within that space of indistinction, the human's ability to act is both very limited and potentially crucial. It is limited because the human has to deal with a spatially and corporeally inscribed past that demands positions radically different from the ones assumed presently.[111] It is limited because the lawscape in the anthropocene leaves its future imprint with every past movement, and in so doing captures future generations in such atmospheric hyperobjects as global warming, extinct biodiversity, exhausted non-renewables, technological everywares, global finance, perpetual population movements and so on. It is potentially crucial because it relies on an ontology of indistinction that works through spatiotemporal assemblages and in so doing, produces new and off-centred possibilities of human action in relation to the assemblage of which one partakes. This would be the grand ecological withdrawal from excessive consumerist practices, resource wastage, or even the main anthropocentric position of utilising the globe as a resource. The way one is expected to act, however, is not necessarily limited to standard legislative strategies. That too, but if there is anything that the lawscape shows, is that there are various spatiolegal

---

109 Braverman, 2007
110 Johnson *et al.*, 2014
111 Gatens and Lloyd, 1999: 81, are more optimistic about the possibility of repositioning of the human in relation to the past. Collectivity is the key: 'Often the political dimensions of collective responsibility involve our taking responsibility for a past in which we did not act – perhaps did not even exist.'

regimes that can operate in tandem, folded within each other, often parallel to each other, without breaking the continuum.

Withdrawal can be in the form of *stasis* as we have seen in Chapter 4, namely a pause that pulsates with revolt, with turning and reorienting. This is reminiscent of at least some discussion on resistance,[112] and especially resistance connected to space. Pile and Keith have shown how the two can be intimately connected.[113] Douzinas refers to *stasis* as resistance and a specific Athenean topology.[114] In her work on Israel and Palestine, Irus Braverman refers to strategies of everyday resistance as well as counter-hegemonic grander gestures.[115] In the context of electronic surveillance, Gary Marx has suggested 11 ways in which resistance is generated.[116] While comparable to the general mobilisation that flows from discourses of resistance, withdrawal is a particular form of resistance. Not all resistance is withdrawal. Respectively, withdrawal is not just resistance but spatial reorientation. Withdrawal constitutes a spatialisation of some of the existing resistance discourse by bringing forth its connection to spatial justice. To repeat, withdrawal is a corporeal move away from the atmosphere. This means, first, that withdrawal needs a position of knowledge with regards to the atmospherics in hand; second, that it always involves bodies in relation to the space they generate; third, that it does not exhaust itself to a resisting presence but expands to reconstruction (of the lawscape) through reorientation[117] – we must remember that a body withdrawing is law withdrawing, since every body embodies the law; and fourth, that it is intimately connected to a spatialised and embodied understanding of justice as desire. Thus, as opposed to resistance, withdrawal actively withdraws from atmospherics and its affective desire, drawn by an ontological desire for spatial justice. Importantly, it cannot be directly translated into conflict because even conflict is often inscribed within the atmospheric conatus. As we have seen, there can be no conflict of atmospheres and yet there exist atmospheres of conflict. This means that conflict is impossible within the specific atmosphere (if a non-conflictual atmosphere); or it might mean that conflict is already inscribed in it (if conflictual); or that conflict emerges as a rupture that allows the atmosphere to become-other atmosphere, thus

112 It is very difficult to generalise resistance. Like rupture, it crops up everywhere and in various guises. See Caygill, 2013
113 Pile and Keith, 1997
114 Douzinas, 2012
115 Braverman, 2007 and 2012
116 Marx, 2003
117 This is often difficult to distinguish from the proliferating discourse of resistance, and especially those forms that posit a reorientation and reconstruction. In Braverman 2007, for example, several instances of what she terms 'everyday tactics of resistance' could be read as withdrawal, since they involve a withdrawal from atmospherics of hegemony, and the opening up of a space of reorientation (which avoids thus being co-opted by atmospherics), showing how 'resistance is in itself a form of power' (2007: 360).

entrapping desire once again. The withdrawing body does not remain captured within the atmospheric becoming-other but moves itself away from the process. It does not just resist the process but reorients the process towards a different lawscaping process. Of course, as we have seen, conflict can open up a crack in the atmosphere: becoming-other atmosphere is not necessarily a smooth process. It opens up the rift between ontology and epistemology, makes an atmosphere speed up, acquire a different viscosity. All this might 'wake up' a somnabulist body and allow it to withdraw by revealing the disparity between desire and atmospheric desire. But the problem is that bodies simply carry on with the atmosphere in its becoming-other, thus moving from an atmospheric regime in which conflict is inscribed within, to conflictual atmospheres, and either stay there, attuned to the conflict for conflict's sake, or follow on to another, different but same, atmospheric regime without ever withdrawing from atmospherics. Withdrawal from an atmosphere lands a body on the lawscape, with its different margins for reorientation. And while it is always the same lawscape, this same lawscape might also be a different one.

There is an irresistible law of bodies coursing through the lawscape, and this law is withdrawal. Heidegger writes: 'once we are drawn into the withdrawal, we are ... caught in the draft of what draws, attracts us by its withdrawal.'[118] Building on Heidegger, a new theorisation of assemblages generally responding to the name object oriented ontology as pioneered by Levi Bryant and along such thinkers as Graham Harman, Timothy Morton, Patricia Clough and others, has given us a way of apprehending what they call 'objects', which, as I showed earlier in the context of atmosphere, are what I have here called bodies. So humans are objects; communities are objects; packs of wolves are objects, air, vegetables, other planets, ideologies, fears, pleasures: all objects. All are part of a constellation that holds them together while allowing them to carry on being *unified* as Harman writes.[119] Harman's work on objects from a Heideggerian point of view consistently resists to objects being either *undermined* as non-fundamental and static, or *overmined* through objections based on relationality or purely residing in the mind.[120] Harman's project is to show how objects are ontologically unified yet withdrawn from each other. Withdrawal, in that sense, is an ontological quality. Levi Bryant, also working on withdrawal, writes: 'withdrawal is not an accidental feature of objects arising from our lack of direct access to them, but a constitutive feature of all objects regardless of whether they relate to other objects.'[121] Objects withdraw from each

---

118 Heidegger, 1976: 16
119 Harman, 2011
120 These are terms Harman, 2011 uses.
121 Bryant, 2011: 32

other, indeed 'absolutely from every relation',[122] and so also from humans, since they are neither reducible to the sum of their relations (they are like Leibniz's monads), nor however are they ever fully revealed (in terms of what we can call here for simplification purposes, essence). Levi Bryant has put together Harman's withdrawal and Luhmann's autopoietic closure in order to show how every object is self-othering through its very withdrawal.[123] Every object is

> withdrawn in a dual sense. On the one hand, objects are withdrawn from other objects in that they never directly encounter these other objects ... On the other hand, objects are withdrawn even from themselves as the distinction through which operations are possible, the endo-structure of objects, withdraws into the background, as it were.[124]

This is where the concept of spatial justice comes in. Right in the middle of this double ontological withdrawal, one notices that everything is underlined by a desire: I want to be where you are, exactly there, exactly then. In order to do this, a body must withdraw. Which means that withdrawal is a desire (to move, to pass), as well as an inherent quality of bodies/objects. This can mean only one thing: *the gesture of withdrawal, indeed the desire to withdraw, is ontological: it is part of the conative quality of every body*. Withdrawal is more than a mechanism for spatial justice. It is an ontological reality, hardwired in bodies as a matter of course, of quotidian survival, of conative becoming. Withdrawal is the autopoietic reality par excellence: the autopoietic body has to carry on being and becoming, and can only do this through this double gesture of withdrawal. Spatial justice as ontological withdrawal follows the double movement. On the one hand, a body withdraws from the atmospherics of other bodies (through the desire that is ontologically driven and tautologous with conative existence); on the other, a body also withdraws from its own body.[125] The latter withdrawal has a side effect. Bryant again: 'objects simultaneously withdraw and are self-othering.'[126] Withdrawing from oneself is not mere catatonia but proper schizophrenia.

Self-othering generates the possibility of proliferating lawscapes. One needs to self-other in order to generate a different lawscape that will change the register. One needs to be bilingual. Withdrawal from oneself, from one's previous lawscape, and movement towards the next lawscape. But how much of this is ontological, namely part of the continuum, and

---

122 Harman, 2005: 76
123 Bryant, 2011
124 Bryant, 2011: 160
125 The self splits into two, observer and observed, in order to operate. This is the Luhmannian distinction between description and self-description, or indeed politics of identity and ipseity as I have defined it in Philippopoulos-Mihalopoulos, 2007a
126 Bryant, 2011: 135

how much of this is the result of a necessary rupture of the continuum? In order to generate different lawscapes, do we rely on the possibility of transformation or do we construct it? To put it differently, is the desire to move alive if one is deprived of the possibility of change? Can we maintain the hope of transformation in the continuum? Is it worth carrying on trying? I have already stated that withdrawal *must* be seen as a separate materiality, for otherwise the desire for withdrawal will die away, and with it the body itself. This is, once again, the ontological/epistemological rift that Timothy Morton was referring to.[127] For Harman, this is the difference between what he calls 'real' and 'sensual' objects:

> 'sensual objects would not even exist if they did not exist for me, or for some other agent [human or nonhuman] … The real objects that withdraw from all contact must somehow be translated into sensual caricatures of themselves, and these exaggerated profiles are what must serve as fuel for the causal relations that are impossible between concealed real things.'[128]

I have doubts whether the distinction between real and sensual is ontological, and whether indeed manages to avoid phenomenology. I would, however, suggest that the sensual is the necessary rupture of the ontology of the continuum, which needs to be perpetuated in order for the conatus of every body to carry on. A body needs to fake it. A body needs to dissimulate the fact that cannot touch other bodies, that cannot change the lawscape to a different one altogether, a lawless lawscape, a lawless spatiality of justice. A body needs its drugs that will convert the abhorrent reality of a machine that carries on regardless, into sensuality, sweet ruptures, breaks on the skin of the world.

There is another instance in which the rupture, however illusionary, is necessary: when making a judgment that aims at spatial justice from within the lawscape, the body of the judge (lawmaker, citizen, *homo sacer*) withdraws from the falling body of the law. The space of justice opens up just here, at the edge of destruction and the beginning of reconstruction – the liminality of de- and reterritorialising waves of legality. In the same way, spatial justice might imagine itself to be alegal, even illegal. This is all part of the necessary epistemological rupture of the lawscaping continuum. Spatial justice cannot be worked out except through some form of lawscape. Only the lawscape can determine the direction of the withdrawal. Unlike the desire to be in a specific space, withdrawal is potentially multidirectional, lost in the openess of space. As we have seen, there are atmospheres, themselves direct results of withdrawal, that appear fairer, closer to spatial justice. Yet, there is no a priori measure but only the

127 Morton, 2013
128 Harman, 2011: 74–5

conditions of the specific lawscape that includes the atmospherics and its outside. This is constantly obvious but it became even more obvious when I asked my students to 'map' the various affects (legal affects, but also the police, the religious authorities, the City of London corporation representatives, the business people on their lunch break, the occupiers, their cold, their boredom, their enthusiasm, their fears, their ideas about justice and so on) that were circulating in front of St Paul's Cathedral, in London, during the large 'Occupy' movements in 2011 across the globe (see also, Chapter 1). This anarchic, spontaneous tribe that embodied pure molecular politics of emergence, had created a tight lawscape of spatial divisions in which spatial justice, their own withdrawal by taking stance, would emerge and be practised. The space before the cathedral, and with it the whole City of London, London as a whole and the networked assemblage of the global occupy movement, were reorienting the lawscape – a new lawscape but still within that continuum of lawscape, where bodies move and generate law and space as they do.

However successful a withdrawal might be, itself cannot escape matter. Just as the thing it revolts against, namely the law, withdrawal is materially embodied and emplaced. It is always in the lawscape, yet it *resists* this materiality of the law. But how does one do this? This seems to be the main question of current sociolegal and critical legal theory, whether this is national, regional or international. As Eslava and Pahuja put it, 'how can we revolt against international law when it positions itself as the ultimate law in institutional, normative and doctrinal terms?'[129] How to reform and revolt? This is done through the only means available, namely the space and body of the law. We return to the law, not to conserve it and its institutional fantasies, but to reorient it. So, from within law and within matter. But there is no difference between the two: law has always been material through and through. Yet, as we have seen, law presents itself as immaterial, abstract, universal, non-geographical. Here we have yet another rupture – not in the service of spatial justice but in the service of the conative desire of the law. As we have already seen, this is one of law's greatest tricks: the dissimulation of its material. Spatial nature is both convincing and necessary, for otherwise the law could not claim access to that cudgel of cudgels, impartial, blind, objective justice. And so the myth goes. In that way, law has managed to dissimulate the fact that it is material through and through. That the law is not just the text, the decision, even the courtroom. Law is the pavement, the traffic light, the hoodie in the shopping mall, the veil in the school, the cell in Guantanamo, the seating arrangement at a meeting, the risotto at the restaurant. But we forget.

Spatial justice is disconnected from historicisation and thrown in the space of *here*, namely the space that vibrates with history through its material

129 Eslava and Pahuja, 2012: 22

appearance. Not an abstract history but a history of the *here*. Not an atmospheric, constructed *here* but a *here* that changes through in/visibilisation. Not a history that legitimises atrocities but a history that accepts the need for bodies to be here, exactly where other bodies might also want to be. This means that any historical claim, ethnic identity construction and deep-seated belief about belonging can only be factored in to the extent that they are 'imprinted' on, 'rooted' in, 'embodied' by the very surface on which spatial justice emerges. The link to space is not as simple as land ownership, legal claims to property or right by birth. While spatial justice can be linked to the legal regime of property, its function is sometimes complementary and often oppositional. In her work on Australian aborigines, Sarah Keenan employs the concept of institutional 'holding up',[130] namely the way property ownership is being de facto 'held up' (supported, validated, brought forth) by social institutions, as opposed to, I would argue, mere possession or occupation which can only be relevant if validated. Possession, in the sense of corporeal presence and psychological projection of intent is not automatically held up by society (think of squatters, graffiti writers or illegal immigrants).[131] Spatial justice urges a being-held-up, not by society and its institutions, but by the very spatiality of the relation. In other words, when society holds up ownership, the outcome hardly deviates from standard understandings of legal claims to property, which, in their turn are not necessarily 'objective' but rather serve particular social hierarchies, priorities, power structures and imbalances, as Keenan abundantly shows in her work. But when space itself, in its very own texture, which includes physical arrangement of bodies and objects, existing paths of movement and pause, spatial narratives and potential for disorientation *and* orientation, bears the 'imprint' of such holding up, the outcome relies not on unchallenged power imbalances but on corporeal presences that may or may not be translatable to legal language. This is the meaning of spatial justice: the move to 'possess' space, even though this space seemingly belongs to other bodies, in the conviction that this move will be 'held up' by the space itself, and that is responsibly situated in the assemblage of its emergence.

This is by no means a simple feat. I remember seeing published widely in the media at some point in 2005 a photograph of a graffiti written on a wall in Gaza by an Israeli soldier that said 'this is the only land I know.' Is this a media-stunt or reality of the continuum? We cannot afford to care about the distinction. Lawscaping technologies of architecture, property division, atmospherics of exclusion are hard at work altering the space of the law and making it impossible to distinguish, on the one hand, history as sedimentation on the earth and the bodies on and in it, and on the other, engineered historicisation that inflates supremacist, colonialist,

---

130 Keenan, 2013
131 See also Blomley, 2004, for a thorough analysis of the postcolonial city.

nationalist or otherwise land-bound ideologies. But the fact remains that one is often born and brought up in engineered atmospherics. This, coupled with the very simple fact that one only knows one land, that only one body can ever be at the same space at the same time, makes clear how there can *never* be a 'solution'. How the teleology of the end-result, with its obsessive idea of absence of conflict, is not only part of the problem but *the* problem. Justice must be recast as a conflictual space, full of erupting laws and spreading normativities. Justice away from the law is not a lawless justice. It is certainly a risky, potentially dangerous space, emptied of pillared security and smooth lines. It is also a space of constant reconstruction, rapid concept formation, applied acrobatics of thought and action. A space of justice – and indeed spatial justice – is a space where the law is being erected at every moment as if for the first time. Like a group of nomads that must set up home every time they stop for the night, in the same way the law is re-erected through a repetition that might create difference. No doubt there is legal repetition that simply generates identity, a sameness of application indistinguishable from case to case. In parallel to this, however, there is legal repetition, itself seemingly far from the materiality of a revolt, yet steeped into the materiality of the law, that destroys the law as it stands and builds it from scratch, every time new.[132] This kind of repetition generates the difference of justice.[133] But not always. There is no guarantee, no prescription for the emergence of justice. Luhmann calls justice a contingency formula,[134] since its emergence is always contingent on the conditions, and therefore not predictable. Spatial justice remains contingent, relying on law's materiality. Yet, and perhaps unsurprisingly, this space of justice is nearer than it seems. It appears at various moments, from within the lawscape. Justice is already here, a latent promise that materialises with every movement.

What does this understanding of the concept of spatial justice bring to the discussion? To start with, it forcefully and unapologetically rushes space in. Space and its conceptualisations have greatly changed, and law, both as a discipline and as a social function, must keep up with it. Space is no longer the local background but the radically disorienting factor of law's emergence. Second, the discussion achieves the much delayed link between justice and law in their spatiality. Justice is closer to law for the latter's spatial distribution but also at a distance from law in terms of withdrawal. This schizophrenic connection which I have elsewhere described as 'justice through and despite of the law',[135] must be fleshed out rather than quietly subsumed in a political theory setting. Justice is the movement of *going-*

---

132   Derrida, 1992
133   Deleuze, 2004c; for an elaboration on this, see Philippopoulos-Mihalopoulos, 2011b.
134   Luhmann, 2004
135   Philippopoulos-Mihalopoulos, 2003

against yet through the lawscape in attempting to cross the line of law's normative geometry while being inscribed within it. Spatial justice comes through and despite the law, riding on law's spatial ingestion yet escaping it, withdrawing from it in an immanent violence. The law must be killed, its proud edifice razed to the ground for justice to arise – but then again, only by climbing up the steps, the folded staircase of the lawscape, can justice reorient the lawscape. This is how the present understanding of withdrawal differs from equivalent concepts in Hardt and Negri's withdrawal of the multitude,[136] or Paolo Virno's 'engaged withdrawal',[137] or indeed Judith Butler's critique of traditional politics of withdrawal.[138] Withdrawal is not the end-purpose. It is merely a stage, a rupture in order to return to the lawscape, to the inescapable immanence of the law as institution. Third, the kind of spatial justice discussed here moves away from the unyielding concepts of anthropocentrism and moral necessity. The body and its spatial movement is not human-specific but potentially and simultaneously everything. In that sense, no moral judgment can withstand the constant fluctuation that ensues a description of a space of folds, itself interfolded with itself.

So, the question remains: who after all gets the concert hall seat? When law plays games with itself (or when hall administrators get greedy and double-book seats!), the question of spatial justice emerges. The negotiation is neither prescribed, nor easy. It is not a question of putting one body against another and letting them fight it out. Nor is it a question of simple temporal priority ('I got here first'), although this might count if spatially inscribed. It is a question of creating a breathing space, through a rupture of atmospherics and onto the lawscape, where spatial justice can emerge, where several corporeal movements can be tried out and where bodies might find themselves in need of withdrawal from recognised positions, security of choices or historical belongings. In those moments, the bodies expose their spatiality and its imprints in order to allow the law to listen to it. It is, once again, the law that will listen in and will enable a decision.

One final example: I was fortunate to see an artwork by Ingeborg Lüscher at Hamburger Bahnhof in Berlin a few years ago called *The Other Side: Israel/Palestine*. The artwork consisted of three long horizontal screens situated next to each other. In a series of silent black-and-white relatively brief takes, the faces of approximately 30 Palestinians and Israelis were shown, one at a time. The face appeared on the first screen, only to disappear afterwards and reappear on the second screen and then finally the third. Every time however, the expression was different, as if something had happened to which we were not privy. I then noticed that three plaques

---

136 Hardt and Negri, 2004
137 Virno, 2004
138 Butler, 1993, and 1997

were positioned underneath each screen. The first read: 'Think. Who are you, your name, your origin?' The second: 'Think. What has the other side done to you?' All the participants were asked the same questions. All of them had lost loved ones during the conflicts. Lüscher was filming them while asking these questions, which we could not hear. We were not told who was on which 'side' and, although one could guess, the takes were meant to conflate the sides rather than to keep them as 'sides'. The variation and emotional impact of the facial expressions of the participants were overwhelming. The usual trajectory was one of pride and defiance mixed with pain; this would then change on the second screen to intense pain and increasingly deepening sorrow. So far, so usual: a quest for identity politics of origin which leads to taking sides, and a habitual historicisation through blame attribution. The saving difference was the emotional privileging of the affect and the fact that all faces were inhabiting the same spaces/screen. But the most devastating moment was when the third screen would come alive with the participant's face. It was the screen that betrayed most expectations, and went against most projections of how the participants would react and how their expressions would change. These last screens were a humble triumph against synthesis. They could not be predicted on the basis of the previous screens. They emerged from a withdrawal from the atmospherics of conflict so powerful that it was humbling, so defiant that it was devastating.

The plaque underneath the third screen read 'Think. Can you forgive?'

Chapter 6

The islands

We are folded in the ocean. We are on a ship, dancing in the eye of a storm. On the horizon, a line of land, short and flat: an island. The ship moves inexorably towards the shallows, out of human control and as if drawn by a grand teleology. The landing was never going to be smooth and uneventful. It is violent, vomited, ravaged. We were not looking for a utopia, we were not opening our sails to a favourable wind. We are shipwrecked. We have no choice. We are part of the greater ocean, and this, the lack of choice, is our freedom. And we shall be rewarded. Not just one but two islands will emerge from the surface. One at a time though. The first island is a painstaking, fully normalised, fully legalised lawscape, compulsively striated, its spatiality obsessively invisibilised. This is Robinson Crusoe's island, full of lines of property and propriety. He calls it *Speranza* ('hope' in Italian) because it is the lawscape of escape. Through the lawscape, Robinson wants to escape the island and either leave it completely behind for the pleasures of eighteenth-century European society carried onboard a passing ship; or, failing that, totally invisibilise the island and replace it with the matrix of civilisation, the perfection of the rule, the straightness of anthropocentrism. And then comes another island, also Speranza, but different from the first. This other island is fully given to a mad becoming where all the bodies (island included) become elemental. Is this the second island? You will have to read to the end of the chapter to find out.

One clarification: this is not Defoe's Crusoe. There are many others, each one important in its own right,[1] but I have chosen to focus on Michel Tournier's novel *Vendredi*, which has the double allure of, first, having its title changed to the 'Other', namely Friday (*Vendredi* in French); and second, becaue it has had Deleuze poring over it and writing on it. My role therefore is this: I map the novel and Deleuze's text on the novel from the

---

1   Notably J. M. Coetzee's 1986 *Foe* written from the perspective of Susan Barton, a female character of another novel by Defoe; and Patrick Chamoiseau more recent *L'empreinte à Crusoé*, 2012, in which Crusoe is an adopted identity by an amnesiac cast-away who reads the name inscribed on the island and assumes it is him.

point of view of the law and spatial justice. I map a series of islands, each one a body of repetition, each one moving towards the finale of the book. The text ends up with a paroxysm of erected singularity that unfolds with the emergence of the body of spatial justice.

6.1 The first island

Deleuze's essay on Tournier's *Vendredi ou les Limbes du Pacifique* (in English, the title has been simplified to 'Friday'),[2] maps a self-enclosed, immanent island.[3] Deleuze uses Tournier's *Vendredi* as a diagrammatic text, namely as something that 'does not function to represent, even something real, but rather constructs a real that is yet to come, a new type of reality.'[4] This new type of reality is immanent to the space of its emergence, namely to its own *plane of immanence*, the latter both on the basis of and beyond the control of such reality. We have encountered the plane of immanence in Chapter 2 of this book, and indeed throughout the book in various guises: the continuum, atmosphere, lawscape, surface. They all constitute the parameters of thinking that determine a problem. We posit the problem, indeed a question, in order to come up with a concept, as Deleuze and Guattari ask us to do in *What is Philosophy*.[5] A concept is always an answer to a problem, put in the form of a question. Both problem (here, the lawscape, atmosphere, the continuum) and concept (here, spatial justice as immanent rupture) emerge from within the place of immanence. This is neither a clean-cut epistemological exercise, nor a representation of the way things work. On the contrary, the plane of immanence is a continuous exploration that becomes unsettled with the changing conditions of movement within. Moira Gatens describes it as 'a plane of experimentation, a mapping of extensive relations and intensive capacities that are mobile and dynamic.'[6] For this chapter, the plane of immanence takes the form of a desert island. Its coastline, brimming with movement, pushes out its desert, living the island overflowing with bodies. 'What is deserted is the ocean

---

2   Tournier, 1967 and English translation, 1969. In what follows, I give the translated pages reference followed by the pages of the original (in its folio edition) in brackets, stating whenever I have modified the translation. The novel operates on multiple levels, such as the animistic, the sexual, the intercorporeal, the philosophical, the reflective, the 'perverse'. In many respects, it is a paradigm of what a Deleuzian programmatic novel might be. The novel appears to be a source of inspiration for Deleuze's philosophy.
3   Deleuze was a friend of Tournier's and has referred to the latter's novels in his work. The present essay was first published as Deleuze, 1967, subsequently incorporated as a postface in Tournier's book and as an appendix in Deleuze, 2004b. I have used the latter edition of the essay.
4   Deleuze and Guattari, 1988: 142
5   Deleuze and Guattari, 1994
6   Gatens, 1996b: 165

around it', writes Deleuze in one of his texts on islands.[7] Ready to be populated, the island offers itself to the shipwreck.

Tournier's *Vendredi* is the story of Robinson-meets-Friday. It differs radically, however, from the Defoe original, of which incidentally Deleuze has written that 'one can hardly imagine a more boring novel, and it is sad to see children still reading it today.'[8] Deleuze's dismissal of Defoe's Robinson is on account of the novel's preaching, capitalist, property-obsessed tone that makes 'any healthy reader' dream 'of seeing [Friday] eat Robinson.'[9] In some ways, Tournier's treatment of the topic is Deleuze's dream come true. Although no cannibalism is practised between the two men, there is a certain reciprocal ingestion that characterises the whole novel. Indeed, Tournier's version begins with an initial Defoe-like storyline of Robinson's attempt to master his surrounding space, only quickly to move to a meditation on a becoming-elemental, that is becoming-tree, becoming-Friday and eventually becoming-island. Tournier's Robinson ends up in an assemblage consisting of the insular earth and Friday's body. Rather eccentrically, however, Robinson's sexual becoming couples with the former rather than the latter. Sexual or not,[10] the connection between bodies and space is one of double-capture, a wasp-orchid ingestion. Desire flows and rests, thrusts itself wildly around and forms clots of thick craving. Its movement is in accordance with a law that trammels the whole assemblage. This law, however, does not determine the assemblage any more than being determined by it. The law in *Vendredi* cannot be dissociated from either the body or the space of its appearance. This is the *Vendredi* lawscape, distinctly posthuman, fractal and immanent.

Deleuze's reading of *Vendredi* offers a thesis on the connection between space and body.[11] Both elements are explicitly present in the novel. Let me start with space first. In another difference from the Defoe original, *Vendredi* is flooded by a sense of spatiality and in many respects entirely measured by it. The space of the island is never a mere backdrop but an active force. The two men languidly lull along the movement of the island, to the final point that the space of the island is not different from the bodies that walk it. The island as destination, imagination and geography, hosts these 'heavenly nuptials, multiplicities of multiplicities'[12] of bodies

---

7   Deleuze, 2004a: 11
8   Deleuze, 2004a: 12
9   Deleuze, 2004a: 12
10 Deleuze's essay focuses on sexual perversion, a reading readily offered by the novel itself. While law can be read through desire (indeed one could argue that the two are inseparable,) for reasons of textual consistency, I am obliged to focus on the connection between space and bodies in relation to the law.
11 Although arguably its most explicit focus is the connection between the self and the world. However, unsurprisingly for Deleuze, both are discussed in their material dimension, namely as bodily affects in space.
12 Deleuze and Guattari, 1988: 35

and space flowing together in an elemental dance. Thought, memory, movement, future projections, philosophical enquiries, desire, fear, nostalgia: they are all mediated by the space of the island. Above all, *hope* is the island's main function, which is why Robinson quickly names it *Speranza*: initially envisaged as Robinson's hope to be saved from the elements, the cannibals, and the solitude, it quickly embodies hope to save Robinson from himself and his human nature.

Space does not only appear in its abstract quality as the field in which the story unfolds, nor in its concreteness as just an as-yet-unmapped territory in the ocean. On the contrary, the space of the island is material and immaterial, a manifold body that floats on the ocean. Its extension, namely its physical quality, includes the ocean of the voyage as well as that last vestige of the outside, the shipwreck itself now well ingested in the spatial immanence of the island. As Deleuze writes, 'humans would have to reduce themselves to the movement that brings them to the island.'[13] Human bodies are just another body in relation to the body of the island. The two affect each other, communicate their presence to each other through their motion and rest, as Spinoza taught us.[14] This is the way the continuum becomes animated: through the difference of bodies and their ontological withdrawal from each other, yet at the same time steeped in the indistinction between bodies. The movement *to* the island becomes one with the movement *of* the island, thus fusing imagination with the materiality of geography and the voyage with the desert island. In this sense, the endeavour is not a utopia, at least in its usual sense of successful failure that the voyage towards a *ou*-topos entails. The voyage is immanent to the (eventual) joy of becoming one with the *topos*: it is folded in the coastline of the island. The voyage is no longer anything more than the shipwreck from which one can draw useful tools and useless rituals.[15] The voyage has become a space of the island, simultaneously a potential means of striation, organisation, categorisation, *and* a space from which the origin prepares the island for its teleology, its emergence as pure spatiality. One must not forget the voyage; nor however must one lose oneself in it. The island remains desert and the perfect case of a manifold,[16] namely a field of rapidities and slowness defined internally, by and through itself and its inclusive immanence. The island as manifold includes the space of all possible states

---

13   Deleuze, 2004a: 11
14   Spinoza, 2000: L2
15   As Deleuze writes, 2004a: 9, 'we need only extrapolate in imagination the movement they [the shipwrecked] bring with them to the island. Only in appearance does such a movement put an end to the island's desertedness; in reality, it takes up and prolongs the *elan* that produced the island as deserted.'
16   See earlier, Chapter 2 for space as manifold; also De Landa, 2005, which is a spatially pronounced description of multiplicity.

the island can have, as well as the invariant structure of that space:[17] 'you have to keep enough of the organism for it to reform each dawn.'[18]

Through its gravitational pull, its fully spatialised *here*, the island grounds the continuum of the bodies inhabiting it: Robinson, Friday, the goats Robinson has tamed, the ship, even the island itself are bodies. But so are the log that Robinson keeps and the body of the law that he builds as soon as he lands on the island. Bodies emerge all the time from the folds of the manifold, defined both by their extensive qualities, such as the space they occupy, and by the way they affect and are being affected by other bodies. Bodies form other bodies, arranged not in hierarchy but in a system of movement and rest in which positions change and power balances are being restructured. The island is moving in that direction, eventually to become, as I show below, this most exquisite of bodies, a body without organs.

While space and bodies are explicitly dealt with in Deleuze's text on *Vendredi*, the law is less so. Deleuze anchors law on the concept of perversion. He writes: 'Perversion is a bastard concept, half-juridical, half-medical. But neither law nor medicine are entirely suited to it.'[19] Law (along with medicine) constitute the system in which the pervert introduces his or her desire and tries to make it work as an internal limit. In so doing, the law does not see the pervert as someone who desires but as a threat to a well-established system of desire. Deleuze's understanding of law as *logos* runs throughout his work. It is small surprise, therefore, that direct references to law are limited and dismissive. Still, in his essay on *Vendredi*, the law is constantly and implicitly there. The law determines the space between the self and the Other with regard to desire. Thus, the Other gives depth to the world because she is assumed to be able to access what I cannot: 'I desire nothing that cannot be seen, thought, or possessed by a possible Other.'[20] Her presence determines what is *possible* by delimiting and thus defining the field of what I see, where I move, what I touch. The Other is the authority that subsumes my ontology into a specific epistemological perspective: I can only know through the Other. The Other is the law that operates spatially by rendering things available to me, offering to me the 'possible' thereby excluding the things that are to remain out of reach for me. The law, just as the Other, is 'neither an object in the field of my perception nor a subject who perceives me: the Other is initially a structure of the perceptual field, without which the entire field could not function as it does.'[21] This is the '*a priori Other*', 'the absolute structure' that places things into categories and distributes space according to the organisation of its own structure.

17   Buchanan and Lambert, 2005
18   Deleuze and Guattari, 1988: 199
19   Deleuze, 2004b: 343
20   Deleuze, 2004b: 345
21   Deleuze, 2004b: 346

The presence of the Other manifests itself through a spatial law, that I shall call the *law of the Other*. As I have shown in this book, the law striates the very space on which it operates; it determines distances and propinquities between bodies; it restricts, enables or forces movement; it renders visible bodies with which a connection is possible while obscuring others, thus rendering the connection impossible. The law of the Other is the lawscape: the law society needs in order to carry on functioning according to its pre-given structure. Tournier writes: 'He knew now that man resembles a person injured in a street riot, who can only stay upright while the crowd packed densely around him continues to prop him up.'[22] The law of the Other, the lawscape 'is the structure which conditions the entire field and its functioning, by rendering possible the constitution and application of the preceding categories.'[23] The law of the Other is the great categoriser, finding its breath in the body of the Other. The law emplaces my body on an island of controlled flows, where only the bodies revealed to me by the law can be part of my understanding of my own ontology in relation to others.

And so it is in *Vendredi*. Robinson begins with an obsessive juridification that striates the island. He divides it in excruciatingly delimited spaces that contain resources or are designated for the ritual of log-keeping, or for church-like spiritual concentration. He also divides the island in present and future visibility, thereby keeping the seed stock well hidden while using a small portion for his present needs, deferring thus most of it to a future promise. The time of the lawscape is fulfilled: a society is repeated on the island that harks back to what Robinson had before the shipwreck. Smelted in the rocks of the island and the sinews of Robinson, the law determines the assemblage between the body of Robinson and that of the island, and, furthermore, invisibilises as much as possible the spatiality of the island while fully visibilising its law. The island gives way to an asphyxiating grid of regulating lines, structures and gaps that fiercely reflect the oceanic sun from afar: an absurd human effigy in the midst of the posthuman continuum, a grotesque lighthouse that desperately tries to invisibilise the brutality of space through its own sad brutality. This is an angular, rectilinear lawscape that reterritorialises the desert island and produces capital accumulation on the cutting edge of further deterritorialisation, namely what Deleuze and Guattari define as the outcome of the interplay between *logos* (that reterritorialises, thus striating) and *nomos* (that deterritorialises, thus rendering space smooth).[24] As Tournier's Robinson reflects in his log, 'Robinson is infinitely rich only when he coincides with the whole island.'[25] Robinson is the perfect

---

22 Tournier, 1969 (1967): 40 (38)
23 Deleuze, 2004b: 348
24 This is a simplification, as I show below. Deleuze and Guattari, 1988
25 Tournier, 1969 (1967): 67 (70)

coloniser, involving the body of the island in a circular narcissism that makes the island finally achieve its potential only through the body of Robinson.[26] The colonial assemblage is productive because it is determined by the law of the Other. Robinson constructs the Other, indeed represents the Other in the form of rituals, dressed-up ceremonies, self-imposed limits or even a ban on the consumption of crops, calendar division of time, and a desperate self-imposed obligation for vocal articulation of all thinking. The reader witnesses the collective enunciations of 'acts and statements, of incorporeal transformations of bodies.'[27] Robinson obsessively controls even the way his senses thematise space. Right from the start he codifies everything in a *Chartre de l'Ile*,[28] namely a code of several articles and paragraphs *and* a map of the island, a cartography of movement and a striation, not only of the body of the island but also of his own body, his sensorial involvement, his effluxes and ingestions, his position on the island at any time during the day, his breathing and his vocality. The affects of all the bodies involved becoming orchestrated in a frenzy of cohesion, directionality, teleology.

Robinson needs the law because he is still dependent on the Other as structure. Tournier writes: 'Once again he found that to build, to organize, and to make and abide by laws were sovereign remedies against the dissolving effect of the absence of the Other.'[29] Through normative fortification, he constructs his desire for the society he is missing (which is, according to Deleuze and Guattari, the function of the law: 'so, *that's* what I wanted!'[30]). The Other is present in her absence, haunting the lawscape from within. This is the atmospherics of extreme legal visibilisation, where the law takes over so completely and so inescapably that comes full circle from the *logos* of prison to the illusion of *nomic* freedom. The island offers the same kind of false sense of security that is reproduced as in an anomic lawscape. *Speranza* has become Robinson's shopping mall: the perfect atmosphere of security, where threat is kept out (yet visibly there) through the intricate grid of Western civilisation's mad oceanic outlet. The ocean is striated as a shopping district for the needs of the grid. The land is striated as an affective prison, and the air is partitioned in functionally differentiated activities. Robinson's atmosphere is made to last, future-tending, past-reinscribing yet fully present, in an illusionary elimination of every possible conflict with the continuum. We are in invited to yet another perfect atmospheric rupture, filled with a perverse spatial justice where everything has its place through the lawscape and nothing ever needs to withdraw.

26 Bhabha, 2005
27 Deleuze and Guattari, 1983: 88
28 Tournier, 1969 (1967): 69 (71). The English translation is *Charter*.
29 Tournier, 1969 (1967): 76 (79) modified translation.
30 Deleuze and Guattari, 1988: 125

Yet, things change. Deleuze points out that Tournier's novel 'develops the very thesis of Robinson: the man without Others on his island.'[31] Robinson's *thesis*, his *position* on the island, is one of shifting juxtaposition. Indeed, the absence of the Other manifests itself with an asphyxiating presence that leads to frenetic juridification. But then, the Other is finally ingested in her absence and nothing is left: the Other's presence is finally properly absent. In a beautiful, almost Kierkegaardian passage, Robinson indulges the Nietzschean ecstasy of what he subsequently calls 'a moment of innocence'.[32] Robinson withdraws into a pre-edenic state of atemporality and aspatiality, when he is caught unawares by a pause in his daily routine caused inadvertently by the stopping of his salvaged hourglass. This pause operates like a legal lacuna, an abrupt estoppel of the legal flow, a withdrawal from his own restrictive lawscape. The legislative gap cracks open Robinson's subject/object distinction that he has constructed for himself in relation to the island, and dramatically minimises the distance between them. The body of the law is wounded and leaking. The lawscape seems dizzyingly withdrawn: no more desire for the law. A moment of innocence that supercodes any judgment of guilty or not-guilty pushes Robinson on a new line of flight, radically immanent to his existing assemblage: it exposes him to a painful yet profoundly shaking awareness of his affect, '*these nonhuman becomings of man*', and edges him to let it flow together with the island as a *percept*, or what Deleuze and Guattari have called the '*nonhuman landscapes of nature.*'[33] The 'barrier of the skin which separates the inner from the outer world'[34] collapses, becomes 'common tangency' to recall Michel Serres,[35] and Robinson becomes aware of progressively shedding his human nature when 'he perceived, when he awoke one morning, that his beard, growing in the night, had begun to take root in the earth.'[36]

Robinson is finally free to fall, literally in the depths of the island, deep in its womb, lose time and render visible another cavernous insular space. He spends an unspecified amount of time in a deep dark hole in the ground, a space of intense claustrophobia in the novel that leaves a Robinson blanched out, weak, savage, smooth. Robinson encounters what Reza Negarestani in *Cyclonopedia* refers to as 'chtuloid ethics', namely the *chtonion* (underground, inside the earth) ethical encounter that does not indicate openness to the Other in the Levinasian mode, but is butchered open from the inside. In chtuloid ethics, there is no choice but to tear

---

31   Deleuze, 2004b: 344
32   Tournier, 1969 (1967): 90 (94)
33   Deleuze and Guattari, 1994: 169, original emphasis.
34   Tournier, 1969 (1967): 69 (70)
35   Serres, 2008: 80 'contingency means common tangency: in it the world and the body intersect and caress each other ... Everything meets in contingency, as if everything had a skin.' See Chapter 3.
36   Tournier, 1969 (1967): 130 (138)

oneself apart, thus resisting the atmospheric economy of moral openness towards the construction of the Other. Robinson becomes the true geo-philosopher, conflating epistemology with ontology, leaving behind the phenomenological and throwing his body into the earth: 'the earth cannot be narrated by its outer surface any longer but only by its plot holes, vermicular traces of exhumation.'[37] Robinson emerges in a milky consistency, newborn and healed, one with the body of the island. The withdrawal to the womb is the unfolding of spatial justice. Bodies are now emplaced in a different way. Not in the atmosphere of hyperjuridification but in a posthuman embeddedness: the island is not anthropomorphised. Rather, Robinson is island-morphised. Now, it is the island that is Robinson, co-extensive with him, and not the other way round. Robinson slowly abandons the law of the Other and slides in a new assemblage with the body of the island. The emerging lawscape is an *insular* one, the island being more potent than Robinson. He becomes one with the elements and loses completely the lawscaping presence of the absence of the Other as structure. Robinson passes from the representation of the Other as juridical structure in the form of law's compulsive spatial striation, to a different juridical structure: that of the demise of the Other. Law is no more mediated through the Other but emanates directly from within Robinson's assemblage with the body of the island. The lawscape becomes posthuman. A newfound freedom that is a direct product of Robinson's *encounter* with the island smoothens up the territory by whispering a different law altogether – the law of *becoming other*. This is no longer a law of striation but of continuous, uncontrollable, nomadic perambulation.

6.2 Are we there yet? The second island

Robinson passes from the law as *logos* to the law as *nomos*. The passage seems irreversible. We are saved. We have settled in the space of justice. We have left behind the big bad *logos* and we now sleep with the smooth *nomos*. But are these two laws so distinct? Do they follow such an unproblematic temporality of irreversibility? And where *is* Friday in all this? Remarkably, he appears when Robinson's reorientation is well under way. Friday arrives too late to be able to resuscitate the order of the Other, yet a little too early to slide along the assemblage Robinson-island. This is a clue: the assemblage will change again, the lawscape will be transformed again. Perhaps we have not arrived to the final island after all.

Friday's legality is a different one from that of Robinson, a legality of surface rather than depth, and for this more *profound*.[38] While Robinson

---

37   Negarestani, 2008: 239
38   'is it the case that every event is of this type – forest, battle and wound – all the more profound since *it* occurs at the surface?' (Deleuze, 2004b: 12).

enjoys withdrawing underground in an arborescent move, Friday spreads his body on the island, grass-like, exposed and seemingly without need for shelter, even reversing the available surfaces by replanting trees upside-down, or using the element of wind to make music on an aeolic harp. Not without a certain envy, Robinson parallels him to Aphrodite, perennially emerging and free-floating on the sea spray. Their encounter is a 'joyful' one, if joy is understood according to Rosie Braidotti,[39] herself reading Spinoza as a force that pushes (here, the law) from within and into forceful encounters with other bodies. The encounter encompasses the possibility of violence, of mutual destruction, of cannibalistic ingestion but also of new becomings and different power configurations:

> Heretofore it was only a question of how a particular thing can decompose other things by giving them a relation that is consistent with one of its own, or, on the contrary, how it risks being decomposed by other things. But now it is a question of knowing whether relations (and which ones?) can compound directly to form a new, more 'extensive' relation, or whether capacities can compound directly to constitute a more 'intense' capacity or power.[40]

This is not a synthetic process but an affective one. Bodies (extensive) and the ideas of these bodies (intensive) come together without merging. In exactly the same way, *logos* and *nomos* never merge, nor however can they ever emerge in isolation. Logos is the law of the Other that made Robinson excavate the island, erect edifices dedicated to specific functions and strew fences and limits everywhere. Robinson's law of depth striated the island with cavities of past crops and observatories of future salvation perched on the highest peak. When however, Robinson began to become-island, especially after the advent of Friday, the law of depth (*logos*) gave way to the law of surface (*nomos*). Space is now manifold, emergent and spread out, a playground of nomadic existence where one never gets lost because there is no centre, no measure, and no one direction. The surface has surfaced (especially after a spectacular explosion caused by Friday that destroyed the cave with the provisions) and the island becomes a smooth membrane.

Still, as I mentioned in Chapter 2, the binarism of *logos/nomos* is only impressionistic. To start with, the insular plane of immanence contains the two in overlapping events. Even before Friday's advent, Robinson's law contained spaces of smoothness, Sundays and rest and sleep and allowances for cave retreats and moments of innocence. Likewise, Friday's law striated the space around which he could lie about and enjoy the sun.

---

39 Braidotti, 2006
40 Deleuze, 1988: 126

An example: Friday made full use of every single part of the grand goat that he killed after a battle. The goat in its various manifestations as food, clothing, tools, toys, music instruments could only be made thus by following an intense and ritualistic series of actions in strict order and with a view to (elemental) possession. Rather than the usual fantasy of the native as *of nature*, Friday emerges in the novel as a striator that understands the folds of the surface he is striating. Even at its progressive formation, the assemblage Robinson-Friday-Speranza moves between the two laws and on the two spaces, overlapping, in *glissando* or stuttering or frictionless, resting or rushing. And this brings me to the second point: although they do not merge, to see them as two different things is one way of rupturing the continuum of the lawscape and putting it into categories available for literary, political or legal manoeuvring. Likewise, the attempt to separate deterritorialisation from reterritorialisation is problematic because it assumes that one can take place without the other. But every de- is also reterritorialisation. Every escape from the lawscape lands in the lawscape. Every deterritorialising needs a reorientating reterritorialising. Friday is firmly based on the *here* of the island. He deterritorialises Robinson's relics of structure that manically hold onto the vertical, and brings him from the law of the depth to that of the surface. Yet, Friday also reterritorialises the island according to his own 'minor' lawscape, which is unhierarchical and elemental. Deterritorialisation occurs not in an opposition but from within the structure, *in equal measure invited and invaded by the structure*. The movement is always immanent, always from within: the minor from within the established. The law of surface moves across the law of depth and captures it while luring it.

Back on the island, everything is precipitating towards its final apotheosis. According to Deleuze, Tournier puts the story 'in terms of end, and not in terms of origin.'[41] Right from the start, the novel moves headlong to its edge with a teleology that is foreshadowed as both irresistible and inevitable. By treating the voyage and the shipwreck (the origin) as a space within the island, Tournier is willingly folded in an immanence that 'makes it impossible for him to allow Robinson to leave the island.'[42] The challenge with any manifold surface however is to carry on unfolding it, 'not how to finish the fold, but how to continue it, to have it go through the ceiling, how to bring it to infinity.'[43] Tournier's solution is to bring infinity down, ground it on the immanence of elements and thus liberate it from the desire to transcend. The end of the novel follows Robinson's ends: both are defined in terms of their insular immanence. Although 'these ends represent a fantastic deviation from our world',[44] they remain inalienable to the world that has

---

41   Deleuze, 2004b: 342
42   Deleuze, 2004b: 342
43   Deleuze, 2006: 39
44   Deleuze, 2004b: 343

been progressively constructed by the various movements on the island. The end had been present well before the law of the Other fused with the law of Friday, even before the shipwreck takes place. The end is revealed to Robinson on board the ship by a tarot reader:

> A snake biting its tail is the symbol of that self-enclosed sexuality, in which there is no leak or flaw. It is the zenith of human perfectibility, infinitely difficult to achieve, more difficult still to sustain. It seems you are destined to rise even to these heights, or so the Egyptian tarot cards say. My compliments, young sir.[45]

The tarot image brings to mind the Jungian uroborous snake with its paradoxical flow constitutive of its immanence.[46] Immanence is characterised by *enantiodromia*,[47] Jung's alchemic loan of a vitalistic paradoxical force that flows in two simultaneous and opposing directions. This is the way continuum and rupture work.

Compare the above to what Deleuze writes: 'the power of the paradox therefore is not all in following the other direction, but rather in showing that sense always takes on both senses at once, or follows two directions at the same time.'[48] 'Sense' is understood as both 'meaning' and 'direction', both mind and body. Every direction endows the body that moves with a different meaning. The two directions form a continuum of immanence, or to paraphrase Edgar Morin, a closure that rests on its openness.[49] The plane of immanence simultaneously accommodates smooth and striated spaces, *logos* and *nomos*, state and nomads, Robinson and Friday. The enantiodromic, parallel movement of these directions *is* the abstract machine of the plane of immanence. This is what connects while keeping apart continuum and rupture. The paradox behind Deleuze and Guattari's thinking is not a binarism but a fold.[50] There is nothing outside except for the infinity of internal elongation. Even in Friday's law, the law without the Other, striation persists. Likewise, in Robinson's law, smoothness is always round the corner. In the law of the Other, there is identification with society, with the missing Other, with desire as determined by the structure. In the law without the Other, there is diversity, a multiplicity that 'gives itself a singularity.'[51]

Let me return briefly to Deleuze's paradox. In one of his treatments of *Alice in Wonderland*, Deleuze plays with the various levels of common sense/

---

45 Tournier, 1969 (1967): 12 (12)
46 Jung, 1963
47 From Greek *enanti* which means 'opposite' and 'across', and *dromos* which means 'route', 'path', 'flow'.
48 Deleuze, 2004b: 88
49 '*L'ouvert s'appuie sur le fermé*', Morin, 1986: 203
50 Deleuze and Guattari, 1988: 8ff.
51 Deleuze, 2004b: 87

good sense. Schematically, common sense is about identification with the Other, whereas good sense is singular thinking. Common sense is shared sense but also (self-)imposed sense, a frame through which the path is revealed, the decision is sanctioned. Society, in its presence or absence, is embodied in common sense and imposes the law of the Other through the unity of the 'I'. To put it in the terminology I have been using in this book, common sense is of the lawscape that tends towards or has achieved the atmospheric, namely the fully normalised. This lawscape is based on externally determined identities and maintains itself through a centralised understanding of belonging. Good sense on the other hand is the direction from the singular to the regular with a teleology and in some ways a responsibility to stretch the former over the latter. This means that good sense requires a differential emplacement of the body, a withdrawal from the atmosphere that would allow a body to reorient the lawscape towards the emergence of singularity. Even though I have crudely simplified the intricacies of the two senses, I must stress that the two are not opposing: 'the force of paradoxes is that they are not contradictory; they rather allow us to be present at the genesis of the contradiction.'[52] The genesis is multiple and explosive, its effects trickling in the way one sense flows into each other. There is no final resolution and no transcendence but simply a repetition of such emergence,[53] from atmosphere to lawscape and then, inevitably, back again, but to a different orientation. The Deleuzian paradox is the space where the superlative body emerges, the *Body without Organs*: no organism, just flow and flow that stops the flow, organs that throw themselves in space, on the surface, and then move against each other, molecular explosions, violence of continuum and rupture, clashes of Otherness and elements. Nothing is thrown away; all is recycled on the ruptured continuum.

6.3 Repetition: the double island

The lawscapes of Robinson and Friday are both necessary, as they stand within yet exceeding each other, in order to materialise the grand teleology of the novel. But this grand teleology is not the island emerging from Robinson's legal lacuna that dragged him into the depths of the island and his assemblage with Friday's lawscape. His withdrawal from the register of

---

52 Deleuze, 2004b: 86
53 'Good sense could not fix any beginning, end, or direction, it could not distribute any diversity, if it did not transcend itself toward an instance capable of relating the diverse to the form of a subject's identity ... Conversely, this form of identity within common sense would remain empty if it did not transcend itself toward an instance capable of determining it by means of a particular diversity, which would begin here, end there, and which one would suppose to last as long as it is necessary to assure the equalization of its parts' (Deleuze, 2004b: 90).

the striated lawscape was necessary. His repositioning allowed for the lawscape to reorient itself and one could even say enabled the emergence of a 'different' lawscape, a second island indeed. It does not stop here though, and with this I depart from Deleuze's reading of *Vendredi*. The teleology of the novel forces us to gaze upon yet another island: an island which is the *repetition of the second island*. This island is simultaneously the second island *and* its excess, in the same way that the second island *withdraws* from its repetition and thus enables it to transform into something else. Thus Robinson: 'Now I have been transported to that other Speranza, I live perpetually in a 'moment of innocence'. Speranza is no longer an uncultivated land that must bear fruit, nor Friday a savage towards whom I bear the duty of civilising. Both demand my undivided attention, a careful contemplation, an ecstatic surveillance, for I think – nay, I am certain – that every moment I am seeing them for the first time.'[54] The other island is not merely a smooth space of nomadic movement. It is that too, but it also a space that demands surveillance, contemplation, vigilance – in short, striation: 'you don't reach the Body without Organs, and its place of consistency, by wildly destratifying it.'[55]

The repetition of the second island is not a third island but a *double* that 'is not a replica of things. It is, on the contrary, the new upright image in which the elements are released and take possession of themselves again, having become celestial and forming a thousand capricious elemental figures.'[56] The repetition of the second island is neither Robinson's nor Friday's: '[Friday] indicates *another*, supposedly true world, an irreducible double which alone is genuine, and in this other world, a double of the Other who no longer is and cannot be.'[57] Friday 'indicates' it but is not it, nor can he provide for it. The emergence of repetition is not simply a product of dialectic interpenetration or a synthesis. In Chapter 2, I referred to the time of the lawscape, itself always double, always rippling across the continuum of difference, always *repeating*. The *double* island repeats the *second* island and in that way, difference emerges. But this is not a fool-proof calculation. The emergence is violent, tremor-like and uncontrollable, entirely contingent on the reorientation.

Let me sum up before I proceed to the final topos of the argument: Robinson's law, the striating law of the Other, folds in the law of Friday, the nomic law of smooth space, in an assemblage that generates its very own lawscape, the second island. In its turn, the second island withdraws (in the novel Friday and Robinson withdraw from each other in a respectful, undemanding indifference), only for its *double* to emerge. The double island is

---

54 Tournier, 1969 (1967): 205 (220) translation modified.
55 Deleuze and Guattari, 1988: 161
56 Deleuze, 2004b: 351, translation modified.
57 Deleuze, 2004b: 355

the elemental space of legal teleology, overflowing with the elemental luminosity of spatial justice. Justice emerges from within the lawscape in 'capricious elemental figures.'[58] This is the connection between lawscape and spatial justice – the latter emerging from the conflicts of the former but not necessarily and not always. No formula can be followed, and no engineering can take place, in striation or in smoothness, that guarantees the emergence. Only withdrawal allows a reorientation of the lawscape towards the teleology of justice: justice requires that the reorientated lawscape put itself 'in terms of end, and not in terms of origin.'[59] But even this orientation is counter-intuitive. For the moment of emergence of spatial justice is that of a movement *away* from the goal: '*withdrawal, freaks*'.[60]

Spatial justice takes place in a maelstrom of withdrawal. In *Vendredi*, bodies withdraw for the islands to emerge. So, allow me to add to the previous definitions of spatial justice this: *spatial justice is the movement of withdrawal of the lawscape with the aim of a possible emergence of a space of a different, repeated legality: the lawless legality of justice*. My aim is to emphasise the immanence of justice, namely its self-enclosed generation, necessarily based on a connection of withdrawal. Spatial justice is immanent to law, flowing along the legal orientation towards justice, yet overcoded by the withdrawal of the law. Spatial justice is the rupture of the lawscape, inside of which its very own self-rupture is folded. Spatial justice retains the lawscape within, in withdrawal and perennial movement, like the empty square of the Deleuzian chessboard: 'there is no structure without the empty square, which makes everything function.'[61] The double island orients everything on account of its empty space, a space of withdrawal within. Deleuze carries on by urging us to keep moving the square: 'today's task is to make the empty square circulate.'[62] *The space of withdrawal is always there but needs to be constantly flowing, for otherwise justice becomes frozen in the regime, a pillar amidst other pillars.* Just as justice cannot be disengaged from the lawscape in its paradoxical flow of the logic and the nomic, in the same way there is no telling how much of either needs to be withdrawn for the empty square to follow the lines of escape and keep on moving. Withdrawal is a revolutionary, dangerous move that takes risks by allowing spaces to discover their immanent legality.

The double island is the product of multiple encounters, indeed multiple judgments between bodies, a series of judgments that might lead

---

58 Deleuze, 2004b: 341
59 Deleuze, 2004b: 342
60 'But the revolutionary knows that escape is revolutionary – *withdrawal, freaks* – provided one sweeps away the social cover on leaving, or causes a piece of the system to get lost in the shuffle. What matters is to break through the wall' (Deleuze and Guattari, 1983: 277).
61 Deleuze, 2004b: 61
62 Deleuze, 2004b: 84

ontologically or teleologically to the island's infinitely repeated singularity. It is the space of *here* into which the law throws itself, the luminosity of 'erected' spaces, the singularity of 'erected' times: 'each day stands separate and erect, proudly affirming its own intrinsic value ...They so resemble each other as to be superimposed in my memory, so that I seem to be ceaselessly reliving anew the same day.'[63] Repeated *anew*, in difference: these are the quotidian (every-day-the-same), banal, micro-judgments that huddle together and gather force towards the emergence of spatial justice. The space of justice is the space of 'second origin', which is 'more essential than the first, since it gives us the law of repetition, the law of the series'[64] that does 'not add a second or third time to the first, but carr[ies] the first to the "nth" power.'[65]

The double island, the space in which spatial justice emerges, is then the *desert island* par excellence. It is uncharted, unreachable, closed, withdrawn, 'a sacred island.'[66] To maintain this sacredness, the island must remain desert yet open to shipwrecks and castaways arriving: 'far from compromising it, humans bring the desertedness to its perfection and highest point.'[67] But this is not the end. The fold has to keep on being unfolded to infinity. For these humans, just as everything else, *must* withdraw from the desert island for the 'highest point of desertedness', namely the world without law, to keep on emerging. Lawscapes also withdraw. Whether oppressive machines of striating law or smooth distributors of nomadic law, lawscapes remain objects, bodies of connection and singularity, and as such withdraw.[68] Sometimes withdrawn lawscapes become atmosphere. But at other times, they become pure becoming. This is the final twist to the story: in the novel, spatial justice emerges as a double island. Yet, this emergence is not spatial justice anymore. It is fully spatialised and fully embodied, but it is somehow other than spatial justice. It is pure im/materiality, without the image of law, that is to say without any representation of the law.[69] The emerging justice is one of absolute singularity, meteoric, free-floating,

---

63 Tournier, 1969 (1967): 204 (219)
64 Deleuze, 2004a: 13
65 Deleuze, 2004c: 2
66 Deleuze, 2004a: 13
67 Deleuze, 2004a: 10
68 See Chapter 5, p. 000.
69 I have chosen here not to focus on Deleuze's return to perversion in his text on Tournier because I feel that it deals with a different kind of legality from the one I am interested here. Still, it is indicative that the second (for me, double) island for Deleuze, to some extent following Lacan and de Sade, is the embodiment of a perversion that comes out of an aberrant *withdrawal* from structure. Logos sees this as harm to another (see Deleuze, 2004b: 358), and this is according to Deleuze a basic legal misinterpretation. However, perversion, this bastard 'half juridical, half medical' (Deleuze, 2004b: 343) concept reveals a world without the Other yet with its own immanent and teleological structure – once again, a legality of the middle.

inhuman, elemental. There is no longer any lawscape to land on, nothing to catch the sedimentation of the law. The air has been recovered and released from any enclosure, the elements are running wild. The lawscapes have withdrawn, the bodies too. There is no point in space to which one can reorient oneself. The emerging justice is alone in the vastness left behind when both space and law withdraw. This justice augurs the end of the lawscape and the meteoric buoyancy of justice. This is no longer a spatial justice. This is an *elemental justice* that comes *after the end* of the era of the Anthropocene, when the inhuman eye will look back to a human withdrawal that has led nowhere. This is life without humans. This is justice without the lawscape. But this kind of justice can only be written in texts that repeat themselves, on the same island, with the same bodies, every time different.

Bibliography

Agamben, G., *The Highest Poverty: Monastic Rules and Form of Life*, trans. A. Kotsko, Stanford: Stanford University Press, 2013
Ahmed, S., *The Cultural Politics of Emotion*, Edinburgh: Edinburgh University Press, 2006
Aime, M., Allovio, S. and Viazzo, P.P., *Sapersi muovere. Pastori transumanti a Roaschia*. Roma: Meltemi, 2001
Albertson Fineman, M. and Grear, A. (eds), *Vulnerability: Reflections on a New Ethical Foundation for Law and Politics*, Farnham: Ashgate, 2013
Almond, P., 'The Dangers of Hanging Baskets: "Regulatory Myths" and Media Representations of Health and Safety Regulation', *Journal of Law and Society*, 36 (3), 352–75, 2009
Althusser, L., 'Ideology and Ideological State Apparatus (Notes Towards an Investigation)', Lenin and Philosophy and Other Essays, *New York Monthly Review Press*, 85–126, 2001
Amin, A., 'Collective Culture and Urban Public Space', *City* 12(1), 5–24, 2008
Amin, S., *Global History: A View from the South*, Cape Town: Pambazuka Press, 2011
Amin, A., *Land of Strangers*, Cambridge: Polity, 2012
Amin, A. and Thrift, N. *Cities: Reimagining the Urban*, Cambridge: Polity Press, 2002
Anderson, B., 'Affective Atmospheres', *Emotion, Space and Society* 2, 77–81, 2009
Andreini, B. *et al.*, Patent No. US 8544217 B2 (October 1, 2013), Washington DC: US Patent and Trademark Office, 2013
Appadurai, A., *Modernity at Large: Cultural Dimensions of Globalisation*, Minneapolis: University of Minnesota Press, 1996
Arendt, H., 'Collective Responsibility', in J.W. Bernauer (ed.), *Amor Mundi: Explorations in the Faith and Thought of Hannah Arendt*, Dordrecht: Martinus Nijhoff, 1987
Arias, S., 'The Geopolitics of Historiography from Europe to the Americas', in B. Warf and S. Arias (eds), *The Spatial Turn: Interdisciplinary Perspectives*, New York: Routledge, 2009
Arsic, B.,'Thinking Leaving' , in I. Buchanan and G. Lambert (eds), *Deleuze and Space*, Edinburgh: Edinburgh University Press, 2005
Arvidsson, M., 'Embodying Law in the Garden: An Autoethnographical Account of an Office of Law', *Australian Feminist Law Journal* 39, 25–49, 2014
Azuela, A., 'Mexico City: The City and Its Law In Eight Episodes, 1940 – 2005', in A. Philippopoulos-Mihalopoulos (ed.), *Law and the City*, London: Routledge, 2007

Babbage, C., *The Ninth Bridgewater Treatise*, London: John Murray, 1838
Bachmann, G. and Beyes, T., 'Media Atmospheres: Remediating Sociality, in M. Doulis and P. Ott (eds), *Remediate: At the Borders of Film, Internet and Archives*, Munich: Wilhelm Fink Verlag.
Baraban, R. and Durocher, J., *Successful Restaurant Design*, Hoboken, NJ: John Wiley, 2010
Barad, K., *Meeting the Universe Halfway: Quantum Physics and the Entanglement of Matter and Meaning*, Durham, NC: Duke University Press, 2006
Barad, K., 'Matter Feels, Converses, Suffers, Desires, Yearns and Remembers: Interview with Karen Barad', in R. Dophijn and I. van der Tuin (eds), *New Materialism: Interviews and Cartographies*, Ann Arbor, MI: Open Humanities Press, 2012
Barr, O., 'A Moving Theory: Remembering the Office of Scholar', *Law Text Culture* 14(1), 40–54, 2010
Battersby, G., 'Equitable Fraud', *Legal Studies* 15(35), 1995
Bauman, Z., *Liquid Love*, Cambridge: Polity, 2003
Beard, J., *The Political Economy of Desire: International Law, Development and the Nation State*, New York: Abingdon, 2007
Béland, D. and Lecours, A., 'The Politics of Territorial Solidarity: Nationalism and Social Policy Reform in Canada, the United Kingdom and Belgium', *Comparative Political Studies* 38(6), 676, 2005
Bell, V., 'Declining Performativity: Butler, Whitehead and Ecologies of Concern', *Theory, Culture & Society* 29(2), 107–23, 2012
Benjamin, W., *The Arcades Project*, trans. H. Eiland and K. McLaughlin, Cambridge, MA: Harvard University Press, 2002
Bennett, J., *The Enchantment of Modernity: Crossings, Energetics, and Ethics*, Princeton, NJ: Princeton University Press, 2001
Bennett, J., *Vibrant Matter: A Political Ecology of Things*, Durham, NC: Duke University Press, 2010
Bentley, L. and Flynn, L. (eds), *Law and the Senses: Sensational Jurisprudence*, London: Pluto Press, 1996
Bergmann, S., 'Theology in its Spatial Turn', *Religion Compass* 1(3), 353–79, 2007
Berlant, L., *The Female Complaint*, Durham, NC: Duke University Press, 2008
Berlant, L., *Cruel Optimism*, Durham, NC: Duke University Press, 2011
Bhabha, H., *The Location of Culture*, London: Routledge Classics, 2005
Bhandar, B., 'The Conceit of Sovereignty: Toward Post-Colonial Technique', in H. Lessard, R. Johnson, and J. Webber (eds), *Storied Narratives: Narratives of Contact and Arrival in Constituting Political Communities*, Vancouver: UBC Press, 2009
Bigot, D., 'Security, Exception, Ban and Surveillance', in D. Lyon (ed.), *Theorizing Surveillance: The Panopticon and Beyond*, Cullompton, Devon: Willan Publishing, 2006
Black, C., *The Land is the Source of the Law: A Dialogic Encounter with Indigenous Jurisprudence*, London: Routledge, 2011
Blackman, L., *Immaterial Bodies*, London: Sage, 2012
Blacksell, M., Watkins, C. and Economides, K., 'Human Geography and Law; A Case of Separate Developments in Social Sciences', *Progress in Human Geography* 10(3), 371–96, 1986
Blanchot, M., *The Space of Literature*, trans. A. Smock, Lincoln, NE: University of Nebraska Press, 1982

Blandy, S., 'Gating as Governance: The Boundaries Spectrum in Social and Situational Crime Prevention', in A. Crawford (ed.), *International and Comparative Criminal Justice and Urban Governance*, Cambridge: Cambridge University Press, 2011

Blok, A., 'Rams and Billy-goats: A Key to the Mediterranean Code of Honour', *Man* 16, 427–40, 1981

Blomley, N., *Law, Space and the Geographies of Power*, New York and London: Guilford Press, 1994

Blomley, N., 'Landscapes of Property', *Law and Society Review* 32(3), 567–612, 1998

Blomley, N., 'From 'What' to "So What?": Law and Geography in Retrospect', in J. Holder and C. Harrison (eds), *Law and Geography: Current Legal Issues Volume 5*, Oxford: Oxford University Press, 2002

Blomley, N., *Unsettling the City: Urban Land and the Politics of Property*, New York and London: Routledge, 2004

Blomley, N., 'The Spaces of Critical Geography', *Progress in Human Geography* 32(2), 285–93, 2008

Blomley, N., Delaney, D. and Ford, R. (eds), *The Legal Geographies Reader*, Oxford: Blackwell, 2001

Boano, C., 'Violent Spaces: Production and Reproduction of Security and Vulnerabilities', *The Journal of Architecture* 16(1), 37–55, 2011

Bohm, D., *Wholeness and the Implicate Order*, London: Routledge, 2002

Böhme, G., *Atmosphäre*, Frankfurt: Suhrkamp, 1995

Böhme, G., 'Atmosphere as the Subject Matter of Architecture', in P. Ursprung (ed.), *Herzog and deMeuron: Natural History*, Montreal: Lars Muller and Canadian Centre for Architecture, 398–406, 2005

Boltanski, L. and Chiapello, E., *The New Spirit of Capitalism*, London: Verso, 2007

Bonomi, A. and Albruzzese, A. (eds), *La Città Infinita*, Milano: Bruno Mondadori, 2004

Bonta, M. and Protevi, J., *Deleuze and Geophilosophy: A Guide and Glossary*, Edinburgh: Edinburgh University Press, 2004

Borch, C., 'Foam Architecture: Managing Co-isolated Associations', *Economy and Society* 37(4), 548–71, 2008

Borch, C., 'Foamy Business: On the Organizational Politics of Atmospheres', in W. Schinkel and L. Noordegraaf-Eelens, *In Medias Res: Peter Sloterdijk's Spherological Poetics of Being*, Amsterdam: Amsterdam University Press, 2011a

Borch, C., The *Politics of the Senses: Crowd Formation through Sensory Manipulation*, unpublished manuscript, 2011b

Borch, C., *The Politics of Crowds: An Alternative History of Sociology*, Cambridge: Cambridge University Press, 2012

Borch, C., 'The Politics of Atmospheres: Architecture, Power, and the Senses', in C. Borch (ed.), *Architectural Atmospheres: On the Experience and Politics of Architecture*, Basel: Birkhäuser, 2014

Bottomley, A., '*Shock to Thought: Encounters (of a Third Kind) with Legal Feminism*', *Feminist Legal Studies* 12(1), 29–65, 2004

Bottomley, A., 'From Walls to Membranes: Fortress Polis and The Governance of Urban Space in 21st Century Britain', *Law and Critique*, 18(2), 171–206, 2007a

Bottomley, A., '*A Trip to the Mall: Revisiting the Public/Private Divide*', in H. Lim and A. Bottomley (eds), *Feminist Perspectives on Land Law*, London: Routledge, 2007b

Bottomley, A., 'They Shall Be Simple in their Homes ... The Many Dimensions of Co-operative Housing', in A. Bottomley and S. Wong, *Changing Contours of Domestic Life, Family and Law: Caring and Sharing*, Oxford: Hart Publishing, 2009

Bottomley, A. and Moore, N., 'Blind Stuttering: Diagrammatic City', *Griffith Law Review* 17(2), 559–576, 2008

Bottomley, A. and Moore, N., 'Matters of Ownership: A "People's Port" for Dover?', *Northern Ireland Legal Quarterly* 64(3), 365–82, 2013

Braidotti, R., *Metamorphoses: Towards a Materialist Theory of Becoming*, Cambridge: Polity Press, 2002

Braidotti, R., *Transpositions: on Nomadic Ethics*, Cambridge: Polity Press, 2006

Braidotti, R., 'Between The No Longer and the Not Yet: Nomadic Variations on The Body', *Server Donne*(www.women.it/cyberarchive/files/braidotti.htm), 2007

Braidotti, R., 'Locating Deleuze's Eco-Philosophy: between Bio/Zoe-Power and Necro-Politics', in R. Braidotti, C. Colebrook and P. Hanafin (eds), *Deleuze and the Law: Forensic Futures*, Basingstoke: Palgrave Macmillan, 2009

Braidotti, R., *The Posthuman*, Cambridge: Polity, 2013

Brasier, R., 'Concepts and Objects', in L. Bryant, N. Srnicek and G. Harman (eds), *The Speculative Turn: Continental Materialism and Realism*, Melbourne: re.press, 2011

Braverman, I., 'Powers of Illegality: House Demolitions And Resistance in East Jerusalem', *Law and Social Inquiry* 32(2), 333–72, 2007

Braverman, I., 'Loo Law: The Public Washroom as a Hyper-Regulated Space', *Hastings Women's Law Journal* 20(1), 45–71, 2009

Braverman, I., 'Checkpoint Watch: Bureaucracy and Resistance at the Israeli/ Palestinian Border', *Social & Legal Studies* 21(3), 297–320, 2012

Braverman, I., 'Animal Frontiers: A Tale of Three Zoos in Israel/Palestine', *Cultural Critique* 85, 122–62, 2013

Braverman, I., Blomley, N., Delaney, D. and Kedar, A. (Sandy), 'The Expanding Spaces of Law: A Timely Legal Geography', *Buffalo Legal Studies Research Paper Series*, Paper No. 2013-32, 2013

Brennan, T., *The Transmission of Affect*, Ithaca, NY and London: Cornell University Press, 2004

Brenner, N., 'The Limits to Scale? Methodological Reflections on Scalar Structuration', *Progress in Human Geography* 25(4), 591–608, 2001

Brighenti, A. 'On Territory as Relationship and Law as Territory', *Canadian Journal of Law and Society/ Revue Canadienne Droit et Société* 21(2), 65–86, 2006

Brighenti, A., *Visibility in Social Theory and Social Research*, Basingstoke: Palgrave, 2010

Brighenti, A. (ed.), Urban Interstices: The Aesthetics and Politics of Spatial in-Betweens, Aldershot: Ashgate, 2013

Bryant, L., *The Democracy of Objects*, Michigan: Open Humanities Press, 2011

Bryant, L., Srnicek N. and Harman, G. (eds), *The Speculative Turn: Continental Materialism and Realism*, Melbourne: re.press, 2011

Buchanan, I. and Lambert, G. (eds), *Deleuze and Space*, Edinburgh: Edinburgh University Press, 2005

Butler, C., 'Géographie Critique du Droit et Production de l'Espace: Théorie et Méthode Selon l'oeuvre d'Henri Lefebvre', in P. Forest (ed.), *Géographie du Droit: Épistémologies, Dévelopements et Perspectives*, Québec: Presses de l'Université Laval, 2009

Butler, C., *Henri Lefebvre: Spatial Politics, Everyday Life and the Right to the City*, London: Routledge, 2013a

Butler, C., 'Space, Justice and Politics', paper presented at Spatial Justice Symposium, Stradmore Island, Australia, 2013b

Butler, J., *Bodies That Matter: On the Discursive Limits of 'Sex'*, London: Routledge, 1993

Butler, J., *The Psychic Life of Power: Theories in Subjection*, New York: Fordham University Press, 1997

Butler, R. and Parr, H. (eds), *Mind and Body Spaces: Geographies of Illness, Impairment and Disability*, London: Routledge, 1999

Cacciari, M., *L'Arcipelago*, Milan: Adelphi, 1997

Calvino, I., *Le Città Invisibili*, Milan: Arnoldo Mondadori, 1993

Canetti, E., *Crowds and Power*, trans. C. Stewart, New York: Farrar, Straus and Giroux, 1984

Carr, H., 'Utopias, Dystopias and the Temporalities of Social Housing: A Case Study of the Spa Green Estate London UK', *Australian Feminist Law Journal* 38(109), 2013

Caygill, H., *On Resistance: A Philosophy of Defiance*, London: Bloomsbury, 2013

Chandler, D., 'The World of Attachment? The Post-humanist Challenge to Freedom and Necessity', *Millennium – Journal of International Studies* 41, 516–35, 2013

Chatterjee, B., 'Cybercities: Under Construction', in A. Philippopoulos-Mihalopoulos (ed.), *Law and the City*, London: Routledge, 2007

Chatwin, B., *The Songlines*, London: Penguin, 1988

Chea P. and Grosz, E., 'The Body of the Law' , in P. Chea, D. Fraser and J. Grbich (eds), *Thinking Through the Body of the Law*, New York: New York University Press, 1996

Chettiparamb, A., 'Fractal Spatialities', *Environment and Planning D: Society and Space*, 31(3), 680–92, 2013

Choy, T., *Ecologies of Comparison: An Ethnography of Endangerment in Hong Kong*, Durham, NC: Duke University Press, 2011

Chryssostalis, J., 'Athens: The Boundless City and the Crisis of Law' , in A. Philippopoulos-Mihalopoulos (ed.), *Law and the City*, London: Routledge, 2007

Cloatre, E., 'TRIPS and Pharmaceutical Patents in Djibouti: An ANT Analysis of Socio-Legal Objects', *Social and Legal Studies* 17(2), 263–281, 2008

Cloatre, E. and Wright, N., 'A Socio-legal Analysis of an Actor-world: The Case of Carbon Trading and the Clean Development Mechanism', *Journal of Law and Society* 39(SI), 76–92, 2012

Clough, P.T., 'The Affective Turn: Political Economy, Biomedia and Bodies', in M. Gregg and G.J. Seigworth (eds), *The Affect Theory Reader*, Durham, NC: Duke University Press, 2010

Colebrook, C., 'Legal Theory after Deleuze', in R. Braidotti, C. Colebrook and P. Hanafin (eds), *Deleuze and Law: Forensic Futures*, Basingstoke: Palgrave Macmillan, 2009

Colebrook, C., *Death of the PostHuman: Essays on Extinction, Vol. 1*, Ann Arbor, MI: Open Humanities Press, 2014

Collinge, C., 'The Differance between Society and Space: Nested scales and the Returns of Spatial Fetishism', *Environment and Planning D: Society and Space* 23, 189–206, 2005

Connolly, W., *A World of Becoming*, Durham, NC: Duke University Press, 2011
Connolly, W., 'The "New Materialism" and the Fragility of Things', *Millennium – Journal of International Studies* 41, 399–412, 2013
Connor, S., 'Michel Serres' Five Senses', in D. Howes (ed.), *Empire of the Senses: The Sensual Culture Teader*, New York: Berg, 2005
Coole, D. and Frost, S., *New Materialisms: Ontology, Agency and Politics*, Durham, NC: Duke University Press, 2010
Cooper, D., 'Talmudic Territory? Space, Law, and Modernist Discourse', *Journal of Law and Society* 23(4), 529–48, 1996
Cooper, D., *Governing out of Order: Space, Law and the Politics of Belonging*, London: Rivers Oram Press, 1998
Corbin, A., *The Foul and the Fragrant*, trans. A. Sheridan, Leamington Spa: Berg, 1986
Cover, R., 'Violence and the Word', *Yale Law Journal* 95, 1986
Crang, M. and Thrift, N., *Thinking Space*, London: Routledge, 2000
Cresswell, T., *On the Move: Mobility in the Modern Western World*, London: Routledge, 2006
Crowe, N., *Nature and the Idea of a Man-Made World: An Investigation into the Evolutionary Roots of Form and Order in the Built Environment*, Cambridge: MA: MIT Press, 1997
Crutzen, P.J. and Stoermer, E.F., 'The Anthropocene', *IGBP Newsletter* 41(17), 17–18, 2000
Cunningham, D., 'The Concept of Metropolis: Philosophy and Urban Form', *Radical Philosophy* 133(September/October), 13–25, 2005
Cunningham, D., 'Spacing Abstraction: Capitalism, Law and the Metropolis', *Griffith Law Review* 17(2), 454–69, 2008
Darwin, C., *The Expression of Emotions in Man and Animals*, Chicago: University of Chicago Press, 1965
Davies, M., 'Feminism and the Flat Law Theory', *Feminist Legal Studies* 16, 281–304, 2008
Davoudi, S., 'Understanding Territorial Cohesion', *Planning, Practice and Research* 20(4), 433–41, 2005
de Certeau, M., *The Practice of Everyday Life*, trans. S. Rendall, Berkeley: University of California Press, 1984
De Landa, M., *Intensive Science and Virtual Philosophy*, London: Continuum, 2002
De Landa, M., 'Space: Extensive and Intensive, Actual and Virtual', in I. Buchanan and G. Lambert (eds), *Deleuze and Space*, Edinburgh: Edinburgh University Press, 2005
Delaney, D., *Race, Place, and the Law, 1836–1948*, Austin: University Texas Press, 1998
Delaney, D., 'Beyond the World: Law as a Thing of this World', in J. Holder and C. Harrison (eds), *Law and Geography: Current Legal Issues Volume 5*, Oxford: Oxford University Press, 2002
Delaney, D., *The Spatial, The Legal and the Pragmatics of World-Making*, London: Routledge, 2010
Delaney, D. and Leitner, H., 'The Political Construction of Scale', *Political Geography* 16(2), 93–7, 1997
Deleuze, G., 'Une Theorie d'Autrui (Autrui, Robinson et le Pervers)', *Critique* 241, 503–25, 1967
Deleuze, G., *Cinema 1: The Movement-Image*, London: Athlone Press, 1986

Deleuze, G., *Spinoza: Practical Philosophy*, trans. R. Hurley, San Francisco: City Light Books, 1988
Deleuze, G., *Cinema 2: The Time Image*, trans. H. Tomlinson and R. Galeta, London: Athlone Press, 1989
Deleuze, G., *Expressionism in Philosophy: Spinoza*, trans. M. Joughin, New York: Zone Books, 1990
Deleuze, G., 'The Fold: Leibniz and the Baroque: The Pleats of Matter', *Architectural Design Profile* 102, 1993
Deleuze, G., *Negotiations*, trans. M. Joughin, New York: Columbia University Press, 1995
Deleuze, G., *Essays Critical and Clinical*, trans. D. Smith and M. Greco, London: Verso, 1997
Deleuze, G., *Desert Islands and Other Texts, 1953–1974*, trans. M. Taormina, Los Angeles: Semiotext(e), 2004a
Deleuze, G., *The Logic of Sense*, trans. M. Lester with C. Stivale, London: Continuum, 2004b
Deleuze, G., *Difference and Repetition*, trans. P. Patton, London: Continuum, 2004c
Deleuze, G., *The Fold: Leibniz and the Baroque*, trans. T. Conley, Continuum: London, 2006
Deleuze, G., *Cours Vincennes: Sur Spinoza*, 17.02.1981, Cours Vincennes University Lecture, www.webdeleuze.com/php/sommaire.html, trans. L. Lambert, in L. Lambert, 'A Sunflower Seed Lost in a Wall is Capable of Shattering that Wall', *The Funambulist Pamphlets* 01: Spinoza, 73–75, 2013
Deleuze, G. and Guattari, F., *Anti-Oedipus*, trans. R. Hurley, M. Seem and H. R. Lane, Minneapolis: University of Minnesota Press, 1983
Deleuze, G. and Guattari, F., *A Thousand Plateaus: Capitalism and Schizophrenia*, trans. B. Massumi, London: Athlone Press, 1988
Deleuze, G. and Guattari, F., *What is Philosophy?*, trans. H. Tomlinson and G. Burchell, New York: Columbia University Press, 1994
Deleuze, G. and Parnet, C., *Dialogues II*, New York: Columbia University Press, 2007
Della Rocca, M., *Spinoza*, London: Routledge, 2008
De Marchi, V., *Fame d'Erba. Etnografia dei Pastori Vaganti del Triveneto*. Tesi di laurea specialistica in Antropologia culturale e Etnografia, Università degli studi di Siena, 2009a
De Marchi, V., 'Pastori del Duemila. Sguardi, Voci e Strategie dei Vaganti del Triveneto', in A. Malacarne (ed.), *Transumanze. Sulle Tracce degli Ultimi Pastori del Triveneto*, Feltre: Agorà Libreria Editrice, 2009b
De Marchi, V., 'I Pastori Transumanti: Risorsa o Depauperamento del Territorio?', in *Le Dolomiti Bellunesi*, Belluno: Rassegna delle sezioni bellunesi del CAI, 2010
Dean, M., Governmentality: Power and Rule in Modern Society, London: Sage, 2010
Dereck, G., '"In Another Time Zone, The Bombs Fall Unsafely": Targets, Civilians and Late Modern War', *Arab World Geographer* 9(88), 2006
Deriu, D. and Kamvasinou, K., 'Critical Perspectives on Landscape: Introduction', *The Journal of Architecture* 17(1), 1–9, 2012
Derrida, J., *Positions*, trans. A. Bass, London: Athlone Press, 1972
Derrida, J., 'Force of Law: The "Mystical Foundation of Authority"', trans. M. Quaintance, in D. Cornell, M. Rosenfeld and D. Gray Carlson (eds), *Deconstruction and the Possibility of Justice*, New York: Routledge, 1992

Derrida, J., *Adieu to Emmanuel Levinas*, trans. P. Brault and M. Naas, Stanford, CA: Stanford University Press, 1999
Derrida, J., 'On Cosmopolitanism', in *On Cosmopolitanism and Forgiveness*, London: Routledge, 2001
d'Souza, R., *Interstate Disputes over Krishna Waters: Law, Science and Imperialism*, Hyderabad: Orient Longman, 2006
de Sutter, L., *Deleuze: La Pratique du Droit*, Paris: Michalon, 2008
Dikeç, M., 'Justice and the Spatial Imagination', *Environment and Planning A* 33, 1785–1805, 2001
Dikeç, M., 'Police, Politics, and the Right to the City', *GeoJournal* 58, 91–98, 2002
Doel, M., *Poststructuralist Geographies: The Diabolical Art of Spatial Science*, Edinburgh: Edinburgh University Press, 1999
Dolphijn, R. and van der Tuin, I., *New Materialism: Interviews and Cartographies*, Ann Arbor: Open Humanities Press University of Michigan Library, 2012
Dorian, M., 'Utopia on Ice', *Cabinet: A Quarterly of Art and Culture* 47, 25–32, 2012
Dorsett, S. and McVeigh, S., 'Conduct of Laws: Native Title Responsibilty, and Some Limits of Jurisdictional Thinking', *Melbourne University Law Review* 36, 470–93, 2012
Douglas, N., 'The Overall Sale Experience', *Socialist Review* 304(March), 2006
Douzinas, C., 'Stasis Syntagma: The Names and Types of Resistance', in M. Stone, I. Wall and C. Douzinas (eds), *New Critical Legal Thinking: Law and the Political*, Abingdon: Routledge, 2012
Drobnick, J., 'Volatile Effects: Olfactory Dimensions of Art and Architecture', in D. Howes (ed.), *Empire of the Senses*, Oxford: Berg, 2005
Ellem, B. and Shields, J., 'Rethinking "Regional Industrial Relations": Space, Place and the Social Relations of Work', *Journal of Industrial Relations*, 41(4), 536–60, 1999
Erkip, F., 'The Rise of the Shopping Mall in Turkey: The Use and appeal of a Mall in Ankara', *Cities* 22(2), 89–108, 2005
Eslava, L. and Pahuja, S., 'Beyond The (Post)Colonial: Twail and The Everyday Life of International Law', *Journal of Law and Politics in Africa, Asia and Latin America –Verfassung und Recht in Übersee (VRÜ)* 45(2), 195–221, 2012
Esposito, R., *Bios: Biopolitics and Philosophy*, trans. and introduced T. Campbell, Minneapolis: University of Minnesota Press, 2008
Esposito, R., *Immunitas: The Protection and Negation of Life*, trans. Z. Hanafi, Cambridge: Polity Press, 2011
Fainstein, S., *The Just City*, Ithaca, NY and London: Cornell University Press, 2010
Fake Industries Architectural Agonism, 'Should "We Patent Architectural Knowledge', *The Shape of Law: Special Issue* 4(38), 2014
Fall, J., *Drawing the Line: Nature, Hybridity and Politics in Transboundary Spaces*, Aldershot: Ashgate, 2005
Fanon, F., *The Wretched of the Earth*, trans. C. Farrington, New York: Grove Press, 1963
Farinelli, F., *Geografia*, Torino: Einaudi, 2003
Fernandes, J.K., 'Panjim: Realms of Laws and Imagination', in A. Philippopoulos-Mihalopoulos (ed.), *Law and the City*, London: Routledge, 2007
Finnegan, R., *Communicating: The Multiple Modes of Human Interconnection*, London: Routledge, 2002

Ford, R., 'Law's Territory (A History of Jurisdiction)', in N. Blomley, D. Delaney and R. Ford (eds), *The Legal Geographies Reader*, Oxford: Blackwell, 200–16, 2001
Foucault, M., *Power/knowledge: Selected Interviews and Other Writings 1972–1977*, Brighton: Harvester Press, 1980
Foucault, M., 'Of Other Spaces', trans. J. Miskowiec, *Diacritics* 16, 22–7, 1986
Foucault, M., *The Birth of Clinic*, London: Routledge Classics, 2003a
Foucault, M. *Society Must be Defended: Lectures at the Collège de France*, 1975–1976, (ed.), M. Bertani and A. Fontana, trans. D. Macey, New York: Picador, 2003b
Foucault, M., *Abnormal: Lectures at the Collège De France 1974–1975*, trans. G. Burchell, London: Verso, 2003c
Foucault, M., 'The Language of Space', trans. G. Moore, in S. Elden and J. Crampton (eds), *Space, Knowledge and Power: Foucault and Geography*, Aldershot: Ashgate, 2007
Gagnol, L. and Afane, A., 'When Injustice is Spatial. Pastoral Nomadism and the Territorial Imperative in Niger's Sahara region', trans. C. Hancock, *Spatial Justice/Justice Spatiale* 2(October), 2010
Gatens, M., 'Through a Spinozist Lens: Ethology, Difference, Power', in P. Patton (ed.), *Deleuze: A Critical Reader*, Oxford: Blackwell, 1996a
Gatens, M., *Imaginary Bodies: Ethics, Power and Corporeality*, London: Routledge, 1996b
Gatens, M. and Lloyd, G., *Collective Imaginings: Spinoza, Past and Present*, London: Routledge, 1999
Gibas, P., 'Uncanny Underground: Absences, Ghosts and the Rhythmed Everyday of the Prague Metro', *Cultural Geographies* 20, 485–99, 2013
Goodrich, P., 'First We Take Manhattan: Microtopia and Grammatology in Gotham', in A. Philippopoulos-Mihalopoulos (ed.), *Law and the City*, London: Routledge, 2007
Goodrich, P., *Legal Emblems and the Art of Law*, Cambridge: Cambridge University Press, 2014
Goodrich, P., Douzinas, C. and Hachamovitch Y., 'Politics, Ethics and the Legality of the Contingent', in C. Douzinas, P. Goodrich and Y. Hachamovitch (eds), *Politics, Postmodernity and Critical Legal Studies*, London: Routledge, 1994
Goodwin Dunleavy, P., 'Professionalism – A Mock Trial: People of Georgia v. the Modern Lawyer', *Journal of Southern Legal History* 3, 257–96, 1994
Goonewardena, K., 'The Urban Sensorium: Space, Ideology and the Aestheticization of Politics', *Antipode* 37(1), 46–71, 2005
Grabham, E., 'Shaking Mr Jones: Law and Touch', *International Journal of Law in Context* 5(4), 343–53, 2009
Graham, N., *Lawscape: Property, Environment, Law*, London: Routledge, 2011
Grear, A., 'Challenging Corporate 'Humanity': Legal Disembodiment, Embodiment and Human Rights', *Human Rights Law Review* 7(3), 511–43, 2007
Grear, A., 'The Vulnerable Living Order: Human Rights and the Environment in A Critical and Philosophical Perspective', *Journal of Human Rights and the Environment* 2(1), 23–44, 2011
Grear, A., 'Human Rights, Property and the Search for Worlds Other', *Journal of Human Rights and the Environment* 3(2), 173–95, 2012
Greed, C., *Women and Planning*, London: Routledge, 1994
Greenfield, A., *Everyware: The Dawning Age of Ubiquitous Computing*, Boston, MA: New Riders, 2006

Gregg, M. and Seigworth, G.J. (eds), *The Affect Theory Reader*, Durham, NC: Duke University Press, 2010

Gregory, D., *Geographical Imaginations*, Oxford: Blackwell, 1993

Gregory, D., 'From a View to Kill: Drones and Late Modern War', *Theory, Culture and Society* 27(7–8), 190–4, 2011

Gregory, D., Martin R. and Smith, G. (eds), *Human Geography: Society, Space and Social Science*, Basingstoke: Palgrave, 1994

Grosz, E., *Space, Time and Perversion*, London: Routledge, 1995

Guattari, F., *Three Ecologies*, trans. I. Pindar and P. Sutto, New Brunswick, NJ: Athlone, 2000

Haldar, P., 'In and Out of Court: On Topographies of Law and the Architecture of Court Buildings', *International Journal for the Semiotics of Law* 7(2), 185–200, 1994

Halsey, M., *Deleuze and Environmental Damage: Violence of the Text*, Aldershot: Ashgate, 2006

Halsey, M., 'Majesty and Monstrosity: Deleuze and the Defence of Nature', in A. Philippopoulos-Mihalopoulos (ed.), *Law and Ecology*, London: Routledge, 2011

Haraway, D., *The Haraway Reader*, London: Routledge, 2004

Haraway, D., *When Species Meet*, Minneapolis: University of Minnesota Press, 2008

Hardt, M. and Negri, A., *Empire*, Cambridge, MA: Harvard University Press, 2001

Hardt, M. and Negri, A., *Multitude: War and Democracy in the Age of Empire*, New York: Penguin, 2004

Harman, G., *Guerrilla Metaphysics: Phenomenology and the Carpentry of Things*, Chicago: Open Court, 2005

Harman, G., *Prince of Networks: Bruno Latour and Metaphysics*, Melbourne: re.press, 2009

Harman, G., *The Quadruple Object*, Winchester: Zero Books, 2011

Harvey, D., *Social Justice and the City*, Oxford: Blackwell, 1973

Harvey, D., *The Condition of Postmodernity. An Enquiry into the Origins Cultural Change*, Oxford: Blackwell, 1990

Harvey, D., *Justice, Nature and the Geography of Difference*, Oxford: Blackwell, 1996

Harvey, D., *Spaces of Hope*, Edinburgh: Edinburgh University Press, 2000

Harvey, D., *Spaces of Capital: Towards a Critical Geography*, London: Routledge, 2001

Harvey, D., 'The Right to the City', in L. Lees (ed.), *The Emancipatory City: Paradoxes and Possibilities*, London: Sage, 2004

Hay, A., 'Concepts of Equity, Fairness and Justice in Geographical Studies', *Transactions of the Institute of British Geographers*, 20(4), 500–508, 1995

Hayles, K., *My Mother Was a Computer: Digital Subjects and Literary Texts*, Chicago: Chicago University Press, 2005

Heidegger, M., *What is Called Thinking?*, trans. J. Glenn Gray, New York: Harper Perennial, 1976

Hemmings, C., 'Invoking Affect: Cultural Theory and the Ontological Turn', *Cultural Studies* 19(5), 548–67, 2002

Herzogenrath, B. (ed.), *An [Un]Likely Alliance: Thinking Environment[s] with Deleuze| Guattari*, Newcastle-upon-Tyne: Cambridge Scholars Publishing, 2008

Herzogenrath, B. (ed.), *Deleuze|Guattari and Ecology*, Basingstoke: Palgrave Macmillan, 2009

Holder, J. and Harrison, C. (eds), *Law and Geography*, Oxford: Oxford University Press, 2003

Holmqvist, C., 'Undoing War: War Ontologies and the Materiality of Drone Warfare', *Millennium – Journal of International Studies* 41, 535–52, 2013
Hou, J. (ed.), *Insurgent Public Space*, London: Routledge, 2010
Howes, D., 'Hyperesthesia, or, the Sensual Logic of Late Capitalism', in D. Howes (ed.), *Empire of the Senses*, Oxford: Berg, 2005
Hubbard, P., 'Thinking Spaces, Differently?', *Dialogues in Human Geography* 2(1), 23–6, 2012
Hubbard, P., Kitchin R., Bartley, B. and Fuller, D., *Thinking Geographically: Space, Theory and Contemporary Human Geography*, London: Continuum, 2002
Hyde, A., *Bodies of Law*, Princeton, NJ: Princeton University Press, 1997
Ingold, T., *Lines: A Brief History*, London: Routledge, 2007
Ingraham, C., *Architecture, Animal, Human: The Asymmetrical Condition*, London: Routledge, 2006
Irigaray, L., *L'Oubli de l'Air chez Martin Heidegger*, Paris: Les Editions de Minuit, 1983
Irigaray, L., '*La Mystérique*', in *Speculum of the Other Woman*, trans. G. Gill, Ithaca, NY: Cornell University Press, 1985a
Irigaray, L., *This Sex which is not One*, trans. C. Porter with C. Burke, Ithaca, NY: Cornell University Press, 1985b
Irigaray, L., *Marine Lover: Of Friedrich Nietzsche*, trans. G. Gill, New York: Columbia University Press, 1991
Jensen, O., 'Negotiation in Motion: Unpacking a Geography of Mobility', *Space and Culture* 13, 389, 2010
Jeyaraj, P., 'The Instagram Lawscape', *Westminster Law Review* 3(1), 2013 www.westminsterlawreview.org/wlr17.php
Johnson, E., Morehouse, H., Dalby, S., Lehman, J., Nelson, S., Rowan, R., Wakefield, S. and Yusoff, K., 'After the Anthropocene: Politics and Geographic Inquiry for a New Epoch', *Progress in Human Geography*, 1–18, 2014
Jones, M., 'Limits to Thinking Space Relationally', *International Journal of Law in Context* 6(3), 243–55, 2010
Joseph, J. 'Resilience as Embedded Neoliberalism: A Governmentality Approach', *Resilience: International Policies, Practices and Discourses* 1(1), 38–52, 2013
Joyce, R.J., *Competing Sovereignties*, London: Routledge, 2013
Jung, C., 'Mysterium Coniunctionis: An Inquiry into the Separation and Synthesis of Psychic Opposites in Alchemy', *Collected Works* 14, trans. R.F.C. Hull, London: Routledge & Kegan Paul, 1963
Kamenka, E. and Tay, A.E.S. (eds), *Justice*, New York: St Martins Press, 1980
Kaplan, C., *Questions of Travel*, Durham, NC: Duke University Press, 1996
Karki, M.M.S., 'Nontraditional Areas of Intellectual Property Protection', *Journal of Intellectual Property Rights*, 10(3), 499–506, 2005
Karplus Y. and A. Meir, 'The Production of Space: A Neglected Perspective in Pastoral Research', *Environment and Planning D: Society and Space* 31(1), 23–42, 2013
Karremans, J., Stroebe, W. and Claus, J., 'Beyond Vicary's Fantasies: The Impact of Subliminal Priming and Brand Choice', *Journal of Experimental Social Psychology* 42(6), 792–8, 2006
Kedar, A. (Sandy), 'On the Legal Geography of Ethnocratic Settler States: Notes Towards A Research Agenda', in J. Holder and C. Harrison (eds), *Current Legal Issues: Law and Geography*, Oxford: Oxford University Press, 402–39, 2003

Keenan, S., 'Subversive Property: Reshaping Malleable Spaces of Belonging', *Social and Legal Studies* 19(4), 423–39, 2010

Keenan, S., 'Property as Governance: Time, Space and Belonging in Australia's Northern Territory Intervention', *Modern Law Review* 76(3), 464–93, 2013

Kerruish, V., 'At the Court of the Strange God', *Law and Critique* 13(3), 2002

Kierkegaard, S., *Either/Or*, trans. W. Lowrie, Princeton, NJ: Princeton University Press, 1949

Kierkegaard, S., *Repetition and Philosophical Crumbs*, trans. M.G. Piety, Oxford: Oxford World Classics, 2009

Kisner, M., *Spinoza on Human Freedom: Reason, Autonomy and the Good Life*, Cambridge: Cambridge University Press, 2011

Kitchin, R. and Dodge, M., *Code/Space: Software and Everyday Life*, Cambridge, MA: MIT Press, 2011

Koskenniemi, M., *From Apology to Utopia. The Structure of International Legal Argument*, Helsinki: Finnish Lawyers' Pub. Co., 1989

Kotsakis, A., 'Heterotopias of the Environment: Law's Forgotten Spaces', in A. Philippopoulos-Mihalopoulos (ed.), *Law and Ecology*, London: Routledge, 2011

Koziak, B., 'Homeric *Thumos*: The Early History of Gender, Emotion, and Politics', *The Journal of Politics* 61, 1068–91, 1999

Lakhani, D., *Subliminal Persuasion: Influence and Marketing Secrets They Don't Want you to Know*, Hoboken, NJ: Wiley, 2008

Lakoff, G. and Johnson, M., *Philosophy in the Flesh*, London: HarperCollins, 1999

Lambert, L., *Weaponized Architecture: The Impossibility of Innocence*, New York: DPR, 2011

Lambert, L., *The Funambulist Pamphlets 01: Spinoza*, New York: Punctum Books, 2013a

Lambert, L., *The Funambulist Pamphlets 06: Palestine*, New York: Punctum Books, 2013b

Lash, S., *Critique of Information*, London: Sage, 2001

Latour, B., *We Have Never Been Modern*, trans. C. Porter, Cambridge, MA: Harvard University Press, 1993

Latour, B., *The Making of Law: An Ethnography of the Conseil D'État*, Cambridge: Polity Press, 2009

Latour, B., 'Some Experiments in Art and Politics', *E-flux* 23(3), 2011

Law, J., 'After ANT: Complexity, Naming and Complexity', in J. Law and J. Hassard (eds), *Actor Network Theory and After*, Oxford: Blackwell, 1999

Law, L., 'Defying Disappearance: Cosmopolitan Public Spaces in Hong Kong', *Urban Studies* 39(9), 1625–1645, 2002

Layard, A., 'Shopping in the Public Realm: The Law of Place', *Journal of Law and Society* 37(3), 412–41, 2010

Layard, A., 'Property Paradigms and Place-making: A Right to the City; A Right to the Street?', *Journal of Human Rights and the Environment* 3(2), 254–72, 2012

Leander, A., 'Technological Agency in the Co-Constitution of Legal Expertise and the US Drone Program', *Leiden Journal of International Law* 26, 811–31, 2013

Le Bon, G., *The Crowd: A Study for the Popular Mind*, New York: Viking, 1960

LeBuffe, M., 'Spinoza's Normative Ethics', *Canadian Journal of Philosophy* 37(3), 371–92, 2007

Lefebvre, A., *The Image of Law: Deleuze, Bergson, Spinoza*, Stanford: Stanford University Press, 2008

Lefebvre, H., *The Production of Space*, trans. D. Nicholson-Smith, Oxford: Blackwell, 1991

Lefebvre, H., *Writings on Cities*, trans. E. Kofman and E. Lebas (eds), Oxford: Blackwell, 1996

Leibniz, G.W., *Discourse on Metaphysics and Other Essays*, trans. D. Garber and R. Ariew, London: Hackett Publishing, 1992

Lenco, P., *Deleuze and World Politics*, London: Routledge, 2012

Lesley, E., 'Snow Shakers', in F. Candlin and R. Guins (eds), *The Object Reader*, London: Routledge, 2009

Levinas, E., *Otherwise than Being: Or Beyond Essence*, trans. A. Linghis, Dordrecht: Kluwer, 1978

Lewis, R., 'Veils and Sales: Muslims and the Spaces of Postcolonial Fashion Retail', *Fashion Theory: The Journal of Dress, Body & Culture* 11(4), 423–41, 2007

Little, J., *Gender, Planning and the Policy Process*, Oxford: Pergamon, 1994

Llewelyn, J., *Margins of Religion: Between Kierkegaard and Derrida*, Bloomington: Indiana University Press, 2009

Lloyd, G., *Spinoza and the Ethics*, London: Routledge, 1996

Loizidou, E., 'The Body Figural and Material in The Work of Judith Butler', *Australian Feminist Law Journal* 28, 29–51, 2008

Longhurst, R., *Exploring Fluid Boundaries*, London: Routledge, 2000

López, G., 'Before the Law', *The Shape of Law: Special Issue* 4(38), 2014

Lorimer, H., 'Herding Memories of Humans and Animals', *Environment and Planning D: Society and Space* 24, 497–518, 2006

Lorimer, J., 'Multinatural Geographies for the Anthropocene', *Progress in Human Geography* 36(5), 593–612, 2012

Löw, M., 'The Constitution of Space: The Structuration of Spaces Through the Simultaneity of Effect and Perception', *European Journal of Social Theory* 11(1), 25–49, 2008

Lowe, V., 'The Politics of Law-Making: Are the Method and Character of Norm Creation Changing?', in M. Byers (ed.), *The Role of Law in International Politics*, Oxford: Oxford University Press, 2000

Luhmann, N., 'The Morality of Risk and the Risk of Morality', *International Review of Sociology* 3, 87–101, 1987

Luhmann, N., 'Closure and Structural Coupling', *Cardozo Law Review* 13, 1419–42, 1992

Luhmann, N., *Risk: A Sociological Theory*, trans. R. Barrett, New York: Aldine de Gruyter, 1993

Luhmann, N., *Social Systems*, trans. J. Bednarz, Jr., Stanford, CA: Stanford University Press, 1995

Luhmann, N., *Observations on Modernity*, trans. W. Whobrey, Stanford, MA: Stanford University Press, 1998

Luhmann, N., *Theories of Distinction: Redescribing the Descriptions of Modernity*, ed. and introduced by W. Rasch, trans. J. O'Neil *et al.*, Stanford, CA: Stanford University Press, 2002

Luhmann, N., *Law as a Social System*, trans. K. Ziegert, (ed.), F. Kastner, R. Nobles, D. Schiff and R. Ziegert, Oxford: Oxford University Press, 2004

Luhmann, N., *Theory of Society, Volume 1*, trans. R. Barrett, Stanford, CA: Stanford University Press, 2012

Luhmann, N., *Theory of Society, Volume 2*, trans. R. Barrett, Stanford, CA: Stanford University Press, 2013

Lyell, C., *Principles of Geology*, London: Penguin, 1997

Lyotard, J-F., *The Differend: Phrases in Dispute*, trans. G. van den Abbeele, Minneapolis: University of Minnesota Press, 1988

Lyotard, J.-F., *Libidinal Economy*, trans. I. Hamilton Grant, London: Athlone Press, 1993

McAuslan, P., 'Law And The Poor: The Case Of Dar Es Salaam', in A. Philippopoulos-Mihalopoulos (ed.), *Law and the City*, London: Routledge, 2007

McCraty, R., Atkinson, M. and Trevor Bradley, R., 'Electrophysiological Evidence of Intuition: Part 1. The Surprising Role of the Heart', *Journal of Alternative and Complementary Medicine* 10(1), 133–43, 2004

McCormack, D., 'Geography and Abstraction: Towards an Affirmative Critique', *Progress in HumFan Geography* 36(6), 715–34, 2012

McDonough, T. (ed.), *Guy Debord and the Situationist International: Texts and Documents (October Books)*, Cambridge, MA: MIT Press, 2004

McVeigh, S. and Pahuja, S., 'Rival Jurisdictions: The Promise and Loss of Sovereignty', in C. Barbour and G. Pavlich (eds), *After Sovereignty: On the Question of Political Beginnings*, London: Routledge, 2010

MacMurray, R., 'Trademarks or Copyrights: Which Intellectual Property Right Affords Its Owner the Greatest Protection of Architectural Ingenuity?', *Northwestern Journal of Technology and Intellectual Property* 3(1), 111–120, 2005

Mahmud, T., 'Geography and International Law: A Postcolonial Mapping', *Santa Clara Journal of International Law* 5(2), 525–61, 2007

Mahmud, T., 'Law of Geography and the Geography of Law: A Postcolonial Mapping', *Washington University Jurisprudence Review*, 3(1), 64–106, 2010a

Mahmud, T., '"Surplus Humanity" and Margins of Law: Slums, Slumdogs, and Accumulation by Dispossession', *Chapman Law Review* 14(1), 2010b

Malabou, C., *What Should we do with Our Brain?*, New York: Fordham Press, 2008

Manderson, D. (ed.), 'Legal Spaces', *Law Text Culture* 9 (Special Issue), 2005a

Manderson, D., 'Proximity and the Ethics of Law', *UNSW Law Journal*, 28(3), 697–720, 2005b

Marder, M., *Plant Thinking: A Philosophy of Vegetal Life*, New York: Columbia University Press, 2013

Marlin-Bennett, R., Embodied Information, Knowing Bodies, and Power', *Millennium – Journal of International Studies* 41, 601, 2013

Marston, S., Jones, J. P. III and Woodward, K., 'Human geography without scale', *Transactions of the Institute of British Geographers*, 30(4), 416–32, 2005

Martin, A., 'Trademark Infringement Lawsuit between DPP & Jay-Z Now Settled', *WrestleView*, 2007

Marx, E., 'Vi Sono Pastori Nomadi nel Medio Oriente Arabo?', in U. Fabietti and P. C. Salzman (eds), *Antropologia delle Societa Pastorali Tribali e Contadine*, Pavia: Ibis, 1996

Marx, G.T., 'A Tack in the Shoe: Neutralizing and Resisting New Surveillance', *Journal of Social Issues* 59(2), 369–90, 2003

Massey, D., *Space, Place and Gender*, Cambridge: Polity, 1994

Massey, D., 'Geographies of Responsibility', *Geografiska Annaler, Series B, Human Geography, Special Issue: The Political Challenge of Relational Space* 86(1), 5–18, 2004

Massey, D., *For Space*, London: Sage, 2005
Massey, D., *World City*, Cambridge: Polity, 2007
Massumi, B. (ed.), *The Politics of Everyday Fear*, Minneapolis: University of Minnesota Press, 1993
Massumi, B., *Parables for the Virtual: Movement, Affect, Sensation*, Durham, NC: Duke University Press, 2002
Massumi, B., 'National Enterprise Emergency: Steps towards an Ecology of Powers', *Theory, Culture and Society* 26(6), 153–85, 2009
Maturana, H. and Varela, F., *Autopoiesis and Cognition*, Dordrecht: Reidel Publishing, 1972
Mawani, R., 'Insects, War, Plastic Life', in B. Bhandar and J. Goldberg-Hiller (eds), *Plastic Materialities: Legality, Politics, and Metamorphosis in the Work of Catherine Malabou*, Durham, NC: Duke University Press, 2014
Meillassoux, Q., *After Finitude*, trans. R. Brassier, London: Continuum, 2008
Menezes, S., 'Smelly Rights: Copyright in Perfume', *Patent Baristas*, 2010
Merelle Ward, A., 'Letter from Amerikat: Happy Bunny Day!', *The IPKat*, 2011
Merleau-Ponty, M., *The Visible and the Invisible*, trans. A. Lingis, Evanston, IL: Northwestern University Press, 1968
Merleau-Ponty, M., *Phenomenology of Perception*, trans. C. Smith, London: Routledge, 1995
Merriman, P., Jones, M., Olsson, G., Sheppard, E., Thrift, N. and Tuan, Y. F., 'Space and Spatiality in Theory', *Dialogues in Human Geography* 2(1), 3–22, 2012
Michulak, M., 'The Rhizomatics of Domination', in B. Herzogenrath (ed.), *An [Un]Likely Alliance: Thinking Environment[s] with Deleuze|Guattari*, Newcastle-upon-Tyne: Cambridge Scholars, 2008
Mieville, C., *The City and The City*, London: Pan Macmillan, 2011
Miller, D., *Social Justice*, Oxford: Oxford University Press, 1979
Milovanovic, D., 'Diversity, Law and Justice: A Deleuzian Semiotic View of "Criminal Justice"', *International Journal for the Semiotics of Law* 20, 55–79, 2007
Minton, A., *Ground Control: Fear and Happiness in the Twenty-first Century City*, London: Penguin, 2009
Mirowski, P. and Nik-Khah, E., 'Markets Made Flesh: Performativity, and a Problem in Science Studies, Augmented with Consideration of the FCC Auctions', in D. MacKenzie, F. Muniesa and L. Siu (eds), *Do Economists Make Markets? On the Performativity of Economics*, Princeton, NJ: Princeton University Press, 2007
Mitchell, D., *The Right to the City*, New York: Guilford Press, 2003
Mitchell, P., 'Geographies/aerographies of Contagion', *Environment and Planning D: Society and Space* 29, 533–50, 2011
Mohr, R., 'Law and Identity in Spatial Contests', *National Identities* 5(1), 53–66, 2003
Moore, N., 'Icons of Control: Deleuze, Signs, Law', *International Journal for the Semiotics of Law* 20, 33–54, 2007
Moore, N., 'Get Stupid: Film and Law via Wim Venders and Others', *Cardozo Law Review* 31(4), 1195–1216, 2010
Moore, N., 'The Perception of the Middle', in L. de Sutter and K. McGee (eds), *Deleuze and Law*, Edinburgh: Edinburgh University Press, 2012
Moore, N., 'Diagramming Control', in P. Rawes (ed.), *Relational Architectural Ecologies: Architecture, Nature, Subjectivity*, London: Routledge, 2013

Moore, N. and Bottomley, A., 'Law, Diagram, Film: Critique Exhausted', *Law and Critique* 23(2), 163–82, 2012

Moran, L., 'The Queen's Peace: Reflections on the Spatial Politics of Sexuality in Law', in J. Holder and C. Harrison (eds), *Law and Geography: Current Legal Issues Volume 5*, Oxford: Oxford University Press, 2002

Moran, L., Skeggs, B., Tyrer, C. and Corteen, K., *Sexuality and the Politics of Violence*, London: Routledge, 2003

Morin, E., *La Méthode II: La Connaissance de la Connaissance*, Paris: Seuil, 1986

Morin, M.E., 'Cohabitating in the Globalised World: Peter Sloterdijk's Global Foams and Bruno Latour's Cosmopolitics', *Environment and Planning D: Society and Space* 27, 58–72, 2009

Morton, T., *The Ecological Thought*, Cambridge, MA: Harvard University Press, 2010

Morton, T., *Hyperobjects: Philosophy and Ecology after the End of the World*, Minneapolis: University of Minnesota Press, 2013

Motha, S., 'Guantanamo Bay, "Abandoned Being" and the Constitution of Jurisdiction', in S. McVeigh (ed.), *The Jurisprudence of Jurisdiction*, London: UCL Press, 2006

Muecke, S., *Joe in the Andamans and Other Fictocritical Stories*, Sydney: Local Consumption Publications, 2008

Mulcahy, L., 'Architects of Justice: the Politics of Courtroom Design', *Social & Legal Studies* 16(3), 383–403, 2007

Mulcahy, L., *Legal Architecture: Justice, Due Process and the Place of Law*, London: Routledge, 2010

Munro, I. and Jordan, S., '"Living Space" at the Edinburgh Festival Fringe: Spatial Tactics and the Politics of Smooth space', *Human Relations* 66(11), 1497–1525, 2013

Murray, J., 'Nome Law: Deleuze and Guattari on the Emergence of Law,' *International Journal for the Semiotics of Law* 19(2), 127–51, 2006

Murray, J., *Deleuze and Guattari: Emergent Law*, London: Routledge, 2013

Mussawir, E., *Jurisdiction in Deleuze: The Expression and Representation of Law*, London: Routledge, 2011

Nancy, J.L., *Los Angeles ou La Ville au Loin*, Paris: Fayard, 1999

Nancy, J.L., *Church, State, Resistance, Keynote Address*, trans. V. Voruz and C. Perrin, in H. de Vries and L. Sullivan (eds), *Political Theologies*, Bronx, NY: Fordham Press, 2006

Nancy, J.L., *The Creation of the World or Globalization*, trans. D. Pettigrew and F. Raffoul, New York: State University of New York Press, 2007

Negarestani, R., *Cyclonopedia: Complicity with Anonymous Materials*, Melbourne: re.press, 2008

Negarestani, R., 'Differential Cruelty', *Angelaki: Journal of the Theoretical Humanities* 14(3), 69–84, 2009

Negarestani, R., 'Solar Inferno and the Earthbound Abyss', in *Pamela Rosenkranz: Our Sun*, Venice and Milan: Istituto Svizzero di Roma and Mousse Publishing, 2010

Negarestani, R., 'Drafting the Inhuman: Conjectures on Capitalism and Organic Necrocracy', in L. Bryant, N. Srnicek and G. Harman (eds), *The Speculative Turn: Continental Materialism and Realism*, Melbourne: re.press, 2011

Neimanis, A., 'Alongside the Right to Water, a Posthumanist Feminist Imaginary', *Journal of Human Rights and the Environment* 5(1), 5–24, 2014

Nietzsche, F., *Thus Spoke Zarathustra*, trans. G. Parkes, Oxford: Oxford University Press, 2005
Nelken, D. (ed.), *Using Legal Culture*. London: Wildy, Simmonds & Hill Publishing, 2012
Neocleous, M., 'Air Power as Police Power', *Environment and Planning D: Society and Space*, 31, 578–93, 2013
Nolan, J., *The Therapeutic State*, Albany: New York University Press, 1998
Nussbaum, M., 'Secret Sewers of Vice', in S. Bandes (ed.), *The Passions of Law*, New York and London: New York University Press, 2000
Nymann Eriksen, N., *Kierkegaard's Category of Repetition*, Berlin: Walter de Gruyter, 2000
Odenwald, S., *Patterns in the Void: Why Nothing is Important*, Boulder, CO: Westview Press, 2007
Oosterman, A. and Cormier, B. (eds), *The Shape of Law: Special Issue* 4(38), 2014
Orford, A. (ed.), *International Law and Its Others*, Cambridge: Cambridge University Press, 2009
Pallasmaa, J., 'Space, Place and Atmosphere: Peripheral Perception in Existential Experience', in C. Borch (ed.), *Architectural Atmospheres: On the Experience and Politics of Architecture*, Basel: Birkhäuser, 2014
Pearson, Z., 'Spaces of International Law', *Griffith Law Review* 17(2), 489–514, 2008
Persaud, S., 'Only Connect: Luhmann and Bioethics', in A. La Cour and A. Philippopoulos-Mihalopoulos, *Luhmann Observed: Radical Theoretical Encounters*, Basingstoke: Palgrave, 2013
Philippopoulos-Mihalopoulos, A., 'Suspension of Suspension: Setting for the Improbable', *Law and Literature* 15(3), 345–70, 2003
Philippopoulos-Mihalopoulos, A., *Absent Environments: Theorising Environmental Law and the City*, London: Routledge, 2007a
Philippopoulos-Mihalopoulos, A., 'Brasilia: Utopia Postponed', in A. Philippopoulos-Mihalopoulos (ed.), *Law and the City*, London: Routledge, 2007b
Philippopoulos-Mihalopoulos, A., 'In the Lawscape', in A. Philippopoulos-Mihalopoulos (ed.), *Law and the City*, London: Routledge, 2007c
Philippopoulos-Mihalopoulos, A., 'Fear in the Lawscape', in J. Priban (ed.), *Liquid Society and Its Law*, Aldershot: Ashgate, 2007d
Philippopoulos-Mihalopoulos, A., *Niklas Luhmann: Law, Justice, Society*, London: Routledge, 2009
Philippopoulos-Mihalopoulos, A., '"…the sound of a breaking string": Critical Environmental Law and Ontological Vulnerability', *Journal of Environmental Law and Human Rights* 2(1), 5–22, 2011a
Philippopoulos-Mihalopoulos, A., 'Repetition or the Awnings of Justice', in O. Ben-Dor (ed.), *Law and Art: Justice, Ethics and Aesthetics*, London: Routledge, 2011b
Philippopoulos-Mihalopoulos, A., 'The Autopoietic Fold: Critical Autopoiesis between Luhmann and Deleuze', in A. la Cour and A. Philippopoulos-Mihalopoulos (eds), *Luhmann Observed: Radical Theoretical Encounters*, Basingstoke: Palgrave Macmillan, 2013a
Philippopoulos-Mihalopoulos, A., 'The World Without Outside', *Angelaki* 8(4), 165–77, 2013b
Philippopoulos-Mihalopoulos, A., 'Critical Autopoiesis and the Materiality of Law', *International Journal for the Semiotics of Law* 27(2), 2014

Philippopoulos-Mihalopoulos, A. and FitzGerald, S., 'From Space Immaterial: The Invisibility of the Lawscape', *Griffith Law Review* 17(2), 438–54, 2008

Pile, S., *The Body and the City: Psychoanalysis, Subjectivity and Space*, London, Routledge, 1996

Pile, S. and Keith, M. (eds), *Geographies of Resistance*, London: Routledge, 1997

Pirie, G.H., 'On Spatial Justice', *Environment and Planning A* 15, 465–73, 1983

Plato, *The Republic*, trans. T. Griffith and ed. G. Ferrari, Cambridge: Cambridge University Press, 2000

Pottage, A., 'Biotechnology as Environmental Regulation', in A. Philippopoulos-Mihalopoulos (ed.), *Law and Ecology: New Environmental Foundations*, London: Routledge, 2011

Pottage, A., 'The Materiality of What?', *Journal of Law and Society* 39(1), 167–83, 2012

Pred, A., *The Past Is Not Dead: Facts, Fictions, and Enduring Racial Stereotypes*, Minneapolis: University of Minnesota Press, 2004

Prior, J.H. and Cusack, C.M., 'Ritual, liminality and transformation: secular spirituality in Sydney's gay bathhouses', *Australian Geographer* 39(3), 271–81, 2008

Prior, J.H., Crofts, P. and Hubbard, P., 'Planning, Law, and Sexuality: Hiding Immorality in Plain View', *Geographical Research* 51(4), 354–63, 2013

Proust, M., *Time Regained: Volume VI – In Search of Lost Time*, trans. A. Mayer and T. Kilmartin, New York: Modern Library, 1983

Pruit, L., 'The Rural Lawscape: Space Tames Law Tames Space', in I. Braverman, N. Blomley, D. Delaney and A. Kedar (eds), *The Expanding Spaces of Law: A Timely Legal Geography*, Stanford, CA: Stanford University Press, 2014

Pue, W., 'Wrestling with Law: (Geographical) Specificity vs. (Legal) Abstraction', *Urban Geography* 11, 566, 1990

Razack, S., 'The Space of Difference in Law: Inquests into Aboriginal Deaths in Custody', *Somatechnics* 1(1), 87–123, 2011

Riello, A., *Dronology: A Symbolic Form of Our Time*, London: Antonio Riello Art Practice, 2014

Rilke, R.M., 'The Eight Elegy: dedicated to Rudolf Kassner', *Ahead of All Parting: The Selected Poetry and Prose of Rainer Maria Rilke*, ed. and trans. S. Mitchell, New York: The Modern Library, 1995

Rose, G., *Feminism and Geography*, Minneapolis: University of Minnesota Press, 1993

Rose, G., 'Performing Space', in D. Massey, J. Allen and P. Sarre (eds), *Human Geography Today*, Cambridge: Polity, 1999

Rose, N., *Governing the Soul: The Shaping of the Private Self*, London: Routledge, 1989

Rose, N., 'Governing Advanced Liberal Democracies', in A. Barry, T. Osborne and N. Rose, *Foucault and Political Reason*, London: UCL Press, 1996

Rousseau, J.J., *Politics and the Arts*, trans. A. Bloom (ed.), Glencoe, IL: Free Press, 1960

Rowan, R., 'Geographies of the Kettle: Containment, Spectacle and Counter-Strategy', *Critical Legal Thinking*, 2010

Ruddick, S., 'Domesticating Monsters: Cartographies of Difference and the Emancipatory City', in L. Lees (ed.), *The Emancipatory City*, London: Sage 2004

Sacco, R., 'Legal Formants: A Dynamic Approach to Comparative Law (Installment I of II)', *The American Journal of Comparative Law* 39(1), 1–34, 1991

Saldanha, A., *Psychedelic Whiteness: Rave Toursim and the Viscosity of Race in Goa*, Minneapolis: University of Minnesota Press, 2007
Saramago, J., *Cain*, trans. M.J. Costa, New York: Houghton Mifflin Harcourt, 2011
Sarat, A., '"… The Law is All Over": Power, Resistance and the Legal Consciousness of the Welfare Poor', *Yale Journal of Law and Humanities* 2(2), 343–80, 1990
Sarat, A., Douglas, L. and Merrill Umphrey, M. (eds), *The Limits of Law*, Stanford, CA: Stanford University Press, 2005
Sarat, A. and Kearns, T.R., 'Beyond the Great Divide: Forms of Legal Scholarship and Everyday Life' , in A. Sarat and T. R. Kearns (eds), *Law in Everyday Life*, Ann Arbor: Michigan University Press, 1995
Sassen, S., *Territory, Authority, Rights: From Medieval to Global Assemblages*, Princeton, NJ: Princeton University Press, 2006
Schaal, C., 'The Registration of Smell Trademarks in Europe: Another EU Harmonisation Challenge', *InterLawyer: Lex E-Scripta*, 2003
Schmitt, C., *The Nomos of the Earth in the International Law of Jus Publicum Europaeum*, trans. G.L. Ulmen, New York: Telos Press, 2006
Schmitz, H., *System der Philosophie 1: Der liebliche Raum*, Bonn: Bouvier, 1967
Schmitz, H., *System der Philosophie 3: Der Gefuhlsraum*. Bonn: Bouvier, 1969
Schmitz, H., *Der unerschöpfliche Gegenstand: Grundzüge der Philosophie*, Bonn: Bouvier, 1995
Sennett, R., *Flesh and Stone: The Body and the City in Western Civilization*, New York: Norton, 1970
Serres, M., *Genesis*, trans. G. James and J. Nielson, Ann Arbor: University of Michigan Press, 1995
Serres, M., *The Five Senses: A Philosophy of Mingled Bodies*, trans. M. Sankey and P. Cowley, London: Continuum, 2008
Shamir, R., 'Suspended in Space', *Law and Society* 30(2), 231–57, 1996
Sharpe, A., 'Structured Like a Monster: Understanding Human Difference Through a Legal Category', *Law and Critique* 18(2), 207–28, 2007
Sharpe A., *Foucault's Monsters and the Challenge of Law*, London: Routledge, 2010
Sifleet, J., 'Trade Dress – Another Way to Protect Against Copycats', *Smart Fast*, 2005
Shoemaker, A., *Aboriginal Australians: First Nations of an Ancient Continent*, London: Thames & Hudson Ltd, 2004
Shoshan, M., 'In the Name of Peace: Another Civic, An Other Law', in Oosterman, A. and Cormier, B. (eds), *Volume: The Shape of Law*, 4(38), 2014
Sloterdijk, P., *Sphären I: Blasen: Mikrosphärologie*, Frankfurt-am-Main: Suhrkamp, 1998
Sloterdijk, P., *Sphären II: Globen: Makrosphärologie*, Frankfurt-am-Main: Suhrkamp, 1999
Sloterdijk, P., *Sphären III: Schäume: Plurale Sphärologie*, Frankfurt-am-Main: Suhrkamp, 2004
Sloterdijk, P., 'Atmospheric Politics', in B. Latour and P. Weibel (eds), *Making Things Public*, Cambridge, MA: MIT Press, 2005
Sloterdijk, P., 'The Nomotop: on the Emergence of Law in the Island of Humanity', *Law & Literature* 18(1), 1–14, 2006
Sloterdijk, P., 'Airquakes', *Environment and Planning D: Society and Space* 27, 41–57, 2009
Sloterdijk, P., *In the World Interior of Capital: Towards a Philosophical Theory of Globalization*, Cambridge: Polity Press, 2013

Smith, D., 'Deleuze and the Question of Desire: Toward an Immanent Theory of Ethics', *Parrhesia* 2, 66–78, 2007
Soja, E., *Postmodern Geographies*, London: Verso, 1990
Soja, E., *Thirdspace*, Oxford: Blackwell, 1996
Soja, E., *Postmetropolis*, Oxford: Blackwell, 2000
Soja, E., 'Taking Space Personally', in B. Warf and S. Arias (eds), *The Spatial Turn: Interdisciplinary Perspectives*, New York: Routledge, 2009
Soja, E., *Seeking Spatial Justice*, Minneapolis: University of Minnesota Press, 2010
Solnit, R., *Wanderlust: A History of Walking*, London: Penguin, 2000
Spectorsky, A.C., *The Exurbanites*, New York: Berkeley Publishing, 1955
Spencer Brown, G., *Laws of Form*, London: George Allen and Unwin, 1969
Spinoza, B., *Ethics, Treatise on the Emendation of the Intellect and Selected Letters*, trans. E. Shirley, (ed.) and introduced S. Feldman, New York: Hackett Publishing, 1992
Spinoza, B., *Ethics*, trans. G.H.R. Parkinson, Oxford: Oxford University Press, 2000
Spinoza, B., *Theological-Political Treatise*, trans. M. Silverthorne and J. Israel (ed.), Cambridge: Cambridge University Press, 2007
Spinoza, B., *Treatise on the Emendation of the Intellect (On the Improvement of the Understanding)*, trans. R.H.M. Elwes, Slough: Dodo Press, 2009
Sreepada, S., 'The New Black: Trademark Protection for Color Marks in the Fashion Industry', *Fordham Intellectual Property, Media & Entertainment Law Journal* 19, 1131, 2009
Stäheli, U., 'The Outside of the Global', *New Centennial Review* 3(2), 1–22, 2003
Starr, D. and Bennett, G.G.,'Trademark Protection of Color Marks in the United States', *China Intellectual Property* 30, 21–5, 2009
Steinberg, P., 'Of Other Seas: Metaphors and Materialities in Maritime Regions', *Atlantic Studies*, 10(2), 156–69, 2013
Stewart, C., 'Atmospheric Attunements', *Environment and Planning D: Society and Space*, 29, 445–53, 2011
Stivale, C.J., 'The Literary Element in "Mille Plateaux": The New Cartography of Deleuze and Guattari', *SubStance* 13, 20–45, 1984
Stone, C., *Should Trees Have Standing? Toward Legal Rights for Natural Objects*, Los Altos: William Kaufmann, 1974
Stramignoni, I., 'Francesco's Devilish Venus: Notations on the Matter of Legal Space', *California Western Law Review* 41, 147–240, 2004
Strawson, J. (ed.), *Law after Ground Zero*, London: Glasshouse Press, 2002
Suthersanen, U., *Design Law: European Union and United States of America*, Andover: Sweet & Maxwell, 2010
Tarde, G., *The Laws of Imitation*, trans. Elsie Clews Parsons, New York: H. Holt, 1903
Tarde, G., *Penal Philosophy*, trans. R. Howell, Montclair: Patterson Smith, 1968
Tarde, G., *Monadology and Sociology*, trans. T. Lorenc, Melbourne: re.press, 2012
Ta kale, A.R., 'Kettling and the Fear of Revolution', *Critical Legal Thinking*, 2012
Thrift, N., *Non-Representational Theory: Space|Politics|Affect*, London: Routledge, 2007
Tournier, M., *Vendredi ou Les Limbes du Pacifique*, Paris: Gallimard, 1967
Tournier, M., *Friday*, trans. N. Denny, Baltimore, MD: John Hopkins University Press, 1969
Trigg, D., 'Place Becomes the Law', *Griffith Law Review* 17(2), 546–58, 2008
Turner, S. and Manderson, D., 'Socialisation in A Space of Law: Student Perform-

ativity at "Coffee House" in a University Law Faculty', *Environment and Planning D: Society and Space* 25, 761–82, 2007

Turpin, E., 'Who Does the Earth Think It Is, Now?', in E. Turpin and A. Arbor (eds), *Architecture in the Anthropocene: Encounters Among Design, Deep Time, Science and Philosophy*, London: Open Humanities Press, 2013

Urry, J., *Mobilities*, Cambridge: Polity, 2007

Valverde, M., *Law's Dream of a Common Knowledge*, Princeton, NJ: Princeton University Press, 2003

Valverde, M., 'Seeing Like a City: The Dialectic of Modern and Premodern Ways of Seeing in Urban Governance', *Law & Society Review* 45(2), 277–312, 2011

Valverde, M., *Everyday Law on the Street*, Chicago: University of Chicago Press, 2012

Vaughan-Williams, N., 'The Shooting of Jean Charles de Menezes: New Border Politics?', *Alternatives* 32, 177–195, 2007

Vaver, D., 'Recent Trend in European Trademark Law: of Shape, Sense and Sensation', *Trademark Reporter* 95, 895–913, 2005

Vergunst, J., 'Rhythms of Walking: History and Presence in a City Street', *Space and Culture* 13, 376–388, 2010

Verona, M., *Dove Vai Pastore? Pascolo Vagante e Transumanza nelle Alpi Occidentali agli Albori del XXI Secolo*, Scarmagno: Priuli e Verlucca, 2006

Virilio, P., *L'Insécurité du Territoire*, Paris: Éditions Galilee, 1993

Virno, P., *The Grammar of the Multitude*, Los Angeles: Semiotext(e), 2004

Vismann, C., 'Starting from Scratch: Concepts of Order in No Man's Land', in B. Huppauf (ed.), *War, Violence and the Modern Condition*, Berlin: Walter de Gruyter, 1997

Vismann, C., *Files: Law and Media Technology*, Stanford, CA: Stanford University Press, 2008

Wall, I., 'The Dis-enclosure of Constituent Power: Tunisia, Agamben and Nancy', *Critical Legal Thinking*, 2011

Wall, I., 'A Different Constituent Power: Agamben and Tunisia', in I. Wall, C. Douzinas and M. Stone (eds), *New Critical Legal Thinking: Law and the Political*, London: Birkbeck Law Press, 2012

Wall, I., Douzinas, C. and Stone, M. (eds), *New Critical Legal Thinking: Law and the Political*, London: Birkbeck Law Press, 2012

Warf, B., 'From Surface to Networks', in B. Warf and S. Arias (eds), *The Spatial Turn: Interdisciplinary Perspectives*, New York: Routledge, 2009

Warf, B. and Arias, S. (eds), *The Spatial Turn: Interdisciplinary Perspectives*, New York: Routledge, 2009

Weizman, E., *The Hollow Land. Israel's Architecture of Occupation*, London: Verso, 2007

Weizman, E., 'Political Plastic', *Collapse VI*, 277, 2010

Weizman, E., *The Least of All Possible Evils: Humanitarian Violence from Arendt to Gaza*, London: Verso, 2012

Wey Gomez, N., *The Tropics of Empire: Why Columbus sailed South to the Indies*, Cambridge, MA: MIT Press, 2008

Whitehead, A., *Process and Reality*, New York: Free Press, 1978

Williams, J., *Stoner: A Novel*, New York: New York Review of Books, 2006

WIPO, 'Smell, Sound and Taste – Getting a Sense of Non-Traditional Marks', *Wipo Magazine* 1(February), 2009

Wolfe, C., *What is Posthumanism?*, Minneapolis: University of Minnesota Press, 2009

Wood, A., 'Recursive Space: Play and Creating Space', *Games and Culture* 7(1) 87–105, 2012

Woodard, B., *On an Ungrounded Earth: Towards a New Geophilosophy*, Brooklyn, NY: Punctum Books, 2013

Xifaras, M., *La Propriété, Étude de Philosophie du Droit*, Paris: Presses Universitaires de France, 2004

Young, A., *Street Art, Public City*, London: Routledge, 2013

Zaino, C. ('The Urban Grocer'), 'Juice Boxes Imitating Fruit', *Lost At E Minor*, 2009

Zumthor, P., *Atmospheres*, Basel: Birkhäuser, 2012

Index

Actual/virtual 7, 46, 82, 87–8, 90, 93, 117, 124–5, 170, 198, 205
Affects 5, 11, 13, 29, 50, 107*ff*, 153, 156, 161, 164, 190, 195, 215, 222 n11, 226
Affective Judgment 171–2, 204
Affective Ontology of Excess 5, 13, 124
Agency 6, 36, 43, 46, 48, 50–1, 175 assemblic 6; contextual 59; of drones 134; human 6, 75, 181; law and 50; legal 21, 51; material 40, 50, 59
Air 6, 13, 44, 58, 87, 89, 106–7, 111, 115, 122–4, 126, 128–33, 135–6, 138–9, 140 n151, 144, 151, 157–8, 161, 165–6, 170–2, 176, 193, 197, 212, 226, 236
Animal 14, 43, 46, 58, 81, 87, 96–7, 100, 111, 113–4, 115, 152–3, 157–9, 167, 169: becoming 100, 158; and human 14, 100, 152, 159, 167; and hybrids 81, 96; law 169
Animality 113–4, 157, 196
Anthropocene 5, 12, 36, 61, 64–5, 210, 236
Anthropocentric/ism 64, 81, 87, 141, 179, 181–2, 210, 220: non– 76, 183
Aspatial/ity 3, 14, 30, 175, 227
Assemblage/assemblic 2, 5–9, 12–4, 22, 30, 46–52, 54–5, 57, 59, 61–3, 70, 78, 80–2, 96–100, 102, 104–6, 119–21, 128, 135, 149, 153–4, 158, 161–2, 167–8, 176, 186, 189, 191, 194–8, 200, 205–6, 209–10, 212, 215–6, 222, 225–8, 232–3
Atmosphere/atmospherics 107*ff*, 151*ff*
Australia 13, 20, 40, 52–4, 65, 88–9, 133, 216
Autopoiesis/autopoietic 6–8, 49–51, 60, 67, 76, 78, 95, 108, 143, 163, 186, 203, 213

Becoming 8, 31, 33, 40–1, 45, 55, 71, 73, 82, 90–1, 100, 119–20, 158, 160–1, 163–5, 167, 169–70, 186, 190, 196, 206, 208, 213, 220, 222, 229, 235: animal 100, 158; atmospheric 164–5, 167, 170, 172–3, 198, 212; being and 8, 186, 206, 213; inhuman 97 n 222; nonhuman 119, 227; other 41, 45, 82, 139 n149, 163, 165, 173, 198, 212, 228
Bennett, Jane 7, 43, 46 n33, 48, 83, 197 n73
Body/Bodies 39*ff*, 107*ff*: and the city 114, 116; desire of 68, 75, 186; human 5–6, 9, 46–7, 58–9, 80, 97, 105, 136 n133m 149, 169, 176, 186, 210, 223; immaterial 5, 59, 70, 149, 194; lawscaping 69, 74–5, 94, 97–8, 103–4, 126; material 1, 5, 51, 59, 70, 136, 149, 176, 194; and mind 34, 231; movement of 5, 34, 47, 68, 96–7, 99, 105, 112; nonhuman 5–6, 9, 44, 46, 59, 69, 80, 154; and objects 12, 16, 49, 69, 125, 213, 216; ontology of 59, 192; posthuman 100, 210; space and 4, 13, 40, 45, 66, 68, 70, 74, 114, 122, 135, 138, 154–6, 164, 170, 215, 222, 224; thought and 118, 154; transhumant 14, 154, 157, 167; withdrawal of 7–8, 49–51, 55, 59, 68, 71, 97, 105–6, 161–2, 167, 169, 175, 198, 211–3, 234; the world 113, 115–7, 191, 227 n35: *See also* Objects; Hyperobjects; Things
Border 56, 72, 115, 130 n101, 142–3, 163, 172, 194, 201,
Boundary/ies 5, 10, 15, 18, 23–5, 28, 46, 48, 54, 72, 81, 89, 93, 97, 100, 102, 115, 118, 130, 132, 142–3. 156, 163, 167, 177–8, 182, 189–90, 208

260 Index

Boundlessness 143, 208
Braidotti, Rosi 3, 60, 62 n90, 77, 80–1, 114 n29, 117, 121, 135, 171 n73, 229
Brennan, Teresa 10, 135, 153, 164, 171, 176, 203

Code/space 42
Colebrook, Clare 6 n3, 64, 65 n101, 78, 79 n152–3, 138, 205–6
Collectivity 7, 14, 46, 53, 61–3, 76, 78, 105–6, 108–9, 118, 120, 122, 125, 130, 137, 144, 160, 176, 179, 181–2, 188, 194, 197, 209–10, 226
Conative 7, 14, 36, 51, 68 n113, 74, 76, 79, 108, 154, 158, 186–7, 190, 196, 206, 213, 215
Conatus 7, 19, 50–1, 59, 76, 108, 197–8, 206, 211, 214
Concrete/Abstract 30–2
Conflict 3, 5–6, 11, 14, 21, 23, 53, 66, 70, 71 n126, 75, 84, 86, 92, 104–5, 123, 130, 145, 151–2, 156–7, 160, 162–5, 170, 175, 177, 186, 188, 192, 196, 198–9, 200, 204, 211–2, 217, 219, 226, 234: atmospheric 163, 165, 170, 211–2; legal 75 n141, 130
Consistency 5, 7 n6, 8, 19, 42–3, 51, 54, 178 n7, 222 n10, 229, 233
Continuum 1–4, 7–14, 26–7. 35, 40, 42, 46–7, 49, 53, 55, 57, 59, 60–3, 68, 70, 73, 76, 80–3, 85, 87–9, 91 n208, 92–4, 112, 114–6, 118, 120, 125, 128, 130–1, 133, 139–40, 142–4, 155, 161, 165, 168, 170–1, 186, 188–9, 192–4, 197–8, 202–4, 207, 209–11, 213–6, 221, 223–4, 226, 231–3: affective 130, 139; Immanence of 9, 92, 231; immaterial 42; lawscape/ lawscaping 13, 68, 83, 85, 87–9, 94, 168, 192–3, 198, 202, 207, 214–5; material 42, 114, 120; ontological 60, 131, 194, 197; ruptured/ and rupture 9–10, 12, 14, 26, 35, 47, 49, 59, 70, 73, 92–3, 116, 128, 130, 139–40, 142, 144, 168, 192–3, 198, 204, 209, 214, 231–2
Corporeal/corporeality 3, 5–6, 11, 24, 29, 39–40, 42, 48, 50, 66, 69, 71, 80, 95, 97, 99, 106–7, 109–113, 115, 118, 134, 138, 140, 153, 155, 165, 167, 174–6, 178, 185, 189, 191, 198, 201–3, 208–11, 216, 218, 221 n2, 226
Critical 3, 15, 17, 23 n49, 25, 29–30, 33, 59–60, 75, 85 n187, 102, 123, 141, 173, 184, 215: geography 17, 23 n49, 30, 123; legal theory 17, 184, 215l scholarship 15, 29, 33
Critique 3, 12, 14, 17, 19–20, 43–4, 104, 109, 141, 180, 204, 218

Deleuze, Gilles 3, 5, 8, 14, 19 n16, 35 n97 and n98, 41, 43, 46, 48, 50, 56–8, 71, 82 n174, 87 n197, 90, 91 n208, 93 n220, 102, 115, 117, 119–20, 138, 149–50, 185 n39, 189 n46, 190, 193, 202 n87, 203–4, 206 n98, 217 n133, 220–4, 225 n23, 227, 228 n38, 229 n40, 230–1, 232 n52 and n53, 233–5
Deleuze, Gilles, and Guattari, Félix 8, 34 n93, 42, 43 n23, 47, 54–6, 80, 87, 95, 97 n222, 99–100, 105–6, 119, 121, 122 n68, 152 n5, 160 n39, 161 n40 and n43, 162, 169, 170 n70, 172 n81, 178 n7, 185 n37 and n40, 186 n41, 187–9, 190 n50, 194, 204–5, 207, 209 n106, 221, 222 n12, 224 n18, 225–7, 231, 233 n55, 234 n60
Desire 1–3, 5–7, 14, 18, 21, 36, 43, 50–1, 56, 63, 68–9, 73–6, 78, 82, 86, 102, 107, 109, 111 n15, 114–5, 117, 121, 122, 129–30, 135–7, 139, 144, 146, 149–50, 152–3, 156, 158, 163, 168, 170–3, 177, 184–7, 190–1, 195–6, 199, 204–6, 211–5, 222–4, 226–7, 230–1
Deterritorialisation 190 n50, 225: *See also* Reterritorialisation
Dialectics/dialectical 4, 11, 14, 46, 80, 160–1, 187
Dissimulate 1–2, 4, 9, 11, 13, 17, 36, 43, 76, 90, 101, 108–9, 126, 129, 137, 144–6, 148, 157, 170–1, 183, 188, 194, 209, 215
Dissimulation 2–4, 8, 10–2, 14, 18, 22–3, 31–3, 36, 52, 84, 90, 107–8, 114, 126, 128–9, 134, 138, 141, 144, 157, 166, 168–73, 209, 215: atmospheric 114, 126, 157, 172; of dissimulation 171 n78; law's 52, 171; reality and 128; self– 4, 7, 20, 137, 145, 157, 162, 170
Drones 133–5

Embodiment/embodied/embody 1, 3, 6, 11, 15, 17, 19, 21, 23, 25, 28–9, 33, 35, 44, 52, 53–5, 65–7, 75, 77–8, 89, 92, 99, 101, 105, 110, 111 n13 115, 117–9, 121–2, 124, 139, 152–3, 155, 161, 175–6, 185, 187, 192–3, 195,

201, 202–3, 205–6, 211, 215–6, 223, 232, 235: *See also* Corporeal/corporeality
Encounter/s 11, 13, 50–1, 56 n71, 68, 81–3, 94, 99, 101–3, 136, 149, 154, 168, 183, 190, 197, 200, 208, 213, 227–9, 234: between bodies 56 n71, 68, 82, 190, 197; corporeal 134; space of 81, 190
Epistemology/ical 4, 7, 8 n11, 10–11, 13, 17, 21, 24, 27–8, 32, 35, 42–3, 46, 52, 59–61, 71–3, 78, 84, 89, 117–8, 165, 175–6, 184, 188, 194, 212, 214, 221, 224, 228: and ontology 10, 28, 35, 61, 212, 228; posthuman 13, 59–61; ruptured 10–1, 32, 59, 194, 214
Ethics 7 n6, 35, 57, 95,192, 203, 206–7: chtuloid 227; morality and 7 n6; posthuman 62 n90; of the situation 63 n95
Ethical 7, 18, 50, 56, 62, 81, 92, 184, 188, 227,
Exterior/Interior 1, 46, 49, 111, 116, 118, 126, 128–32, 138–9, 141–4, 163–5, 193–4: *See also* Outside

Feminism 3, 9, 13, 17–8, 77, 81, 95, 175, 192, 202
Flat Ontology 3, 5, 11, 177
Flow/s 2, 11, 13, 22, 26, 30, 34, 43, 56, 60, 72, 82, 95, 104, 109, 116, 120–1, 146, 153, 155, 159, 177, 204, 206, 225, 227, 231–2, 234
Fold 14, 35, 41, 43, 45, 73, 78, 80, 82, 144, 158, 169, 171 n78, 187, 193, 195, 202 n87, 230–1, 235
Folded 10, 17, 40–1, 45, 52, 54–5, 58, 80, 84, 87–8, 92, 126, 128, 141, 159, 166, 184, 186–7, 189, 192, 211, 218, 223, 230, 234
Folding 10, 36, 40–1, 44–6, 49, 52, 55, 70, 93, 158, 190–1, 207: *See also* Unfolding
Foucault, Michel 15, 41, 66, 74, 75 n138, 80 n164, 86, 111, 132, 138
Fractal/isation 13, 23, 35–6, 79, 87, 125, 132, 192, 194, 222: lawscape 13, 85, 87–8, 93, 195
Friday 14, 220–2, 224, 228–33: *See also Vendredi*
Freedom 34, 39, 48, 62–3, 69, 74, 82, 112, 141–2, 149, 156–8, 186, 195, 205–6, 220, 226, 228

Gated communities 20, 129, 131, 142, 197, 199
Gatens, Moira 3, 8, 9 n14, 10, 35, 43, 111, 206 n97, 221
Gatens, Moira, and Lloyd, Genevieve 8 n8, 10 n18 and n23, 62 n88, 63 n94, 150 n207, 193 n63, 194, 206 n98, 210 n111
Gender 9, 17–8, 20–1, 34, 99, 123, 155 n14, 169 n67, 202
Geography 3–4, 12, 17, 21–3, 26, 28, 33, 95, 122–3, 175, 180, 183, 222: human 39; law and 4, 12, 23, 73, 174, 183–4; legal 30; materiality of 223
Geophilosophy/ical 46, 64
Grosz, Elizabeth 3, 17 n5, 29 n78, 55, 82 n174, 190, 193, 205
Guattari, Félix 3, 70 n120, 80 n162

Haecceity 9, 105–6, 160–1, 206
Harman, Graham 8 n11, 49, 50 n51, 79 n156, 121 n62 and n63, 212, 213 n122, 214
Here 17, 67, 78, 89, 91, 125, 168, 177, 185, 191, 195, 215–6, 224, 230, 235
Here/Now 87, 89
Hyperobjects 77, 193, 210: *See also* Objects

Identity 3, 9, 14, 17–9, 27, 33, 36, 72, 73 n132, 74, 82, 85, 89–91, 93, 95, 105, 130, 140, 142–3, 153, 155, 157, 159–62, 164, 172, 175, 178, 182, 191, 206, 213 n125, 216–7, 219, 220 n1, 232 n53: and haecceity 206; and ipseity 213 n125
Illusion/illusionary 1–2, 7, 10, 13–4, 36, 59, 62–3, 66, 75, 78, 114, 117, 129–31, 138, 143–4, 146, 149–50, 157, 159–60, 166, 168, 171, 181, 193–6, 205, 209, 214, 226: atmospheric 168, 195; of the outside 166, 171, 205; of spatial justice 2, 196; of synthesis 143, 159–60
Immanence/Immanent 1, 8–9, 13, 26, 41–3, 45–6, 51–3, 56–9, 71, 73, 79, 82, 85–6, 92–4, 103, 119, 137, 143, 152–5, 158–9, 161–2, 178 n7, 190, 192–4, 200, 207–9, 218, 221–3, 227, 229, 230–1, 234, 235 n69
Immaterial/ity 5, 25, 34–6, 42–3, 59, 68, 70, 89, 118, 136, 149, 153–4, 170, 194, 215, 235: *See also* Material/ity; Matter

Immobility 162
Indistinction/indistinguishability 5, 7, 9, 12, 36, 50, 61–4, 81, 105, 151, 159, 210, 217, 223
Interior: *See* Exterior/Interior
Invisibilisation/invisibilised 1–4, 6–7, 9, 13, 36, 39, 69, 71 n126, 73–9, 82, 84–90, 93, 98–9, 101–5, 107–8, 130–1, 133, 149, 160, 192, 195, 207, 216, 220: *See also* Visibilisation/visibilise
Invisibility/invisible 76–9, 87, 103–4, 108, 125–6, 130, 134, 169, 191, 193, 204, 207: *See also* Visibility/visible
Irigaray, Luce 43–4, 45 n30, 126 n92, 128, 139 n146
Island 11–2, 14, 43 n24, 86, 129, 201, 220*ff*

Judgment/Judgement 10, 15, 22, 34, 58–9, 83, 92, 109, 134, 171–3, 184, 186, 188, 197, 202–4, 207–9, 214, 218, 227, 234–5: *See also* Affective Judgment
Juridification 67, 89, 225, 227: hyper 228
Jurisdiction 5, 23 n49, 25–6, 29 n77, 31, 33, 52, 66, 82, 92, 132, 154

Kettling 110–1, 142, 144
Kierkegaard, Soren 91–2, 227

Labyrinth 13, 34, 40, 43–4, 59, 71, 76
Latour, Bruno 8, 11, 21, 49, 100 n227, 139
Law in the Lawscape 67*ff*
Law of the Other 225–6, 228–9, 231–3
Lawscape (definition) 66*ff*, 73*ff*
continuum 13, 68, 83, 85, 87–9, 94, 168, 192–3, 198, 202, 207, 214–5; fractal 13, 85, 87–8, 93, 195; law in the 67*ff*; non– 107; ontology of the 13, 76, 118; posthuman 59, 79–81, 103, 114, 151, 154, 210, 228; qualities of 79*ff*; withdrawal of the 4, 68, 83, 107, 128, 134, 234–5
Lefebvre, Henri 19 n17, 27, 31, 34 n92, 40, 45, 56 n71, 71 n126, 72 n128, 74, 179, 181, 200, 202–3
Limit/limitation/limitless 18, 37–8, 46, 48, 114, 143, 159, 161, 224
Line of flight 34, 122, 171–2, 186, 191, 227

Logos/nomos (logic/nomic) 40, 57–8, 129, 143, 157, 171, 173, 186–8, 192, 225–6, 228–9, 231, 234
Luhmann, Niklas 3, 7 n6, 8 n9, 26 n64, 27 n67, 37 n104, 46 n37, 57 n73, 60, 67, 78 n150 and n151, 80, 87–9, 131, 138, 141–3, 171 n76, 194, 207 n103, 208 n104, 213, 217

Manifold 13, 16, 23, 40–5, 55, 59, 66, 70–1, 82, 84, 88, 94, 101, 103, 106, 154–5, 170, 180, 183, 185, 192, 223–4, 229–30
Massey, Doreen 13 n26, 23 n48, 24 n51, 25 n53, 26 n59 and n60, 33, 63
Material/ity 3–6, 8 n11, 10, 12–3, 15–6, 20–1, 23–31, 33–8, 40–3, 46, 47 n38, 48, 50–1, 55, 57, 59, 61, 63, 66–72, 75, 86, 89, 92, 94–5, 97, 99, 101–2, 106, 108 n3, 109, 113–5, 117–20, 124, 126, 128, 131, 136–7, 140, 149, 153–5, 170, 176–8, 192, 194, 199, 201, 209, 214–5, 217, 222 n11, 223, 235: *See also* Immaterial/ity
Matter 21, 44, 46–51, 60, 66, 107–8, 118, 121, 123, 139, 145, 148, 154, 163, 185, 194, 199, 213, 215: and idea 118, 154, 194; non– 145; and space 46, 49, 199
Mobility 47 n39, 162: *See also* Immobility
Morton, Timothy 3, 6 n4, 9 n15, 10, 60 n80, 61 n84, 64 n98, 69, 77, 82, 85 n186, 98, 121, 125, 130 n98, 176 n4, 193, 212, 214
Movement 2–6, 8 n11, 11, 14, 17–8, 20, 29, 33–4, 40, 43, 47, 50, 54, 58, 60, 67–8, 70, 75, 77, 80, 85, 87, 91–2, 94–9, 101–2, 104–6, 111 n13, 112, 114, 119–20, 124, 133, 137, 140 n151, 144, 146, 155–6, 159–62, 164, 172–3, 181, 185–7, 189–91, 195, 197–200, 208–10, 213, 216–8, 221–6, 231, 233–4: atmospheric 137, 172; corporeal 5, 50, 185 n38, 189, 198, 209, 218; nomadic 162, 233; ontological 6, 168; spatial 5–6, 186, 218; of withdrawal 66, 84 n179, 105, 168, 200, 234

Negarestani, Reza 3, 6 n3, 51, 64, 82 n175, 105, 126 n91, 137, 143, 144 n173, 163, 164, n51, 166, 170 n69, 172, 203, 227, 228 n37

Nomad/ic 6–7, 14, 34, 40, 75, 103, 151–2, 155, 156 n19, 159, 161–2, 167 n63, 168–9, 185–7, 202, 204, 217, 228–9, 231, 233, 235

Nomos: *See* Logos/Nomos

Nonhuman 5–6, 9, 44, 46, 54, 59–61, 64–5, 69–70, 79–81, 86–7, 119, 121–2, 124, 135, 152, 154, 199, 210, 214, 227

Object Oriented Ontology (OOO) 3, 5, 7–8, 46, 49 n49, 80, 121, 212

Objects 12, 16, 42 n20, 49–50, 69–70, 98, 121, 125, 130, 137, 171, 179, 212–4, 216, 235: and bodies 12, 16, 49, 69, 125, 213, 216; real 214; sensual 214; *See also* Body/Bodies; Hyperobjects

Occupy Movement 104, 215

Ontological 2, 4–9, 11, 13, 17–8, 24, 29, 32, 35 n97, 43, 46–7, 49–51, 56, 59–62, 69 n117, 70–1, 73, 75, 81–2, 84, 86, 106, 116, 121, 125, 131, 133, 150–1, 161, 168, 188, 194, 197, 203, 211–4, 223: continuum 60, 131, 194, 197; vulnerability 17–8, 60; withdrawal 44, 16, 49, 51, 70, 82, 92, 116, 121, 128, 161, 200 n84, 212–3

Ontology 10–1, 13, 25, 28, 35, 49, 51, 59, 61, 73–6, 78, 85, 93, 98, 118, 120, 124, 131, 155, 161, 177, 192, 203, 210, 212, 214, 224–5, 228: of bodies 59, 192; and epistemology 10, 28, 35, 61, 212, 228; of the lawscape 13, 76, 118; lawscaping 93, 98; material 25, 118, 131; of withdrawal 200 n84: *See also* Affective Ontology of Excess; Flat Ontology; Object Oriented Ontology (OOO):

Other: *See* The Other

Outside 1–2, 8–10, 13, 57, 82, 94, 111, 114–6, 129–31, 136, 138–9, 141–4, 152, 160, 162, 165 n57, 166, 171–2, 175, 187, 193, 202 n87, 204–5, 215, 223: *See also* Exterior/Interior

Parallelism 10, 28, 34–6, 43, 71, 161, 165 n56, 194

Phenomenology/ical 2, 4, 13, 18, 71, 87, 95, 115, 118, 123–4, 131, 151, 154, 159, 163–4, 179, 181, 191, 214, 228

Posthuman/ism 5–6, 11–3, 18, 49, 51, 59–61, 62 n90, 65, 69, 78–81, 87, 100, 102, 103, 105, 114–5, 118, 131, 135, 151, 154, 192, 210, 222, 225, 228: atmosphere 151*ff*; epistemology 13, 59–61; feminism 192; lawscape 59, 79–81, 103, 114, 151, 154, 210, 228

Property 5, 9, 14, 20, 34, 38, 48, 53, 65, 104, 112, 114, 118, 152–8, 160, 162–3, 167, 169, 172, 174 n1, 178, 183, 201–2, 208, 216, 220, 222: intellectual 38, 145, 147–9; law 19–20, 126 n90, 145, 148; lines 34, 112, 126, 153; private 53, 152, 156, 158, 163, 167, 208

Repetition/repeated 12, 14, 24, 47, 52, 56–8, 82, 87–93, 104, 129, 200, 202, 209, 217, 221, 225, 232–5

Responsibility 5, 10, 12, 28, 36, 52, 59, 61–5, 109, 178, 196–7, 203, 206, 210, 232: *See also* in relation to Indistinction/indistinguishability

Reorientation/reorient 2–4, 6, 14, 70, 78–9, 83–4, 94, 103–4, 106, 167–8, 193, 195, 197–200, 202–5, 207–9, 211–2, 215, 218, 228, 230, 232–4, 236

Resistance 4, 26, 28, 50, 75, 77, 103–4, 109, 114, 122, 129, 170, 181, 187–8, 200, 211: *See also* Revolt; Stasis

Reterritorialisation 185, 190, 230: *See also* Deterritorialisation

Revolt 24, 37, 58, 70, 93, 104, 108, 139, 145, 153, 158–9, 162, 165, 181, 185, 187, 189, 207–9, 211, 215, 217

Rhizome/rhizomatic 56, 93, 122, 137, 145, 170, 173, 189–90

Rupture/d 1–2, 4, 6, 9–12, 14, 26–7, 32, 35, 41, 47, 49–51, 59, 68, 70, 73, 82–3, 85, 92–3, 116, 126, 128–31, 139–40, 142, 144, 164–6, 168, 174, 178, 188, 192–5, 203, 205, 207 n103, 209, 211, 214–5, 218, 221, 226, 231–2, 234: atmospheric 126, 164, 192, 218, 226; continuum 9–10, 12, 14, 26, 35, 47, 49, 59, 70, 73, 92–3, 116, 128, 130, 139–40, 142, 144, 168, 192–3, 198, 204, 209, 214, 231–2; and epistemology 10–1, 32, 59, 194, 214; of spatial justice 174*ff*

Saraceno, Tomàs 126–7, 130, 139, 144, 165

Senses/sensorial 13, 16, 59, 80–1, 94–7, 99–100, 109–18, 129, 132, 138, 153–4, 172, 226, 231–2
Serres, Michel 3, 80, 115–8, 120, 122, 124, 190, 227
Shepherds 14, 109, 152, 154–7, 159–61
Singularity/ies 3, 9, 40, 46–7, 49–50, 56 n71, 59, 70, 82, 105–6, 128, 159, 161, 221, 231–2, 235
Skin 7, 10, 59–60, 67, 114–8, 120, 122, 124, 130, 136, 148, 172, 176, 201, 214, 227
Sloterdijk, Peter 11, 86, 122–4, 130–2, 135, 137–40, 142–4, 164
Smooth/smoothness 56–7, 66, 108, 112, 114, 117, 138–9, 145, 155–6, 159, 161–2, 17–3, 185–9, 212, 217, 220, 225, 227–9, 231, 233–5: *See also* Striate/striating/striation; Striated Space
Sociolegal 15, 17, 25, 29–30, 33, 59, 215
Soja, Edward 23 n49, 24 n51, 26 n58, 180–1, 182 n32
Songlines 13, 53
Spatial justice 174*ff*, 220*ff*
Spatial Turn 7, 12, 15*ff*, 174, 176, 178, 183
Spatiotemporal 5, 11, 26, 44, 63, 66–7, 75, 84, 125–6, 142, 154, 160, 176, 181, 210
Spinoza, Baruch 3, 5, 7–10, 19, 34–5, 43, 46–7, 49 n49, 50, 57 n73, 60, 62, 68, 70–1, 105, 111 n15, 118–9, 121, 144 n174, 149–50, 154 n9, 192–4, 197, 206, 223, 229
Stasis 156, 162, 185, 187, 197, 207, 211
Striate/striating/striation 57, 84, 103, 122, 155, 162, 172–3, 185–6, 189, 207, 223, 225–6, 228, 230–1, 233–5: *See also* Smooth/smoothness
Striated Space 112, 185–7, 231
Surface 3, 5, 6 n3, 8–11, 35–6, 38–41, 43, 47, 55, 60, 64, 71, 80, 84, 92–3, 99, 105, 113–5, 117–8, 121, 124, 128, 131, 136, 149, 153, 162–3, 169, 171, 177–8, 189, 192, 194, 200, 208, 216, 220–1, 228–30, 232: law of 229–30; manifold 230; tilted 5, 190, 192, 200

Temporality/temporally 9, 13, 16, 36–7, 39, 55, 87, 92, 105, 108, 125, 147, 152, 176, 178: atemporality 227
Territory 14–5, 53–4, 64, 108, 120, 142, 152 n5, 155, 157–8, 160–2, 172, 175, 193, 203, 223, 228
Territoriality 157–8
The Other 6, 28, 56, 220, 224–9, 231–3, 235 n69: law of the Other 225–6, 228–9, 231–3
Time 9, 12–4, 16–7, 25–7, 29, 36, 67, 87*ff*, 104, 124, 128, 143, 147, 159, 162, 178, 183–4, 186, 197, 202, 225, 233
Things 10, 18, 27, 34–6, 49, 64, 68, 89, 96, 120, 130, 175, 185, 194, 208, 229: between 185; physical 34–5
Tournier, Michel 14, 220–2, 225–7, 230, 231 n45, 233 n54, 235 n63 and 69
Transcend/ental/ence 1, 11, 17, 19, 52, 56–7, 73, 81–2, 85, 92, 94, 168, 192, 197, 230, 232
Transhumance/transhumant 14, 152, 154, 155 n14, 156–65, 167–9, 185, 196, 208–9
Triveneto 12, 14, 43 n24, 151–2, 154, 156, 160, 166, 196

Unfolding 8, 17, 32, 40–1, 52, 55, 59, 80, 93, 124, 193, 228, 230: *See also* Folding
Univocity/univocal 7 n6, 8, 35, 68, 87–8

Vendredi 14, 220–2, 224–5, 233–4
Visibilisation/visibilised 2, 4, 6–7, 9, 13, 36, 39, 73–8, 82, 84–90, 93–4, 98–105, 107–8, 131, 160, 165, 192, 195, 207–8, 216, 226: *See also* Invisibilisation/invisibilised
Visibility/visible 28, 55, 73–9, 87, 97–9, 101, 103–4, 111, 126, 139, 142, 165, 186, 191, 196, 225, 227: *See also* Invisibility/invisible
Vulnerability 18, 23, 60: ontological 17–18, 60

Walking 13, 41, 75, 94, 96–7, 99–100, 102, 106, 138
Withdrawal/Withdraw 1–4, 6–11, 13 n26, 14, 16, 18, 23, 27–8, 31–2, 36, 40, 46–7, 49–51, 55, 59–60, 65, 69–71, 77, 82, 84, 92, 97, 104–9, 112, 116, 121–2, 124–6, 128, 132, 136, 138, 143, 150, 152–3, 156, 161–3, 165–73, 175, 191, 193–4, 196–8, 198*ff*, 226–9, 232–5: atmospheric 14, 128, 165, 167–8, 199–200, 204, 208,

211, 232; of body 7–8, 49–51, 55, 59, 68, 71, 97, 105–6, 161–2, 167, 169, 175, 198, 211–3, 234; corporeal 165, 208; of the law 67, 132, 207, 211, 234; of the lawscape 4, 68, 83, 107, 128, 134, 234–5; movement of 6, 84 n179, 105, 168, 200, 234; ontological 4, 16, 49, 51, 70, 82, 92, 116, 121, 128, 161, 200 n84, 212–3; self- 126, 205, 207, space of 9, 138, 165, 200, 209, 234; of spatial justice 14, 32, 168, 191; transhumant 167–8